Design of Transient Protection Systems

Including Supercapacitor Based Design Approaches for Surge Protectors

Design of Transient Protection Systems

Including Supercapacitor Based Design Approaches for Surge Protectors

Nihal Kularatna

Alistair Steyn Ross

Jayathu Fernando

Sisira James

ELSEVIER

Elsevier
Radarweg 29, PO Box 211, 1000 AE Amsterdam, Netherlands
The Boulevard, Langford Lane, Kidlington, Oxford OX5 1GB, United Kingdom
50 Hampshire Street, 5th Floor, Cambridge, MA 02139, United States

Library of Congress Cataloging-in-Publication Data
A catalog record for this book is available from the Library of Congress

British Library Cataloguing-in-Publication Data
A catalogue record for this book is available from the British Library

ISBN: 978-0-12-811664-7

Publisher: Joe Hayton
Acquisition Editor: Lisa Reading
Editorial Project Manager: Mariana L. Kuhl
Production Project Manager: Surya Narayanan Jayachandran
Cover Designer: Victoria Pearson

Typeset by SPi Global, India

Photographs for cover page- Sadeeshvara Udayanga

Contents

Preface

Surge protection devices and systems are similar to a security-point at the entrance area of a building complex. It allows the easy passage of required electrical signals while potentially dangerous short term high energy transient electrical signals are subjected to a disposal procedure.

Nihal Kularatna

Acknowledgements

The subject presented in this book is a team effort by the four authors. Two authors, Jayathu Fernando and Sisira James, were PhD students in the power electronics research group at The University of Waikato. Nihal Kularatna and his Sri Lankan colleagues are very familiar with electric surge-related disasters in both large industrial electronic systems and consumer electronics that frequently occurred during the tropical electrical storms experienced in Sri Lanka, a small country located on the equatorial belt in the Indian Ocean. After 25 years of experience in Sri Lanka managing various electronic stress facilities, Nihal moved to the University of Waikato to set up a surge test laboratory, assembling a team of researchers whose goal was to develop design, test, and measurement capability in power electronics at the university. Alistair Steyn-Ross joined the research group to provide theoretical and analytical expertise. The resulting long-standing collaborations underpin what has turned out to be a very fruitful and creative enterprise encompassing a broad range of power electronics research projects.

Given this background the team of authors is very grateful to the University's Senior Management and to the School of Engineering for necessary financial resources as well as laboratory space and computing facilities. The team members are very grateful to our former dean of the School of Engineering Professor Bruce Clarkson, and to the former Head of School of Engineering Professor Ilanko Sinniah for helping us with funds and encouragement.

Development of the supercapacitor-assisted surge absorber (SCASA) technique was a major achievement in this subject area, since the team leader strongly believed that supercapacitors have a major role to play in the modern world of power electronic converters and protection systems. After early investigations demonstrated the ability of supercapacitors to safely absorb short-term high-voltage transient surges, WaikatoLink Limited (the University's commercial arm) assisted the team to file a set of international patents. Subsequently the Australian company, Thor Technologies, came forward to commercialise the patented technique.

The authors are very grateful to George Foster-Jones, CEO of Thor Technologies, and to their engineering team led by Willy Nielsen, for their confidence in the SCASA approach to surge suppression. The team is also very grateful to Henry Karpinsky (Senior Engineer, product development) of Thor for his encouragement and for his critique of the endurance testing of the

SCASA based product Smart TVIQ^2. Also we are very thankful to Nidhie Kumar and Tipene Merritt of WaikatoLink Ltd who helped us work with Thor Technologies to enable development of the commercial prototypes of the product.

New PhD students, Sadeeshvara Udayanga and Shanima Silva, are gratefully acknowledged for the specific tasks of cover-page photographs and the calibration document in appendices respectively.

Our senior PhD students, Nimesha Wijesooriya, Thilanga Ariyarathna, Sanduni Malluwawadu and Dilini Jayananda, are gratefully acknowledged for their kind support in manuscript preparation. We thank Dinali Ranmadugala and Gehan Anthonys sincerely for coming forward as a volunteers to help in finalizing and formatting some chapters of the book.

The authors are grateful to Christoph Jehle of TDK-EPCOS AG for organising timely release of copyright permissions. Also we are grateful to James Sutherland of IET publications for his kind advice on reuse of content from IET papers published by the authors. We are very grateful to Thor Technologies for their kind permission to publish limited details on their commercial product.

Nihal Kularatna is always thankful to his beloved wife Priyani, daughters Dulsha and Malsha, sons-in-law Rajith and Kasun, for their unstinting encouragement and appreciation of his technical book writing, while the two little grand-daughters Nethuli and Mineli, keep him entertained during the weekends. Nihal is always grateful to the advice he received during early 1990s from Prof John R Pierce (the Bell Labs engineer who named the transistor) on how to write a technical reference book. Nihal is also sincerely grateful to Sir Arthur C Clarke, who was the patron of Arthur C Clarke Institute for Modern Technologies, for his continuing encouragement for research work during his mid-career of 16 years, which allowed him to start work in this subject area.

Nihal Kularatna
Alistair Steyn-Ross
Jayathu Fernando
Sisira James
School of Engineering
The University of Waikato
Hamilton
New Zealand
25th November 2018

Chapter 1

Background to Surge Protection

1.1 Introduction

At the turn of the century, around 75% of the power generated was processed by power electronics. In the modern scenario of electronics, ultra-large-scale integrated circuits, which progress towards system-on-a-chip (SoC) concepts, are powered by DC power rails as low as 1.2–0.7 V. These two scenarios have made the demands within the power conversion interface complex. Compact ultra-low-DC voltage sources with energy backup, fast transient response, and power management have become mandatory in powering modern electronics. Uninterruptible power supplies and advanced power conditioning with surge protection have become essential in providing clean and reliable AC power.

Modern electronic systems are uniquely vulnerable to power line disturbances because they bring together the high-energy power line and sensitive low-power integrated circuits controlling power semiconductors in the AC-to-AC conversion interfaces as well as the DC–DC converters. The term 'power conditioning' is used to describe a broad class of products designed to improve or assure the quality of the energy source connected to sensitive electronic systems.

Utilities realize that different types of customers require different levels of reliability, and make every effort to supply disturbance-free power. However, normal occurrences, most of which are beyond control and are acts of God, make it impossible to provide disturbance-free power 100% of the time. In addition to these external disturbances, sources within buildings, such as switching of heavy inductive loads, poor wiring, overloaded circuits, and inadequate grounding, can cause electrical disturbances. Many of these power disturbances, particularly transient surges, can be harmful to sensitive electronic equipment with low-voltage DC power supplies. Power disturbances can cause altered or lost data and sometimes equipment damage, which may, in turn, result in loss of production, scheduling conflicts, lost orders, and accounting problems.

There are methods and devices available to prevent these disasters from happening. Protective systems range from those providing minimal protection to those that construct a new power source for critical loads, converting the

Design of Transient Protection Systems. https://doi.org/10.1016/B978-0-12-811664-7.00001-X

standard 'utility-grade power', which may be adequate for most equipment, into the 'electronic-grade power' required by many critical loads. This chapter discusses the protection against transients and surges in power conversion systems required for electronic systems demanding 99.999% or higher reliability. Treatment will be on a low-voltage electronic circuit protection basis only, and hence a discussion on power-engineering aspects is not treated in the book.

1.1.1 Development of Semiconductor Technology

After the invention of the transistor in 1947, followed by the development of integrated circuit concepts in the mid-1950s, an exponential growth of integrated circuits occurred following Moore's law [1]. While silicon and GaAs progressed on similar paths, newer compound semiconductor materials, such as SiGe, GaN, and SiC, were also gradually introduced to cater to high-frequency and/or high-temperature and high-voltage requirements of commercial electronic systems.

With integrated circuits gradually progressing towards SoC concepts with a massively increased number of transistors, the feature size of transistors was gradually dropping towards less than 0.1 μm. For example, companies such as Intel were planning to progress into 22 nm semiconductor processing as early as 2012. Currently many digital components, such as processors and memories, are built using CMOS processes smaller than 15 nm. Fig. 1.1A indicates the Moore's law-based general progress of integrated circuits, while the DC rail voltages are dropping towards sub-1.0 V levels. With the equivalent noise levels increasing within these complex ICs, dropping of the logic levels makes the scenario much more complex [2]. Fig. 1.1B indicates the development of processors similar to the Intel family, and their power supply requirements. Fig. 1.1C indicates the scenario in terms of clock speed, power consumption, and, most importantly, the equivalent impedance of the equivalent processor load. With the processor equivalent impedance dropping below 1 milliohm, while the DC rail voltages are dropping towards sub-1 V levels, if a transient surge voltage appears on the power supply, it could create disastrous consequences.

In summary, this situation is creating a signal integrity crisis, and the chip designers, as well as the power supply designers, are required to pay adequate attention to these situations for reliable product design. Protection against transients and surges therefore becomes a paramount interest.

1.2 Power Line Disturbances

In a single- or three-phase utility power supply, commercial power companies are expected to supply AC power at a nominal RMS voltage with a percentage tolerance such as ±6% with limited amounts of harmonics as per applicable standards. However, practical utility grade power supply carries many

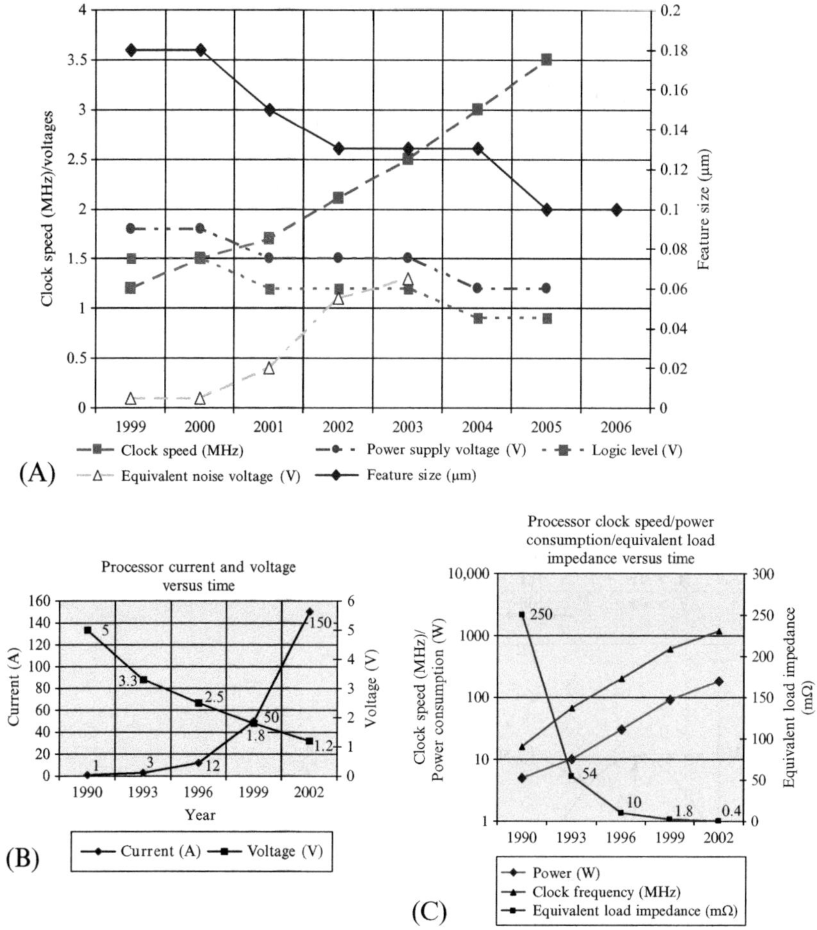

FIG. 1.1 Developments of integrated circuits and processor families: (A) Moore's law driving a signal integrity crisis; (B) development of processors and their power supply requirements; (C) processor speed, power consumption, and equivalent load impedance.

unwanted RMS voltage disturbances, harmonics, noise, and transients. Fig. 1.2A and B depicts the RMS voltage disturbances and transients and noise. Fig. 1.2C indicates the case of a 50 Hz, 230 V RMS power supply, and its corresponding frequency spectrum.

1.2.1 RMS Voltage Fluctuations

1.2.1.1 Voltage Surges

Voltage surges are voltage increases that typically last from about one cycle to one-half of a second. Surges are commonly caused by the switching of heavy

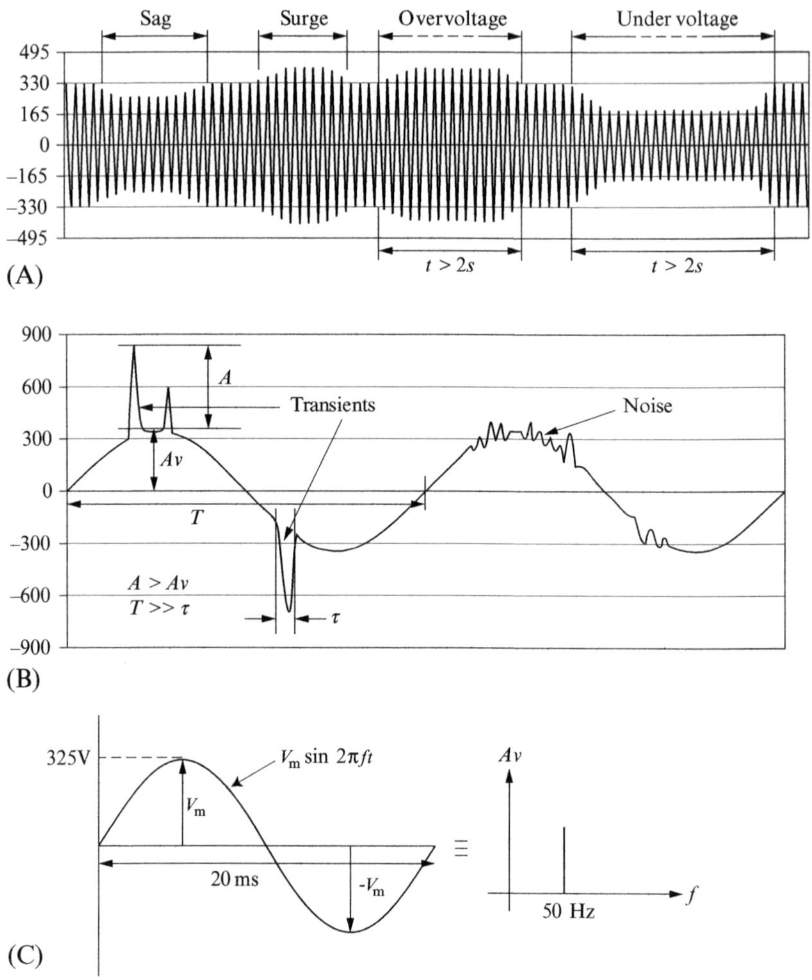

FIG. 1.2 Examples of disturbances on a 230 V RMS, 50 Hz utility AC power supply: (A) RMS voltage fluctuations; (B) transients and noise superimposed on the waveform; (C) ideal waveform and its frequency spectrum.

loads and power network switching. Surges do not reach the magnitudes of sharp spikes, but generally exceed the nominal line voltage by about 20%. This kind of deviation can cause computer data loss, equipment damage, and erroneous readings in monitoring systems. A surge that lasts more than 2 s is typically referred to as an *overvoltage*.

1.2.1.2 Voltage Sags

Voltage sags are undervoltage conditions, which also last from one cycle to one-half of a second. Sags often fall to 20% below nominal voltage and are caused when large loads are connected to the power line. Sags can cause computer data loss, alteration of data in progress, and equipment shutdown. A sag that lasts for more than 2 s is typically referred to as an *undervoltage*.

1.2.2 Transients

As shown in Fig. 1.2B voltage transients are sharp, very brief increases in the supply waveform. These spikes are commonly caused by the on and off switching of heavy inductive loads, such as air conditioners, electric power tools, machinery, and elevators. Lightning can cause even larger spikes. Although they usually last less than 200 μs, typically in the order of tens of microseconds, spikes can be dangerous to unprotected equipment, with amplitudes ranging from about 180% of the peak value to over a few thousands of volts, positive or negative, and sometimes as high as 6000 V or even larger. This high magnitude of sudden voltage variations can wipe out stored data, alter data in progress, and cause electronic-hardware damage. In general, the higher the energy content carried into the circuit by the superimposed transient surge, the higher the potential for damage.

1.2.3 Electrical Noise

Electrical noise is a high-frequency interference in the frequency spectrum of 7000 Hz to over 50 MHz. Noise can be transmitted and picked up by a power cord acting as an antenna, or it can be carried through the power line. These disturbances can be generated by radio frequency interference, such as radio, TV, cellular, and microwave transmission, radar, arc welding, and distant lightning. Noise can also be caused by electromagnetic interference produced by heaters, air conditioners, white goods, and other thermostat-controlled or motor-operated devices.

Although generally nondestructive, electrical noise can sometimes pass through a power supply as if it were a signal and wipe out stored data or cause erroneous data output. Problems result when microelectronic circuitry is invaded by transient, high-frequency voltages collectively called ‘line noise’, which can be grouped into one of two categories: *normal mode* or *common mode* (described later).

1.2.4 Harmonics

Harmonic distortions are usually caused by the use of nonlinear loads by the end users of electricity. Nonlinear loads, a vast majority of which are loads with power electronic devices, draw current in a nonsinusoidal manner. With the increased use of such devices in consumer loads, the presence of distortions in current and voltage waveforms has become a frequent occurrence today. For instance, virtually all modern computer power supplies use a direct rectification mode of AC input (without any step-down transformer) to derive a bulk DC power supply for conversion into lower DC rails using switch-mode techniques. A simple example is illustrated in Fig. 1.3. Because the rectifiers can conduct only when the instantaneous line voltage is higher than the smoothing capacitor voltage, the charging of capacitors takes place only during a small time period, resulting in narrow input current pulses as shown in Fig. 1.3B. Referring to Fig. 1.3C, if any one or several of the loads connected to the distribution bus draws a nonlinear current, terminal voltage at the consumer terminal will be a distorted sinewave. It is quite common to have a flattened-top sinewave due to nonlinear loads such as computer power supplies, UPS rectifiers, or even energy saving lamps. The ultimate result is that the terminal voltage at the consumer end will be a nonsinusoidal waveform with a lot of harmonics.

In view of the widespread use of power electronic equipment connected to utility systems, various international agencies and national agencies have proposed limits on harmonic content injected into the system. More details are available in Kularatna [3, Chapter 9].

1.3 Power Quality Issues

1.3.1 Modes of Transients and Noise

The noise and transient sources superimposed on the utility voltage can come in two different forms, namely *common* mode and *differential* mode. Differential mode is sometimes also referred to as normal mode or transverse mode. In tackling surge protection in power electronic systems, it is very important to deal with both cases. Fig. 1.4 indicates the two cases.

1.3.1.1 Differential Mode (Normal Mode or Transverse Mode)

As the name implies, differential-mode signals appear between the live wire and the accompanying neutral for a case of AC supply input. In the case of a DC power supply, it means the positive or negative rail and its return current path. These two lines represent the normal path of power through the electric circuits, which gives any normal-mode signal a route into sensitive components.

1.3.1.2 Common-Mode Signals

Common-mode (CM) signals are voltage differentials that appear between the ground and either of the two supply lines. Common-mode transients are most often the cause of disruption, because digital logic or analogue signals are either directly or capacitively tied to the safety ground as a zero-voltage data reference point for semiconductors. As a result, transient CM voltage differences as small as 0.5 V can cause that reference point to shift, momentarily ‘confusing’ the

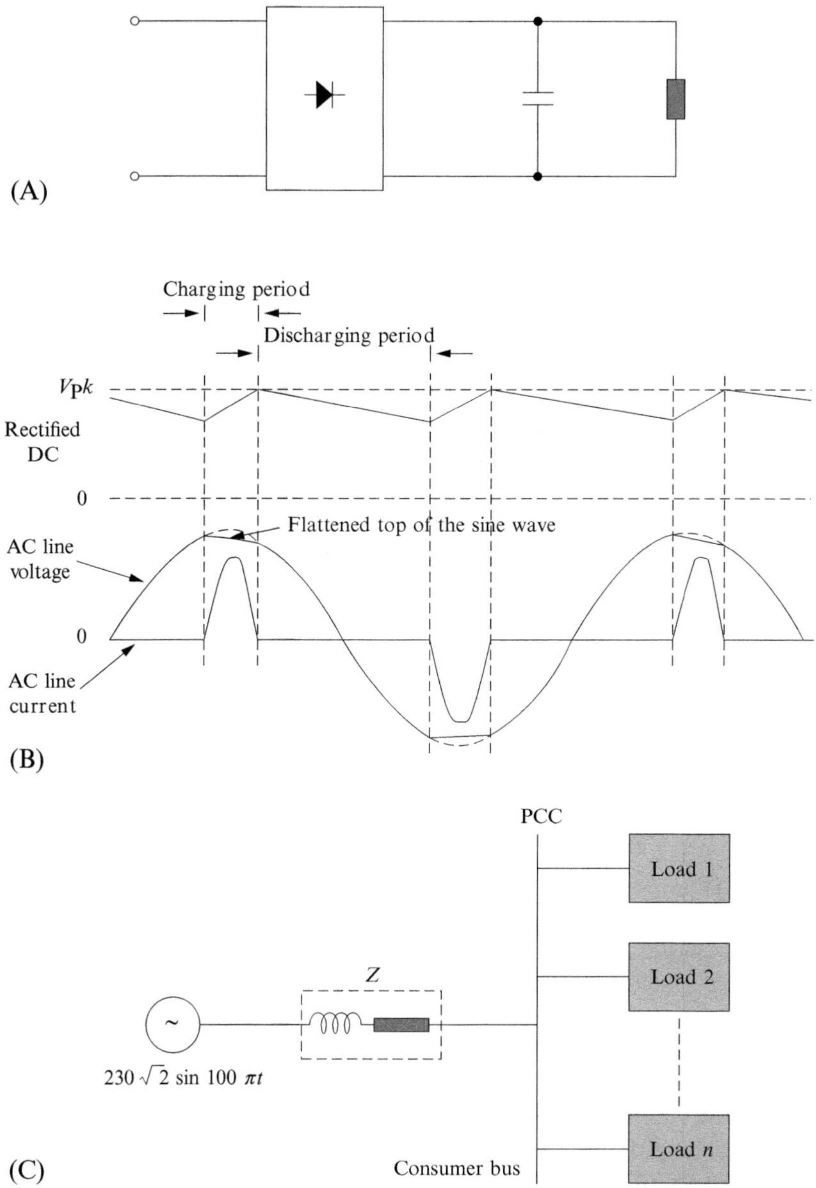

FIG. 1.3 Distortion of the voltage waveform due to nonlinear loads: (A) simple case of bridge rectifier; (B) rectifier input current in relation to input voltage waveform; (C) simplified case of distribution bus, with a 230-VAC 50-Hz sinusoidal generator, and linear bus impedance before the point of common coupling (PCC).

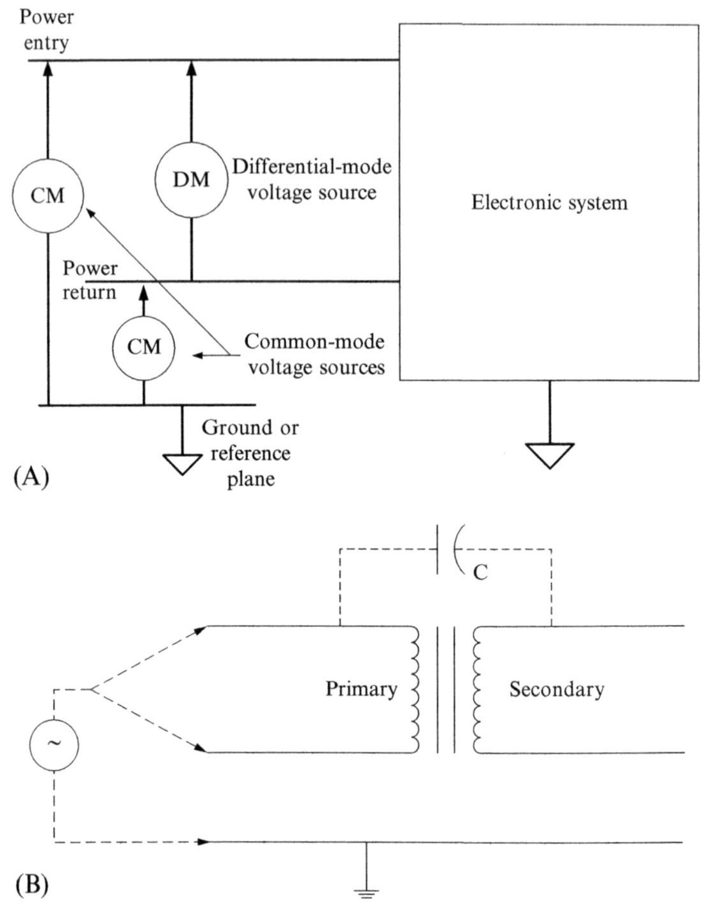

FIG. 1.4 Different modes of signal inputs and the effect of isolation transformer: (A) differential and common modes; (B) common-mode signal coupling towards secondary side of an isolation transformer due to coupling capacitance.

semiconductor chips. Fig. 1.4B indicates the possibility of transferring CM signals due to interwinding capacitances in a transformer with electrical isolation. With the transformer winding process focused on reducing interwinding capacitances, good quality isolation transformer allows you to reduce noise and transients coupled towards load. However, manufacturing isolation transformers with very high CM isolation properties can be expensive.

1.3.2 Development of DC Power Supplies

During the 1970s and early 1980s, most DC supplies were linear types as depicted in Fig. 1.5A. These simple linear power supplies were based on a

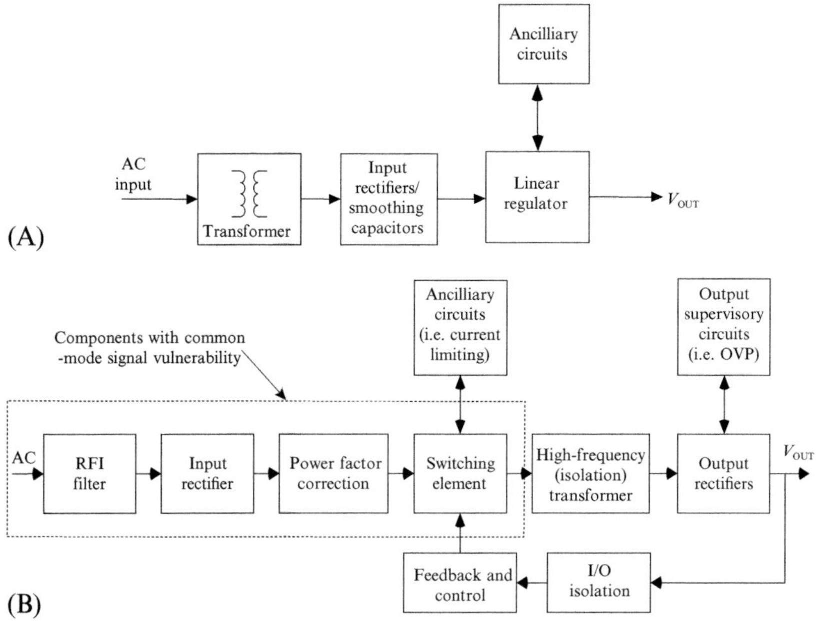

FIG. 1.5 Comparison of linear power supplies and modern SMPS: (A) simple linear power supply with an isolation transformer; (B) modern off-the-line switching power supply and its components vulnerable to common-mode surges.

step-down transformer, which had the advantage of suppressing high-voltage differential- or common-mode transients. Particularly dangerous CM transients were easily suppressed by a reasonably well-designed input transformer with galvanic isolation between the windings and with very low interwinding capacitances. If the step-down transformer is designed to have very low interwinding capacitances, the probability of transferring dangerous levels of CM transients towards the rectifier stage and the linear regulator chips is low.

Fig. 1.5B depicts the situation of modern SMPS designs based on power semiconductors such as BJTs, MOSFETS, or IGBTs where higher VA-rated units carry a power factor correction block as well. In these units the essential (regulatory) condition of ground isolation (for minimizing safety issues) is maintained only by the high-frequency switching transformer, which becomes an integral part of the DC–DC converter stage [3, Chapter 4]. It is important to note that the input bulk DC rail is created by direct rectification of 230 V/50 Hz or 120 V/60 Hz input source creating an unregulated DC rail of 320 or 165 V DC, respectively. Given the complexity of the modern off-the-line SMPS systems, it is undesirable to have any CM signals directly coupled towards the semiconductors. In practice, these systems are much more vulnerable to high-voltage transients, so well-designed surge protection circuits are necessary to protect them. This is particularly true for the protection of components within the dotted lines of Fig. 1.5B.

1.4 Principles of Surge Protection: A Simplified Overview

1.4.1 Levels of Damage or Disruption

Almost all transients and surges are unpredictable and statistical in nature. It is not possible to stop them totally, but in planning engineering facilities every effort is made to minimize their effects. However, there is some part of the transients either induced on circuits or directly conducted into the circuits, which can cause disastrous effects on various end user circuits. A circuit component is likely to fail if it is subjected to an over-current, over-voltage or over-power condition for a finite period of time that is beyond the design specifications of the particular circuit component. Therefore, protecting a circuit against transients is based on two simple principles: (1) limiting the voltage amplitude of the surge at each component to lower values, and (2) diverting the surge-related currents through protection-specific components to absorb the surge energy. In practice a designer is expected to protect a system against superimposed surges as shown in Fig. 1.6. The case is shown for a system where the input supply is 230 V/50 Hz. The voltage spike shown in the figure, which resembles a power surge or a superimposed lightning transient, usually lasts less than a few tens of microseconds, compared to the 20 ms-long AC cycle in this case. In the design of basic power conversion systems, the designers disregard the possibility of these superimposed signals and assume that the input utility source is a sinewave as per specifications considered in the project. In general,

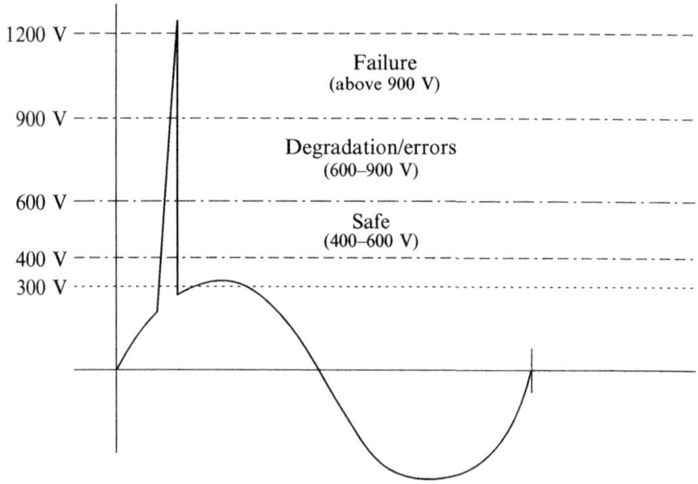

FIG. 1.6 Transient superimposed on AC-input waveforms with their effects.

the possibility of surge damage is related to the height of the spike and the duration of the surge. As indicated in Fig. 1.6, if the surge peak is between 400 and 600 V, the basic circuit blocks are still considered to be safe by design. When the spike amplitude is within 600–900 V, degradation of performance or temporary errors could be expected, while values above this can cause permanent faults.

Because lightning surges and other power system transients are very fast and can have very high amplitudes (many thousands of volts), in order to be effective, surge protection devices (SPDs) need to switch on quickly and handle large amounts of energy in a short time. Normally, upstream circuit breakers or fuses do not have time to trip in the event of the surge diverter activating because the reaction time of circuit breakers and fuses is much slower than the duration of the transient impulse.

1.4.2 Circuit Designer's View to Surge Protectors

All SPD circuits basically work as voltage dividers [4]. Fig. 1.7 illustrates this principal with a series impedance, Z_S, and a load impedance, Z_L. The input impedance, Z_S, always exists due to wiring or transient source impedance. When there is no protection as in Fig. 1.7A, the input surge voltage divides among the two impedances. Z_L is very much greater than Z_S, and hence a higher voltage is developed across the load which causes the damage. When a blocking device with impedance Z_{block} is connected in series as in Fig. 1.7B, its impedance increases due to frequency components of the surge and hence decreases the load voltage. When a shunting SPD component is used as per Fig. 1.7C, it decreases its impedance, Z_{shunt}, during the event of a high-voltage surge to create a shunt current through Z_{shunt} to dissipate the surge energy. If circuit constraints allow, components of Z_S should be selected in order to increase it to dissipate the surge. This will minimize stress on the shunting device.

An ideal SPD must limit the surge energy passed to the load to a minimum level such that the load as well as the SPD are not damaged. The energy passing to the load can be found using the integral of the surge voltage and current, $\int vi\,dt$ [5]. A practical SPD should develop a lower instantaneous voltage across the load (clamping voltage) to limit the current passing to the load, low dynamic

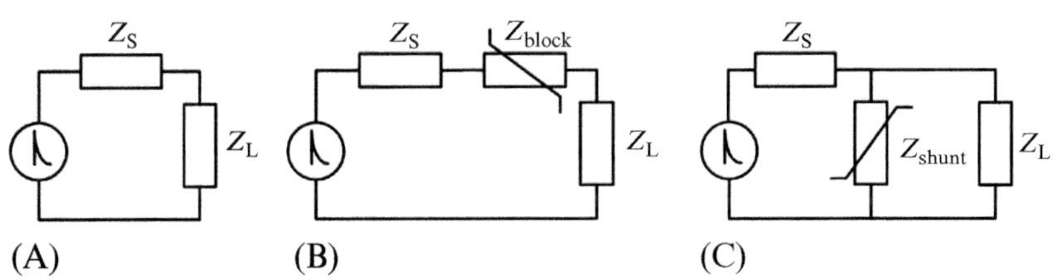

FIG. 1.7 Voltage division with transient/surge protector: (A) without any protection; (B) with series blocking device; (C) with shunting device.

resistance, and fast response time. Longevity, repeatability, board size, cost, reliability, and a safe failure mechanism are also important factors of SPD design. The designs are based on international standards, which are discussed in Chapter 2.

1.4.3 Simplified Calculations

Let us now consider a real life situation with suitable practical circuit parameters. Fig. 1.8A shows a typical AC load connected to a 230 V AC source without the use of any surge protection; here a small series resistance (0.01 Ω) and a series inductance (0.01 μH) are taken as possible values to represent the series effects of the distribution circuit due to its long length. The impedance of the inductor at 50 Hz is

$$X_{s,50\ \mathrm{Hz}} = 2\pi \times 50 \times 0.01 \times 10^{-6} = 314 \times 10^{-8} = 3.14\,\mu\Omega. \quad (1.1)$$

If you pass 10 A RMS through this inductive impedance the voltage dropped across it would be 31.4 μV, and hence negligible. The 0.01 μΩ series resistance would drop only 0.1 μV. This clearly indicates that the distribution line's series resistance and inductive reactance does not have much impact on the transfer of power to a load at 50 Hz.

However, if you take a surge impulse of 10–50 μs, which corresponds to a range of frequencies from 1/10 μs = 100 kHz to 1/50 μs = 20 kHz, the corresponding inductive reactances of Z_s are given by

$$X_{s,100\ \mathrm{kHz}} = 2\pi \times 100 \times 10^3 \times 0.01 \times 10^{-6} = 6.28\ \mathrm{m}\Omega, \quad (1.2)$$

$$X_{s,20\ \mathrm{kHz}} = 2\pi \times 20 \times 10^3 \times 0.01 \times 10^{-6} = 1.25\ \mathrm{m}\Omega. \quad (1.3)$$

Despite the considerable increase in the inductive impedance values, a surge voltage would still force a very large current to pass through this impedance to the load.

In Fig. 1.8B we show some additional components such as a Metal Oxide Varistor (MOV) and an additional series inductance of 1 μH. The MOV shown here, which is in the nonfired state, can be simply considered to be a very large resistance in the order of a few MΩ. (MOVs are discussed in detail in Chapter 3.) If we do an analysis of this circuit by neglecting the effect of the MOV, despite the effect of the added 1 μH inductance, the 50 Hz load current flow will not be blocked. However, if we consider the high-voltage transient surge of several kilovolts (say 5 kV) for a period of 20 μs prior to the firing of the MOV, the circuit still will be equivalent to the one given in Fig. 1.8C where the MOV nearly acts as an open circuit. This circuit will generate frequencies of 50 kHz (1/20 μs = 50 kHz) and its higher harmonics due to the 20-μs transient surge. Let us now estimate the series impedance offered by the total inductance present in the load current's path.

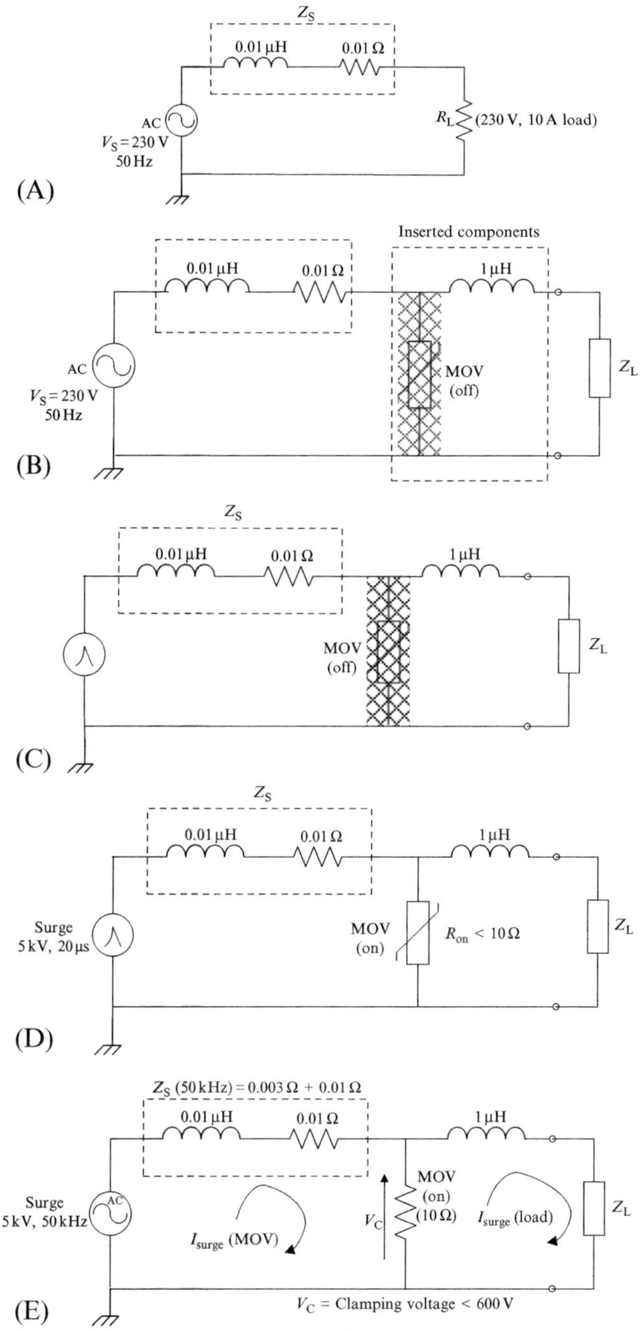

FIG. 1.8 Working of a shunt SPD with a series blocking inductor: (A) without protection prior to the arrival of surge; (B) with protection prior to arrival of surge; (C) at the initial stage of the surge; (D) at the later part of the surge; (E) surge currents in the loops.

$$X_{s,50\ \text{kHz}} = 2\pi(50 \times 10^3)(1.01 \times 10^{-6}) = 0.317\ \Omega. \tag{1.4}$$

Now if we compare the above with the resistance of the 230 V powered load of 10 A, which is 23 Ω, we can see that X_s has some voltage dropping effect on the surge voltage, but still not adequate.

Fig. 1.8D shows the case where the MOV has started to conduct and its resistance has dropped drastically to few tens of ohms or to few ohms (a typical case of 10 Ω is shown in the figure). In this circuit the fired-MOV allows the surge current to pass through easily due to its lowered resistance. If we consider the surge source as a 50-kHz sinewave with a peak amplitude of 5 kV, then as in the discussion related to Fig. 1.8C, the surge source and the fired-MOV now presents the circuit as shown in Fig. 1.8E to the load and the inserted series inductance of 1 μH. The impedance of 1 μH at 50 kHz is 0.314 Ω and this creates a little extra attenuation of possible surge current in the load loop. In case this series inductive component can be made higher, to say 10 μH, to further reduce the surge current in the load. Also at higher and higher harmonics of the surge waveform, the blocking effect will be more significant.

When the MOV is fired into conduction it carries a high surge current and its terminal voltage is maintained close to its clamping voltage which usually amounts to a few hundred volts only. Without going through a detailed analysis, we can see that most of the energy associated with the surge will be absorbed by the MOV. This concept of surge protection will be analyzed in detail in later chapters.

The voltage across the load will be limited to a value less than 600 V due to voltage clamping by the MOV. The energy absorbed by the MOV can be found by calculating $\int vi\,dt$. It is desirable to have a large inductance value for the source impedance, because the greater impedance offered by this during the rising surge would reduce the stress on the MOV. The building wiring becomes a part of this source impedance and would help reduce the surge magnitude for equipment placed far from the mains service entrance of the building.

This concept of series blocking and shunting surge energy is seen in almost all surge protectors. A typical practical protector designed for a three-wire power system is shown in Fig. 1.9. Compared to the case in Fig. 1.8, in

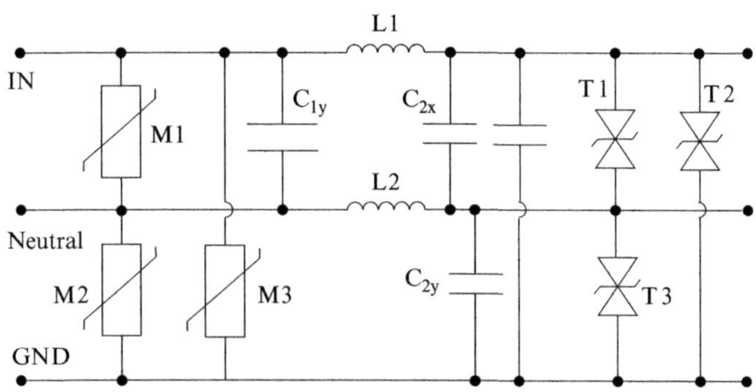

FIG. 1.9 A typical practical surge protector designed for a three-wire system.

Fig. 1.9 capacitors provide additional *near short-circuit paths* to surge currents, giving better protection here. This circuit also provides protection against both common-mode and differential-mode surges.

References

[1] N. Kularatna, Digital and Analogue Instrumentation: Testing and Measurement, Institute of Engineering and Technology, London, 2003 (Chapter 2).

[2] S. Duety, et al., The key for processor performance beyond a GHz is clean power, in: Proceedings of PCIM2000, 2000, pp. 68–74.

[3] N. Kularatna, Power Electronics Design Handbook, Elsevier-Newnes, Burlington, MA, 1998.

[4] J.R. Barnes, Robust Electronic Design Reference Book, vol. 1, Springer Science & Business Media, New York, NY, 2004.

[5] S.J. Goldman, Selecting Protection Devices: TVS Diodes vs. Metal Oxide Varistors, Power Electronics, 36 (2010) pp. 29–33.

Chapter 2

Surge Protection Standards and Practices

2.1 Introduction

Given an overview of power quality issues and background to the simple circuit concepts used in designing surge protection systems in Chapter 1, in this chapter we discuss the standards and practices used in developing and testing surge protector devices (SPD).

Transients discussed in this book are short-duration, high-amplitude (positive or negative), and generally nonrepetitive signals superimposed on the AC or DC power supply inputs. A typical example is shown in Fig. 1.6, with a general indication of what impact it could have on the operation of the circuit powered by the 230 V, 50 Hz AC supply. In general, transients that get induced on the equipment inputs could be due to several different reasons, such as (a) inductive switching, (b) lightning, (c) electrostatic discharge, and (d) electromagnetic pulses. All these are very unpredictable random occurrences. Electrostatic discharge may occur due to build-up of static charge on the human body with voltages as high as 20,000 V. These can generate transients with rise times as fast as 2 kV/ns. Electromagnetic pulses due to nuclear activities where gamma rays can be released could cause transients with rise times around 5 kV/ns, while lightning activity-related transients can generate signals with rise times around 600 V/ns [1]. Given the statistical nature of these different types of transients, SPD developers and manufacturers are guided by a set of standards and practices as summarized in the following sections.

2.2 Surge Protection Standards and Practices

In most countries, the equipment connected to power lines has to be certified by nationally accredited testing laboratories. Most of these national standards are based on the international standards of the International Electrotechnical Commission (IEC). In North America the international standards of the IEC are not valid; instead, standards by the American National Standards Institute (ANSI) and the Institution of Electrical and Electronic Engineers (IEEE) are used,

Design of Transient Protection Systems. https://doi.org/10.1016/B978-0-12-811664-7.00002-1

TABLE 2.1 Standard Test Agencies and Their Recommended Standards for SPDs [2]

Standard Agency	SPD-Related Standards	Remark
IEC	IEC 61000-4-5	Provide a model to simulate surges and then to be able to check if the equipment is able to survive them
	IEC 60364	General rules
	IEC 61643-11	Product standard
	IEC 61643-12n	Selection and application guide
UL	UL 1449	Primary concern is safety
	UL 1449 3rd Ed	Devoted to surge protection manufacturers, defines the parameters as well as the test procedure to qualify an SPD
ANSI/IEEE	IEEE C62.41.1	Risk of transient over voltages to low-voltage networks
	IEEE C62.41.2	Surge environment and types of transients
	IEEE C62.45	Method for testing equipment against transients that are connected to the low-voltage network
	IEEE C62.62	Tests and ratings for SPDs installed on the load side of the service equipment, used to compare SPD performance
	IEEE C62.72	Installation of SPDs in electrical power distribution equipment
NEC	Article 280	Surge arrestor installed on wiring systems over 1000 V
	Article 285	Selection and installation conditions of SPDs
NEMA	LS1 Low-voltage surge protective devices	Provides specific SPD performance. This results can be used to compare actual test results of SPDs

UL, *Underwriters Laboratory Inc.;* IEC, *International Electrotechnical Commission;* ANSI/IEEE, *American National Standard Institute/Institute of Electrical and Electronics Engineers;* NEMA, *National Electrical Manufacturers Association.*

together with the safety standards developed by Underwriters laboratories Inc. [2]. These standards cover the areas such as SPD specifications for different locations, SPD test procedures, and test waveforms. Table 2.1 lists some of these institutions and their standards in relation to SPDs [2, 3].

IEC 61000-4-5 and ANSI/IEEE C62.41 are commonly used by the designers of transient protection systems and they address the most severe transient conditions on both power and data lines caused by lightning strikes and by load switching. Lightning transients may result from a direct strike or induced voltages and currents due to an indirect strike. Switching transients may be the result of power system switching, load changes in power distribution systems, or short-circuit fault conditions. An example is a case of a large air conditioning plant switching off suddenly and pumping inductive energy stored in the motor loads back into the distribution system creating an inductive-surge-related transient.

The IEC 61000-4-5 standard defines a transient entry point and a set of installation conditions [4]. The transient is defined in terms of a generator producing a given waveform and having a specified open circuit voltage and source impedance.

Measurements carried out by the IEEE over a number of years have demonstrated, on a statistical basis, the likely frequency of occurrence, typical amplitudes, and waveshapes to be expected in various locations as a result of artificial and naturally occurring electrical phenomena. These findings are published in IEEE C62.41. The work outlined in the IEC 61000-4-5 and the ANSI/IEEE 62.41 standards provides the basis for the design of AC power-line transient SPDs.

2.2.1 Surge Protector Locations and Categories

When surges propagate through power lines, data lines, or other branched circuits, its energy gets dissipated in several ways. While wire resistance, flashovers, and SPDs in the surge path dissipate part of the energy, the branch circuits divide the energy to smaller quantities depending on the branch impedances and transfer these into individual branches. Due to this fact, facilities are divided into different location categories and specific protection levels are defined based on locations. Fig. 2.1 shows IEEE C62.41 and UL 1449 defined standard location [5, 6]. The standards classify surge protector type based on the potential impact of the transient surge, and location.

2.2.2 ANSI/IEEE C62.41 Categories

2.2.2.1 Category C

This category refers to outside and main service entrances which includes main supply lines, transformers, service connections, and the feeder line to the main service entrance panel, any overhead subfeeder lines or underground lines. Category C surge protector type is applied to protect against externally caused power disruptions such as lightning strike entering a facility via the power line.

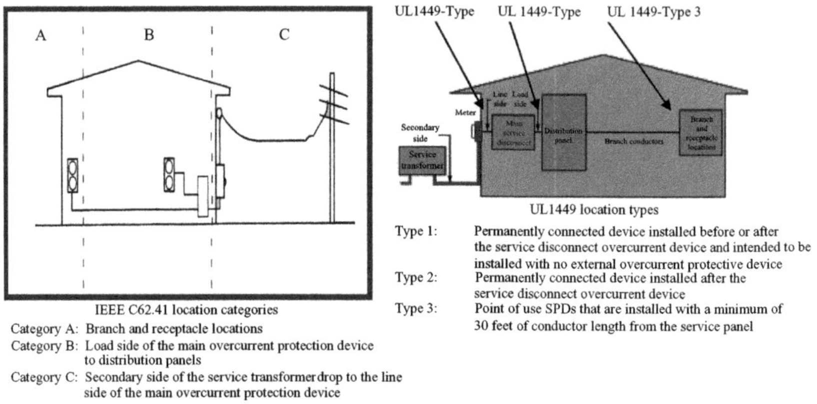

FIG. 2.1 A typical practical surge protector designed for a three-wire system.

2.2.2.2 Category B

This category refers to all major subfeeders, bus systems, and short branch circuits such as distribution panels, industrial buses and feeder systems, heavy appliance circuits, and lighting systems in large buildings that extend almost 10 m from the service entrance. Protection at this location is very effective in suppressing frequently internally generated transients (switching electric motors and motor-driven apparatus), especially for protecting sensitive equipment and equipment which is fed from the substations.

2.2.2.3 Category A

This category refers to any outlets and long branch circuits extending more than 10 m (30 ft) from category B location or 20 m (60 ft) from category C. Surge protection for this location category is applied at the outlets or individual circuit level for individual protection of a specific piece of equipments, such as computers, weighing bridges, measuring equipment, process control equipment, DC power supplies, etc.

2.2.3 UL 1449 Types

The latest UL 1449-2014 4th edition covers enclosed and open-type surge protective devices (SPDs) designed for repeated limiting of transient voltage surges as specified in the standard on 50/60 Hz power circuits not exceeding 1000 V. The 4th edition is also an approved American National Standard for Safety for SPDs and designated as ANSI/UL 1449-2014 (4th edition). The SPD manufacturers recommend their products for different locations based on surge current withstanding capability.

2.2.3.1 *Type 1*

This is a permanently connected, hard wired SPD type that can be used outdoors. These are connected in between the secondary terminal of the supplying transformer and the line terminal of the service equipment overcurrent device as well as the watt-hour meter busbars. Type 1 SPDs can be used without an external overcurrent protection device.

2.2.3.2 *Type 2*

This is a permanently connected SPD type that can be installed on the load side of the service equipment overcurrent device. Type 2 includes the SPDs located at the branch panel and moulded case SPDs.

2.2.3.3 *Type 3*

This type of SPD is intended to be installed indoors at a minimum conductor length of 10 m (30 ft) from the electrical service panel (category B outlets). Typically they are cord-connected, receptacle type, or direct plug-in SPDs.

2.2.3.4 *Type 4*

Component assemblies consisting of one or more Type 5 components together with a disconnect (internal or external) or a means of complying with the limited current tests in UL 1449.

Type 1, 2, 3 component assemblies consist of a Type 4 component assembly with internal or external short circuit protection.

2.2.3.5 *Type 5*

Discrete component surge suppressors, such as metal oxide varistors (MOVs) may be mounted on a printed circuit board, connected by its leads or provided within an enclosure with mounting means and wiring terminations.

When comparing above standards it can be seen that the Type 1 SPDs fall into the category C of IEEE standards and Type 2 consists of both Category B and C while Type 3 is same as category A.

2.2.4 Surge Voltage Waveforms

The IEC 61000-4-5 standard specifies a combination wave consisting of two waveforms, which are shown in Fig. 2.2: the 1.2/50 μs open-circuit voltage waveform and the 8/20 μs short-circuit current waveform. These impulse waveforms are defined by their rise times and half-amplitude duration. For example, an 8/20 s impulse current would have an 8 μs rise time from 10% of the peak current to 90% of the peak current. The 20-μs decay time is measured between half amplitude points.

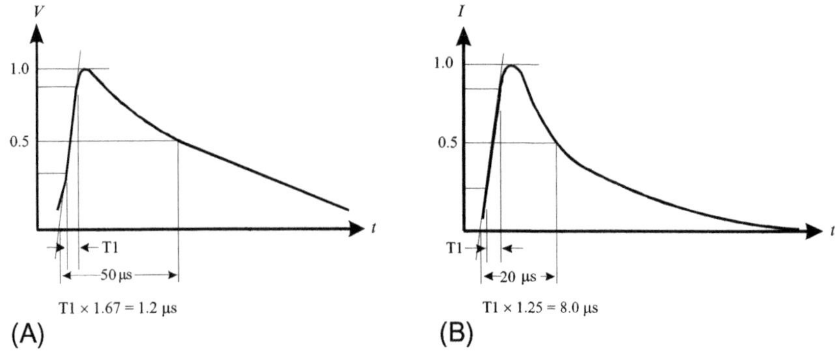

FIG. 2.2 IEC 61000-4-5 combination wave: (A) open-circuit voltage; (B) short-circuit current.

The open-circuit voltage of the combination wave can be represented mathematically as

$$V(t) = A_V V_p (1 - e^{-t/\tau_1}) e^{-t/\tau_2}, \tag{2.1}$$

where $A_V = 1.037$, $\tau_1 = 0.4074$ μs, $\tau_2 = 68.22$ μs, and V_p is the peak voltage. Time t is measured in μs.

The mathematical representation of the short-circuit current of the combination wave is given by

$$I(t) = A_I I_p t^3 e^{-t/\tau}, \tag{2.2}$$

where $A_I = 0.01243$ μs^{-3}, $\tau_1 = 3.911$ μs, and I_p is the peak current.

When connecting the equipment under test (EUT) to a lightning surge simulator, the test leads add some inductance which results in an undershoot in the current delivered to the EUT. The critically damped, unidirectional 8/20 μs combination current waveform presented here does not represent this behaviour. For cross-validation of numerical simulations and the actual tests, the mathematical equation

$$I = 4200 \quad \sin(0.126t) e^{-t/\tau_1} (1 - e^{-t/\tau_2}), \tag{2.3}$$

with $\tau_1 = 28.1$μs and $\tau_2 = 1$μs, is specified in IEEE C62.45 [7]. This equation represents the 3 kA 8/20 μs surge with undershoot. The unit for I is in amperes and t is in microseconds.

The combination wave shown in Fig. 2.2 is also defined in ANSI/IEEE C62.41 along with another standard waveform, which is the 100 kHz ring wave shown in Fig. 2.3. The location categories to which these two standard waves are applicable are indicated in Table 2.2.

2.2.5 Guide on Surge Testing for Equipment Connected to Low-Voltage AC Power Circuits

The IEEE C62.45 (2002) standard provides appropriate surge testing guidelines for equipment survivability, methods of test connection, surge coupling mode

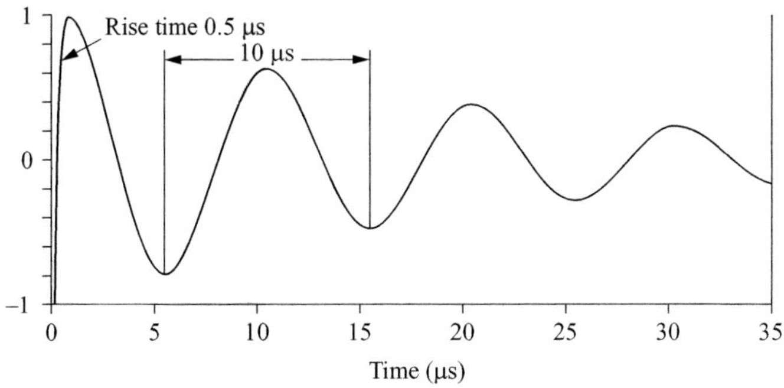

FIG. 2.3 The 100 kHz ring wave.

TABLE 2.2 IEEE C62.41 Location-Based Current/Voltage Waveforms [8]

			0.5 μs × 100 kHz, Ring Wave		1.2 × 5 μs (V) 8 × 20 μs (A), Combination Wave	
Category	*Level of Exposure*	*Voltage (kV)*	*Current (A)*	*Energy (J)*	*Current (kA)*	*Energy (J)*
A1	Low	2	70	0.63	–	–
A2	Medium	3	130	2.34	–	–
A3	High	6	200	5.43	–	–
B1	Low	2	170	1.53	1	9
B2	Medium	4	330	5.94	2	36
B3	High	6	500	13.5	3	81
C1	Low	6	–		3	81
C2	Medium	10	–		5	225
C3	High	20	–		10	900

definitions, testing safety requirements, and various surge suppression techniques. The intent is to provide background information that can help determine if specific equipment or a circuit has adequate surge withstanding capability. The main objective of the document is to call attention to the safety aspects of surge testing. Signal and data lines are not addressed.

Table 2.2 provides the details of standard test waveforms used in different location categories. Effect of external transients in 'A' location is negligible due to the protection provided by the inductance of the building wiring. Category B location is exposed to both internally created and lightning created transients and hence both ring wave and combination wave test parameters are defined. Category C has the highest exposure for lightning and utility switching transients. Combination waveform with high current and voltage, 3–10 kA and 6–20 kV waveform parameters are specified for category C locations.

2.2.5.1 *UL1449 Test Standards*

Clamping Voltage Test (Surge Impulse Test)

Before the 3rd edition of UL1449, impulse tests were performed with a surge waveform of 6 kV/500 A to obtain suppressed voltage rating (SVR), that is the clamping voltage for the SPD. After the third edition, impulse test is performed with a surge waveform of 6 kV/3000 A to obtain the clamping voltage which defined as the voltage protection rating (VPR) of SPD. There is six times more surge energy present in the revised third edition standard waveform and results in a higher clamping voltage. Therefore, VPR values will be numerically higher than former SVR values.

Nominal Discharge Current Test

The UL1449 3rd edition introduced a new SPD test where the device must remain fully operational after being subjected to a total of 15 impulses, 1 min apart. Some consider this to be the 'stress test' for SPDs to measure its durability in the electrical system. It is performed in combination with the surge impulse test above. A 10% deviation in VPR during these tests would result in failure. The value of the nominal discharge current impulse value, I_n, is selected by the manufacturer and can vary from 10 to 20 kA for Type 1 devices and 3, 5, 10, or 20 kA for Type 2 devices. The actual nominal discharge current level achieved by the device must be marked on the label of the device as I_n. Some manufacturers have selected to consolidate the I_n rating for a majority of their Type 1 and Type 2 devices to the highest value of 20 kA. The SPD's and secondary surge arrester's 20-kA nominal-discharge-current rating is a critical requirement in lightning protection systems installation standards [9].

2.3 Circuit Concepts Used in Surge Protection

In general practical circuits developed for surge protection are based on three basic concepts: (1) design the protection circuit separately as an add-on block to the base circuit; (2) attenuate the incoming transient using passive series-connected high impedances, or shunt the surge currents via passive low-impedance circuits (which act as filters for high-frequency components of the surge); (3) use nonlinear devices such as gas discharge tubes, MOVs,

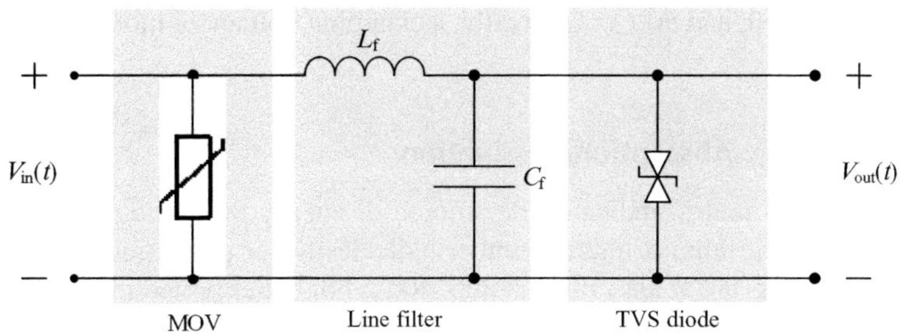

FIG. 2.4 A two-wire category B protection unit with two levels of protection.

transient voltage suppressor (TVS) diodes or TVS thyristors to divert the surge currents and absorb the transient energy.

In Section 1.4.3, we briefly looked at a typical practical protector designed for a three-wire power system which is shown in Fig. 1.9. Let us now consider the two-wire version of this SPD shown in Fig. 2.4.

The combination of line filter and transient suppressor components in this SPD makes it suitable for use in both C62.41 location categories A and B. When designed for category B locations larger components will have to be selected for the transient suppressor components [10]. This circuit uses an MOV for the first level of protection and a TVS diode for the second level of protection. L_f and capacitor C_f form the line filter in the middle for extra protection. The MOV will enter its firing or conduction mode when the transient exceeds the threshold voltage limit thus providing the first level of protection. For very short lived high-voltage transients, the clamping action of the varistors, together with the voltage dropped across the series inductance, holds off the majority of the transient voltage from the output.

For more extended stress conditions, the current in L_f will increase to the point where the output capacitor C_f is charged to a voltage at which the TVS diode is brought into conduction. This diode prevents the output voltage from exceeding its rated clamp value for all stress currents up to the failure point of the diode.

This SPD also prevents voltage transients generated within the driven equipment from feeding back into the supply line. This can be an important advantage when several pieces of equipment in a system are connected to the same supply. Several other SPD circuit concepts are discussed in Billings [10] and Kularatna [11].

2.4 Selection of Surge Protectors

When selecting a surge protector circuit, the following criteria should be considered.

2.4.1 Clamping Voltage

This indicates what voltage will cause the MOV to conduct. A lower clamping voltage provides better protection. There are three levels of protection in the UL rating: 330, 400, and 500 V. Generally, a clamping voltage of more than 400 V is too high.

2.4.2 Energy Absorption/Dissipation

This rating, in joules, indicates the amount of energy the surge protector can absorb before it fails. A higher number indicates better protection. In general, surge protectors are at least rated at 200–400 J. For better protection, a rating of 600 J or more will be required

2.4.3 Response Time

Surge protectors do not function immediately when subjected to a transient surge. The time required to turn 'ON' the SPD and conduct current through the voltage clamping device when exposed to a transient is known as response time. Surge protection devices with longer response times pass the transient voltage to the critical load connected to the power line and will be exposed to the surge for a greater amount of time.

2.4.4 Life Time

The number of transients of specified voltage, rise time, and duration, that the SPD can survive before the threshold voltage changes by more than a specified amount [12] is defined as the life time of an SPD.

References

[1] R.D. Winters, Power supply voltage transient analysis & protection, in: Powercon III, Power Conversion ConferenceTempe, AZ, 1976.

[2] J. Fernando, Surge Protection, CITEL, Technical Report, 2013 (General Catalog 9th edition).

[3] DeGregoria, UL1449 and the 2008 NEC significant changes have been made in the requirements for identifying and evaluating surge protective devices. 2008. Available from: ecmweb.com/power-quality/ul-1449-and-2008-nec. Accessed April 5, 2016.

[4] Semtech International, Surging ideas: TVS diode application note AN96-07. 2002. Available from: manualzz.com/doc/13226289/an96-07-surging-tvs-diode-application-note-protection-pro. Retrieved January 2018.

[5] Transient Voltage Surge Suppressors, Underwriters Laboratories, Northbrook, IL, 2014.

[6] IEEE PC62.41.2-2002/Cor 1/D2, IEEE Draft Recommended Practice on Characterization of Surges in Low-Voltage (1000 V and Less) AC Power Circuits—Corrigendum 1, 2012, 110 pp.

[7] IEEE Std C62.45-2002 (Revision of IEEE Std C62.45-1992), IEEE Recommended Practice on Surge Testing for Equipment Connected to Low-Voltage (1000 V and Less) AC Power Circuits, 2003, pp. 1–85.

[8] Eaton Group, Eatons guide to surge suppression, Eaton Corporation, Technical Report, 2009.

[9] F. Bohac, UL 1449-3rd edition effects on lightning protection systems, GE Industrial Solutions, Technical Report, 2010. Available from: apps.geindustrial.com/publibrary/checkout/UL-1449?TNR=White.

[10] K.H. Billings, Switchmode Power Supply Handbook, McGraw-Hill, New York, NY, 1999.

[11] N. Kularatna, Protection of systems from surges and transients, in: DC Power Supplies Power Management and Surge Protection for Power Electronic SystemsCRC Press, Boca Raton, FL, 2012 (Chapter 9).

[12] Surge and filter protection devices and applications, Rockwell Automation, Inc., Technical Report, 2008. Available from: literature.rockwellautomation.com/idc/groups/literature/documents/wp/4983-wp001-en-d.pdf.

Chapter 3

Components Used in Surge Protection Circuits

3.1 Introduction

A surge protector device (SPD) is like a security area before the entry into a large institution. The transient energy component piggy-backed on the power supply (or telecom signal) entry is absorbed and dissipated within the SPD, while the 50 or 60 Hz energy (or telecom signal) is passed though the SPD unattenuated towards the load side. As we discussed at the end of Chapter 1, in the simple example shown in Fig. 1.8, we combine several simple principles such as (i) voltage divider, (ii) passive filtering, (iii) use of significant nonlinearity of devices, and (iv) energy absorption and dissipation. Typical components used in SPDs are (a) transient-rated nonlinear devices (NLDs), (b) capacitors and inductors. During the period of 2010–17, supercapacitors' ability within R–C loops to absorb and help dissipate transient energy was also documented. This chapter is a discussion on traditional components usable in designing and building SPDs.

3.2 Practical Devices Used in Surge protection Circuits

In developing surge protection circuits, the designer has to ensure that: (1) surge arrestor stages do not alter the normal operation of the power conversion circuit, and (2) components used for surge absorption, diversion, or attenuation should be able withstand the surge. Reliability of the surge protection system is also important, because a very high-level surge might destroy the surge arrestor components.

There are two basic types of transient suppressors:

1. those that attenuate transients and block the propagation to the critical device; and
2. those that divert the transient away from the sensitive load while limiting the residual voltage.

Design of Transient Protection Systems. https://doi.org/10.1016/B978-0-12-811664-7.00003-3

The low-pass filters inserted in series with the critical load attenuate the high-frequency components of the transients and allow the low-frequency power flow without any disruption.

Diverting a transient can be accomplished with a voltage-clamping device or with a crowbar device that switch into a very low impedance mode to short circuit the transient. The designs of these two types, as well as their operation and application, are different [1]. In designing practical surge arrestor circuits, these NLDs are combined with inductors and capacitors.

A voltage clamping device has variable impedance depending on the current flowing through the device or on the voltage across its terminals. These devices exhibit a nonlinear impedance characteristic, that is Ohm's law applicable, but the equation has a variable *R*. The variation of the impedance is monotonic; in other words, it does not contain discontinuities unlike a crowbar device, which exhibits a turn-on action [2].

The most common semiconductor NLDs used are metal oxide varistors (MOVs), thyristors, and avalanche-type back-to-back zener diodes. Gas discharge tubes (GDTs) are also used in some systems. MOVs and avalanche type back-to-back zener diodes are voltage-clamping devices whereas thyristors and GDTs are classified as 'crowbar' devices.

Selection of these components depends on the location, clamping voltage, and the response time. As shown in Table 2.2, category C protection requires higher current capability than category A.

3.2.1 Carbon Block Spark Gap

This is the oldest and most common surge protector type used in power distribution and telecommunication systems. Two carbon blocks 80–150 μm apart [3] are used in these devices. One of the blocks is connected to the system ground and the other one is connected to either the power conductor or the signal cable. The transient energy is dissipated in the arc created between the two electrodes, in the resistance of the carbon blocks and also in the transient source impedance.

3.2.2 Metal Oxide Varistors

MOVs are zinc-oxide (ZnO)-based ceramic semiconductor devices with highly nonlinear current–voltage characteristics similar to back-to-back zener diodes [4]. They are typically formed into a disc shape. The structure of the body consists of a matrix of conductive ZnO grains separated by grain boundaries providing *p–n* junction semiconductor characteristics. Each intergranular boundary displays a rectifying action and presents a specific voltage barrier. When these conduct, they form a low ohmic path to absorb surge energy [5].

Fig. 3.1A shows a typical *V–I* characteristic of an MOV. In order to illustrate the three distinct regions of operation over a wide range of current, a log–log

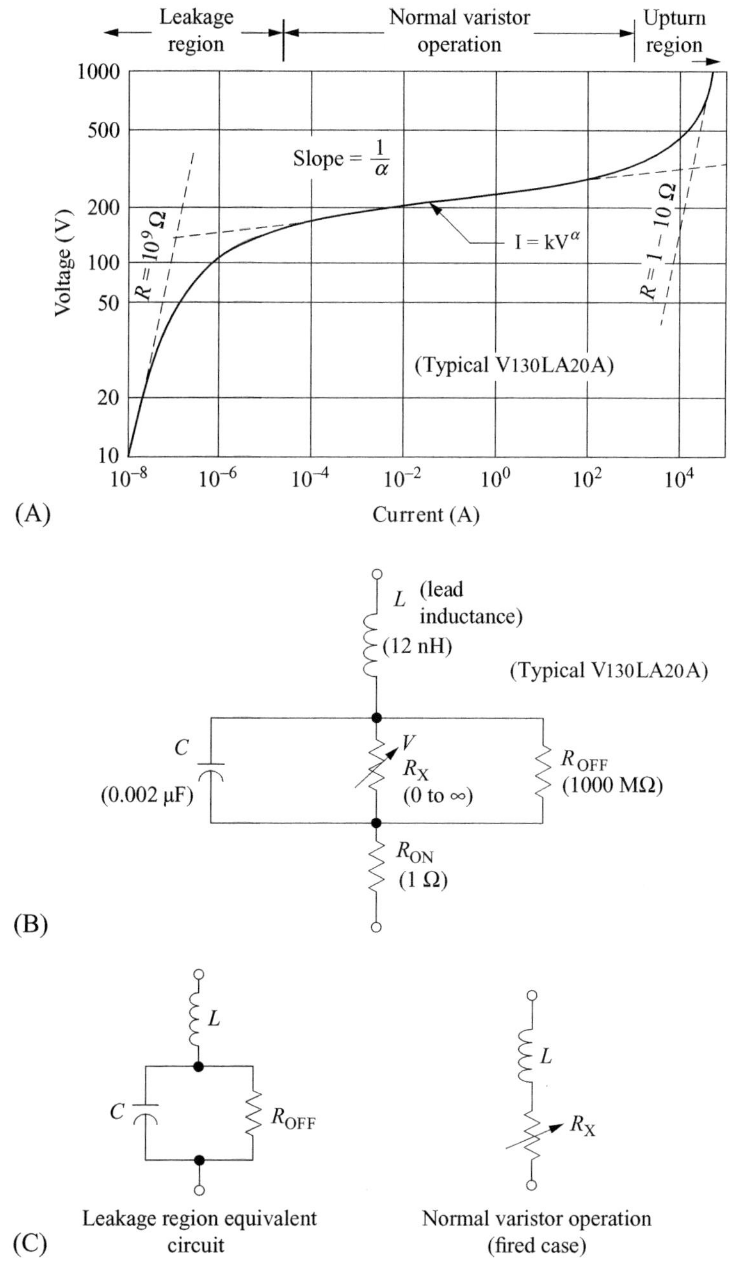

FIG. 3.1 Metal oxide varistor characteristic and equivalent circuits: (A) *V–I* characteristic; (B) generalized equivalent circuit; (C) simplified cases under leakage and conduction [6].

format is used. The basic electrical model for the varistor that could relate to all three regions of operation is shown in Fig. 3.1B [6]. At low currents the *V–I* curve approaches a linear ohmic relationship and shows a significant temperature dependence. Under this condition the MOV is in a very high resistance mode approaching about 1 GΩ or higher. Under this near open-circuit condition, the nonlinear resistance R_X in Fig. 3.1B can be ignored, as R_{OFF} value in parallel will dominate. The resulting equivalent applicable to the leakage region is shown in Fig. 3.1C. R_{OFF} value is dependent on the temperature, but remains in the range of 10–1000 MΩ. It also depends on the frequency in an inversely proportional manner. In the normal varistor operation region or in the conduction mode the value of R_X becomes many orders of magnitudes less than R_{OFF}. The effect of this is for the MOV to absorb much of the transient energy. The equivalent circuit applicable to normal varistor operation is also shown in Fig. 3.1C. At high currents, approaching the maximum rating, the nonlinear resistance is in a low-resistance mode that approximates a short circuit. This is represented by R_{ON} which is the bulk resistance of the zinc oxide grains and would be in the range 1–10 Ω [6].

In the state of the art in MOV modelling, different mathematical models are used for the nonlinear resistance R_X, which plays the major role in surge absorption by the varistor [6, 7]. Two different relationships that govern the operation of this nonlinear resistance are given by Eqs (3.1), (3.2), where i is the current through and v is the voltage across the varistor. The parameters k and α of Eq. (3.1) and parameters B_1, B_2, B_3, and B_4 of Eq. (3.2) are unique to each varistor type.

$$i(v) = kv^{\alpha}, \tag{3.1}$$

$$\log(v) = B_1 + B_2 \log_{10}(i) + B_3 e^{-\log_{10}(i)} + B_4 e^{\log_{10}(i)}. \tag{3.2}$$

The power-law representation given in Eq. (3.1) which represents the MOV's nonlinear resistance in its 'normal varistor operation' region [4, 6, 8] is a versatile model of the MOV since almost all of the surge energies, which is of interest, are transferred when the MOV operates in this region. The parameters k and α of Eq. (3.1) can be experimentally found by curve-fitting for any available MOV as shown in Chapter 5; k is dependent on the device geometry and the exponent defines the degree of nonlinearity in the resistance characteristic and can be controlled by selection of materials and the manufacturing process. A high α-value implies a stronger clamping action. For ZnO technology, typical values lie in the range of 15–30.

The mathematically more complex, highly nonlinear log representation given by Eq. (3.2) represents the MOV's nonlinear resistance from a few microampere (leakage region) to tens of kiloampere in the conducting region [7]. As shown in Chapter 5, modelling of the leakage region can be ignored for the energies to be estimated. The significance of right-hand side coefficients of Eq. (3.2) is provided in Section 4.2.9 of [9], where the characterization of an MOV based on the model of Eq. (3.2) is studied. A weakness of this representation would be the difficulty in estimating the coefficients by a suitable curve-fit for an off-the-shelf MOV. Varistor manufacturers such as *Littelfuse* have provided Simulation Program with Integrated Circuit Emphasis (SPICE)-based models for some of the varistors manufactured by them; in these models the nonlinearity of the MOV characteristic has been represented by extended versions of Eq. (3.2), with up to six terms [10].

High transient energy absorbent capability is achieved by increasing the size of the disc. Typical diameters range from 3 to 20 mm. MOVs turn on in a few nanoseconds and have high clamping voltages, ranging from approximately 30 V to 1.5 kV. Fig. 3.2 compares the behaviour of an MOV with an ideal clamping device, and with a TVS zener diode. Application guidelines on MOVs are given in [11, 12]. Basic theory of operation of ZnO varistors can be found in [4, 13].

Subjecting an MOV to continuous abnormal voltage conditions, rather than short-duration transients, may cause the MOV to go into thermal runaway, resulting in overheating, smoke, and possible fire. To prevent this condition, many modern MOVs include an internal thermal fuse or a thermal cutoff device. Some even extend this capability with an internal indicator that provides a logic output if the device's thermal protection is engaged [14]. An overview comparison of thermally protected MOVs (TPMOV) with traditional MOVs is provided in [15]. These TPMOVs are useful in occasions where continuous AC line overvoltages could occur.

3.2.3 TVS Diodes and Thyristors

TVS diodes and TVS thyristors are two basic types of *p–n* junction devices available for transient surge protection. Both types are available as back-to-back devices (bidirectional devices) suitable for protection against positive or negative surges appearing at power or signal entry inputs. The TVS diode is a clamping device, which suppresses all voltages above its breakdown voltage whereas the TVS thyristor is a crowbar device which switches on when overvoltages rise up to the break-over voltage [16].

Silicon rectifier technology, designed for transient suppression, has improved the performance of regulator-type zener diodes to provide TVS diodes. The major advantage of these diodes is their very effective clamping, which comes closest to an ideal constant voltage clamp [1].

Fig. 3.3 provides an example of a commercial TVS diode device manufactured by ON Semiconductor [17]. As shown in Fig. 3.3A, the *V–I* characteristic curve of the TVS diode is similar to that of a zener diode. Unlike zener diodes, TVS diodes are specifically designed to clamp transient surge pulses, rather than regulate a relatively stable voltage. Like all clamping devices, it automatically resets when the transient drops below the breakdown voltage, but absorbs much more of the transient energy internally than a similarly rated TVS

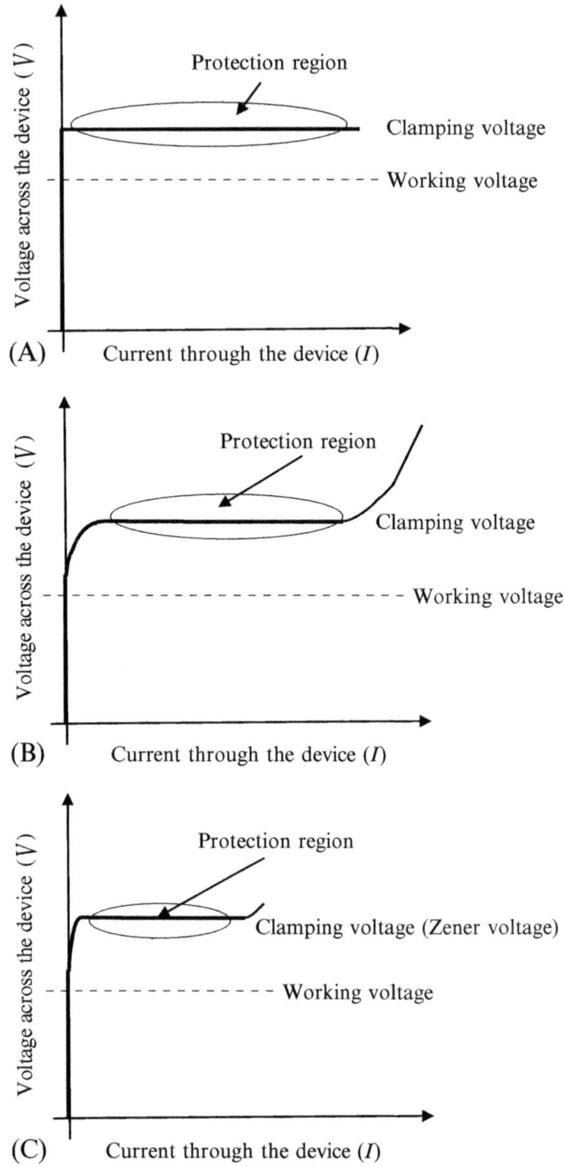

FIG. 3.2 Comparison of clamping type transient protection devices: (A) ideal case; (B) zinc-oxide varistor; (C) TVS zener diode.

thyristor device. A large cross-sectional area is employed for a TVS diode junction, allowing it to conduct high transient currents. These diodes respond almost instantaneously to transient events. Their clamping voltage (which is the voltage measured across the device terminals when the device is fired, and a peak

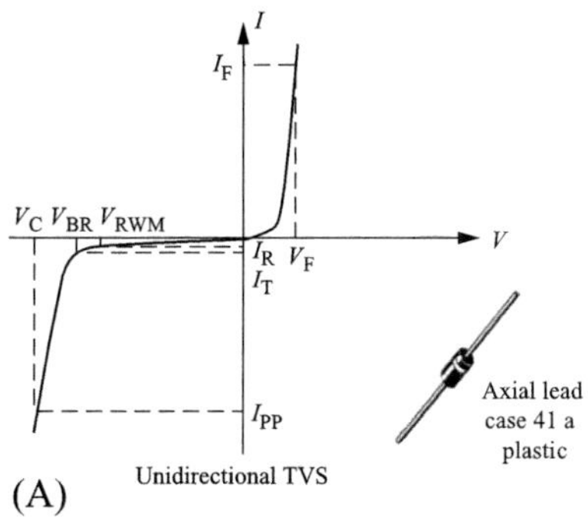

Symbol	Parameter
I_{PP}	Maximum reverse peak pulse current
V_C	Clamping voltage @ I_{PP}
V_{RWM}	Working peak reverse voltage
I_R	Maximum reverse leakage current @ V_{RWM}
V_{BR}	Breakdown voltage @ I_T
I_T	Test current
ΘV_{BR}	Maximum temperature coefficient of V_{BR}
I_F	Forward current
V_F	Forward voltage @ I_F

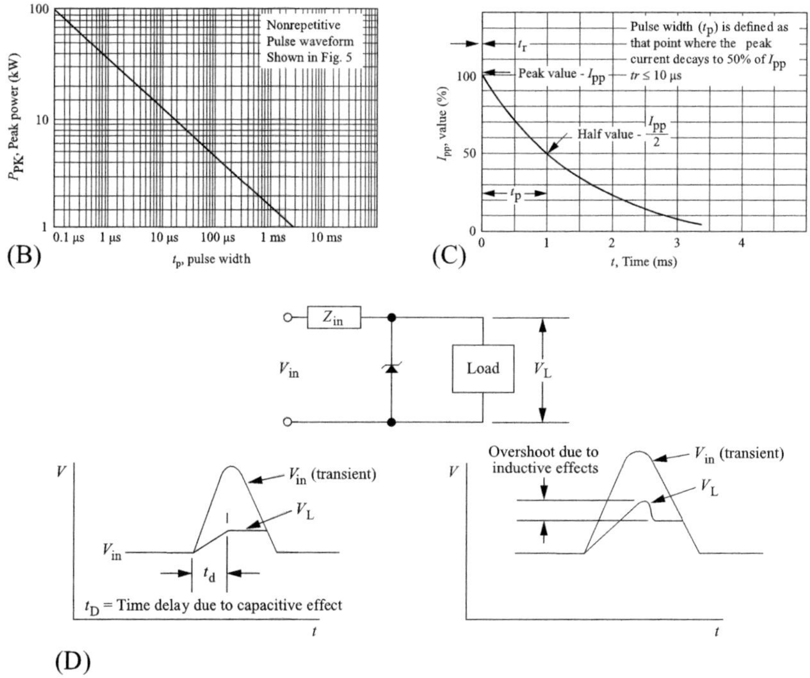

FIG. 3.3 An example of a TVS diode: (A) device behaviour and datasheet parameters; (B) pulse power absorption capability versus pulse width; (C) test pulse used; (D) typical application circuit and voltage at protected load due to device capacitance and lead inductance [17].

pulse current is flowing through the device) ranges from a few volts to a few hundred volts.

A TVS diode can respond to transients faster than other transient protection components such as MOVs and GDTs. Clamping occurs in picoseconds. The fast response time of the TVS diodes means that any voltage overshoot, as shown in Fig. 3.3D, is primarily due to lead inductance. Bidirectional TVS diodes available from several manufacturers have symmetrical response in both positive and negative regions and offer similar clamping to both positive and negative surges. They come in axial leaded and surface mount packages. Datasheet terminology for silicon TVS devices and design calculations are discussed in [18].

When individual excessive surges occur beyond the rating of the TVS diode, it can fail like any other stressed semiconductor component. For the TVS diode, this primarily involves peak pulse power (P_{PP}) and/or peak pulse current (I_{PP}). The basic form of failure mechanism is attributable to excessive heat in the active *p–n* junction of the silicon element [19].

A TVS diode macromodel offering greater accuracy than a standard diode SPICE model is discussed in [20].

TVS thyristors are solid-state devices constructed with four alternating layers of *p*-type and *n*-type materials. The advantage of this device is its low on-state voltage after triggering which provides much higher-current capability for a relatively small chip size compared with a clamp device. TVS thyristors are used extensively in surge protection of communication equipment for the reasons given next.

Because communication equipment with large-scale integrated circuits is very sensitive to surges, its surge protection devices are required to have a high-speed response and a high-surge capability. These devices are also be required to have low capacitance to reduce transmission losses in high-speed wideband transmission systems. Conventional lightning surge protectors, with a GDT, resistors and ceramic varistors, cannot be used in high-speed transmission systems because they cause excessive transmission loss. TVS thyristors have been used for surge protection because of their high per-unit area surge capability. This characteristic enables reduction of device size and hence also its capacitance and transmission loss [21].

TVS thyristors are avalanche triggered components and protect sensitive loads by switching to a low on-state of a few volts, thus providing a 'crowbar' effect with high-current capability. Transition to on-state is initiated at the maximum breakover voltage of the device. Once the TVS thyristors are in the on-state, the current through the device must be interrupted or drop below minimum holding current to restore to nonconduction after the transient has subsided. They are normally not used for protection across DC power or low impedance voltage sources because they may not restore to nonconduction. A TVS diode would be a better choice for such applications.

The surge current capability of a TVS thyristor is determined by structure as well as size. Operating voltage levels begin at 12 V and are available in several increments up through several hundred volts. They do not wear out, but when electrically overstressed fail as a short-circuit [22]. Fig. 3.4 provides an example of a commercial device from ON Semiconductor, with a typical telecommunication outside-plant protection circuit [23].

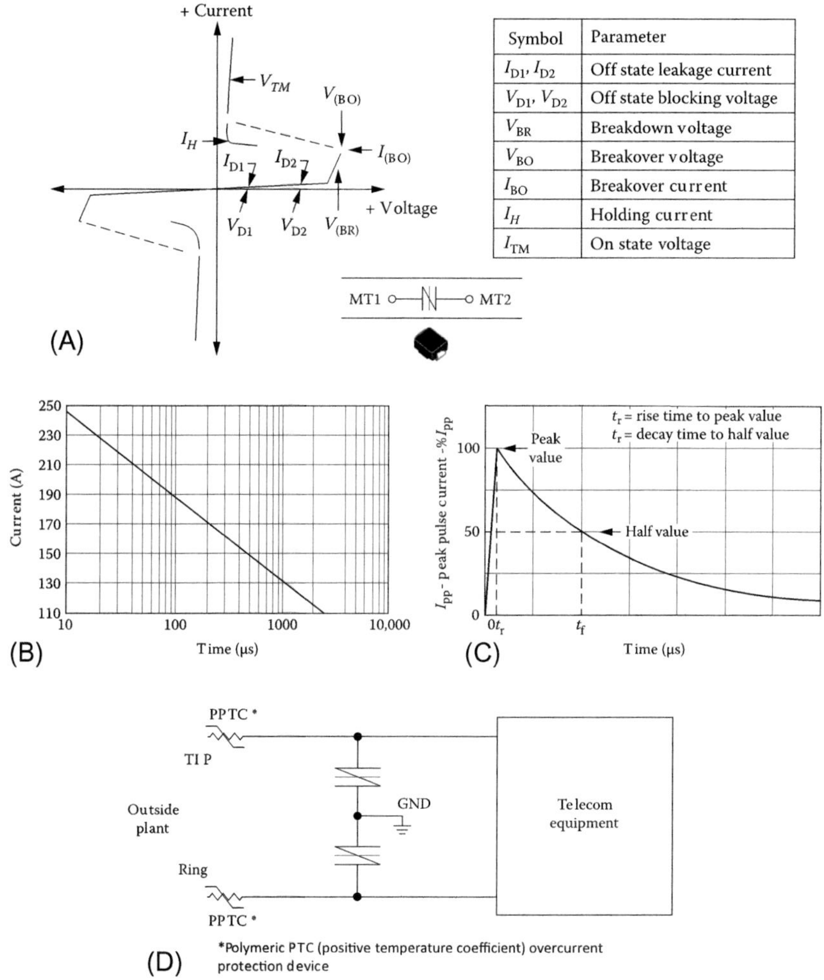

Symbol	Parameter
I_{D1}, I_{D2}	Off state leakage current
V_{D1}, V_{D2}	Off state blocking voltage
V_{BR}	Breakdown voltage
V_{BO}	Breakover voltage
I_{BO}	Breakover current
I_H	Holding current
I_{TM}	On state voltage

FIG. 3.4 An example of a thyristor breakover device: (A) device behaviour and datasheet parameters; (B) pulse power absorption capability versus pulse width; (C) test pulse used; (D) typical application circuit in telecommunication plant with series current limiting device [23].

3.2.4 Gas Discharge Tubes

GDTs (or simply gas tubes) are devices that employ an internal inert gas that ionizes and conducts during a transient event. Because the internal gas requires time to ionize, gas tubes can take several microseconds to turn on or 'fire'. In fact, the reaction time, and firing voltage are dependent on the slope of the transient front. A circuit protected by a gas tube arrestor will typically see overshoot voltages ranging from a few hundred volts to several thousand volts.

The GDT uses specially designed electrodes fitted inside a tube filled with one or more gases under pressure. They are rugged, relatively inexpensive, and have a small shunt capacitance; therefore, they do not limit the bandwidth of high-frequency circuits as much as other nonlinear components [24].

GDTs have extremely high pulse ratings for their size. They are designed for use over a broad voltage spectrum. Impulse striking (firing) voltages typically start at 500 V; these firing voltages are too high to protect microchips but GDTs can be used in conjunction with TVS diodes to provide effective multikiloampere, low clamping level for 5 V signal lines [25]. They are used in telecom network interface device boxes and central office switching gear to provide protection from lightning and alternating current power cross faults on the telecom network [26].

Table 3.1 compares the characteristics of the most widely used TVS devices. References [27, 28] provide more details.

3.3 Capacitors and Inductors

Practically all typical SPDs have LC filters combined with NLDs such as MOVs and bipolar breakdown devices (BBDs) to more effectively divert and absorb the surge energy. If we consider the surge as a 5–100 μs transient event having shapes discussed in Chapter 2, we can model the transient as a combination of sinusoids of frequencies in the range of 10–200 kHz, plus their higher-order harmonics. Now, if we consider the impedance of an inductor given by $2\pi fL$, and the impedance of a capacitor given by $1/2\pi fC$ we can see their value in filtering high-frequency components of a surge superimposed on a 50 or 60 Hz power frequency waveform.

For example, a 10-μH inductor will show an impedance of 3.14 mΩ at 50 Hz, compared with an impedance of 6.28 Ω at 100 kHz. If an inductor of 10 μH is used in the series path of a typical surge protector as in Fig. 1.8, it will impede surge related frequencies 20,000 times more than the 50 Hz power frequency.

A similar argument can be put forward with a capacitor placed in a near-parallel path with the line and neutral of the AC supply.

3.3.1 X and Y Capacitors in Filters

The capacitors used in an electromagnetic interference (EMI) filter stage are of special type [29] and hence referred as X and Y. These capacitors are specially designed considering the safety of the equipment and the user to avoid electric shock. Like normal capacitors, the capacitive reactance of these capacitors is inversely proportional to the frequency. The combination of inductors and capacitors determines the pass band of the EMI filter.

The class X filter capacitors are used across the power lines. Failures of X-type capacitors create a short circuit and cause line fuse or circuit breaker

TABLE 3.1 IEEE C62.41 Location Categories, Frequency of Occurrences, and Surge Waveforms

Suppression Element	Advantages	Disadvantages	Expected Life
Gas tube	Very high current-handling capability	Very high firing voltage	Limited
		Finite life cycle	
	Low capacitance	Slow response times	
	High insulation resistance	Nonrestoring under DC	
MOV	High current-handling	Gradual degradation	Degrades
	Capability	High clamping voltage	
	Broad current spectrum	High capacitance	
	Broad voltage spectrum		
TVS diode	Low clamping voltage	Limited surge current rating	Long
	Does not degrade	High capacitance for low-voltage types	
	Broad voltage spectrum		
	Extremely fast response time		
TVS thyristor	Does not degrade	Nonrestoring under DC	Long
	Fast response time	Narrow voltage range	
	High current-handling capability	Turn-off delay time	

opening. Because these are connected across the live and neutral lines, a failed component does not lead to the risk of an electric shock.

The class Y filter capacitors are normally used in 'line-to-ground' (line bypass) applications. Because the chassis of the equipment is connected to the ground wire, a short-circuited Y capacitor could expose the chassis to the live connection. If there is no proper ground connection, the 'hot' chassis could cause fatal shock. Hence class Y capacitors are designed to open when they fail.

The voltage rating of X and Y capacitors should be more than double the expected voltage in the application. These capacitors are grouped into different classes, X1, X2, X3, Y1, Y2, Y3, and Y4. Type X1 or X2 and Y2 capacitors are used in most end-use equipment. X1, X2, Y1, and Y2 capacitors are impulse tested to 4, 2.5, 8, and 5 kV, respectively.

3.3.2 Inductors

A superimposed surge on a power line or a telecom line input is a short duration, nonrepetitive occurrence with time durations of 1–50 μs. If a simple inductor is connected with a series resistance, and a standard surge voltage waveform is applied to the circuit, it will create a loop current which is governed by the voltage across the inductor given by $L \cdot di/dt$ added to the ohmic drop in the loop. With standard voltage test waveforms of microsecond order rise time, these loop effects allow a designer to use an inductor or a coupled inductor in a surge protector to attenuate the effect of surge propagation towards the load side. If we consider the harmonics of a surge pulse with frequency components in the range of 20 kHz to a few megahertz, an inductor presents a series impedance of $2\pi fL$, where L is the inductance and f is the frequency component of the surge transient pulse, superimposed. Compared to the impedance of the inductor at 50 or 60 Hz line frequency, to the surge components the inductor will show 400–20,000 higher impedance. Given this basic property, coupled with resistors, capacitors, and the NLDs such as MOVs and BBDs, useful surge blocking circuits can be achieved. In almost all cases, inductors are inserted in series and therefore they are expected to carry the line frequency current fed into load, without creating a major voltage drop at 50 or 60 Hz.

In many practical applications, coupled inductors in the form of common or differential mode chokes are practically used in surge protectors. When these are coupled with suitably sized capacitors they work as high-frequency filters to attenuate the propagation of RFI/EMI components across the surge protector.

References

[1] Littelfuse, Transient suppression devices and principles: application note AN9768. (1998) www.digikey.co.nz/Web%20Export/Supplier%20Content/Littelfuse_18/PDF/LF_TransientSuppressionDevices.pdf?redirected=1Retrieved November 2012.

[2] R. Chundru, L. Zhen, D. Pommerenke, K. Keong, L. Cheung-Wei, F. Centola, R. Steinfeld, An evaluation of TVS devices for ESD protection, in: 2011 IEEE International Symposium on Electromagnetic Compatibility (EMC), pp. 62–67.

[3] R.B. Standler, Protection of Electronic Circuits From Overvoltages, Dover Publications, New York, NY, 1989.

[4] L.M. Levinson, H.R. Philipp, The physics of metal oxide varistors, J. Appl. Phys. 46 (1975) 1332–1341.

[5] B. van Beneden, Varistors: ideal solution to surge protection, Power Electron. Technol. 29 (2003) 26–32.

[6] Littelfuse, Littelfuse varistors—basic properties, terminology and theory: application note AN9767.1. (1998). www.digikey.co.nz/Web%20Export/Supplier%20Content/Littelfuse_18/PDF/LF_Varistors.pdf?redirected=1. Retrieved November 2012.

[7] B. Zitnik, M. Babuder, M. Muhr, R. Thottappillil, Numerical modelling of metal oxide varistors, in: Proceedings of the XIVth International Symposium on High Voltage Engineering Tsinghua University, Beijing, China, 2005, pp. 1–6.

[8] G.Z. Zang, J.F. Wang, H.C. Chen, W.B. Su, C.M. Wang, P. Qi, Nonlinear electrical behaviour of the WO3-based system, J. Mater. Sci. 39 (2004) 4373–4374.

[9] S. James, Investigation of Surge Propagation in Transient Voltage Suppressors and Experimental Verification (Ph.D. thesis), University of Waikato, 2014.

[10] Littelfuse, Varistor SPICE models. (2003). www.littelfuse.com/technical-resources/spice-models/varistor-spicemodels.aspx. Retrieved May 2013.

[11] Littelfuse, The ABCs of MOVs: application note AN9311.6. (1999). www.secomtel.com/UpFiles/Attach/0/2007/5/17/141245.pdf. Retrieved November 2012.

[12] D. Tidey, The ABCs of Littelfuse mutilayer suppressors: application note AN9671.2, dev.littelfuse.com/data/en/Application_Notes/an9671.pdf, 1999. Retrieved November 2012.

[13] G.D. Mahan, L.M. Levinson, H.R. Philipp, Theory of conduction in ZnO varistors, J. Appl. Phys. 50 (1979) 2799–2812.

[14] B. Walaszczyk, Multiple protection devices guard against transients, Power Electron. Technol. 28 (2002) 55–65.

[15] D. Dunlap, Thermally protected MOVs resist overvoltage failures, PCIM Power Electronic Syst. 27 (2001) 64–68.

[16] STMicroelectronics, Transil/trisil comparison. (2004). pdf1.alldatasheet.com/datasheetpdf/view/97696/STMICROELECTRONICS/AN574.html. Retrieved June 2013.

[17] O.N. Semiconductor, Transient voltage suppression devices: data sheet for 1N6267A series, (2001). www.icbase.com/file/pdf/add/on/DG-106-001en.pdf. Retrieved November 2012.

[18] M. Clark, Expand the data sheet on silicon transient voltage protection devices, PCIM Power Electron. Syst. 26 (2000) 42–44.

[19] K. Walters, Failure mechanisms for transient voltage suppressors: microNote 135, (2011). www3.microsemi.com/micnotes/135.pdf. Retrieved June 2013.

[20] J. Lepkowski, W. Lepkowski, Evaluating TVS protection circuits with SPICE, Power Electron. Technol. 32 (2006) 44–49.

[21] H. Satoh, Study on increasing the surge capability of a lightning surge protection, semiconductor device, IEEE Trans. Electromagn. Compat. 35 (1993) 311–315.

[22] K. Walters, M. Clark, What is a thyristor surge protective device and how does it work? Available from: www.microsemi.com/sites/default/files/micnotes/105.pdf. Retrieved June 2013.

[23] O.N. Semiconductor, Thyristor surge protectors, data sheet for MMT08B310T3—Rev. 3, (2005). www.onsemi.com/pub_link/Collateral/MMT08B310T3-D.PDF. Retrieved November 2012.

[24] K. Samaras, C. Sandberg, C.J. Saimas, A. Koulaxouzidis, Electrical surge protection devices for industrial facilities—a tutorial review, in: Industry Applications Society 52nd Annual Petroleum and Chemical Industry Conference, 2005, pp. 165–175.

[25] O.M. Clark, R.E. Gavender, Lightning protection for microprocessor based electronic systems, in: Conference Record of the 1990 IEEE, Industry Applications Society Annual Meeting, vol. 2, 1990, pp. 1460–1466.

[26] R. Rosen, W. Simendinger, C. Debbault, H. Shimoda, L. Fleming, B. Stoner, O. Zhou, Application of carbon nanotubes as electrodes in gas discharge tubes, Appl. Phys. Lett. 76 (2000) 1668–1670.

[27] W. Russell, Transients vs electronic circuits: survival of the fittest—part I: the need for transient protection, Powerconv. Intell. Motion 22 (1996) 66–71.

[28] O.M. Clark, Transient voltage suppressor types and applications, Powerconv. Intell. Motion 16 (1990) 19–21.

[29] P.F. Keebler, Chokeless power-line filters with integral surge protection, in: IEEE International Symposium on Electro-magnetic Compatibility (EMC), 2011, pp. 184–189.

Chapter 4

Supercapacitors for Surge Absorption

4.1 Introduction

Surge protectors are typically based on a combination of nonlinear devices such as MOVs and BBDs with inductors and capacitors, as discussed in previous chapters. Within the last 15 years a uniquely new capacitor family, namely electrical double-layer capacitors, commonly referred to as supercapacitors or ultracapacitors, have emerged into the electronic component world. These typically have one million times bigger capacitance for the same volume of an electrolytic or a film capacitor. However, supercapacitors have very low-voltage DC ratings, in the range of 2–4 V only.

If we connect a 1 Ω resistor in series with a capacitance of 1 μF the circuit time constant will be 1 μs. If we apply a step DC voltage source of a specific value to the circuit, the capacitor will reach the source voltage within an approximate time of five time constants. During that time the capacitor will acquire an energy content of $0.5CV^2$ while the loop resistor will dissipate an equal amount of energy.

However, if the DC voltage source persists for a much shorter time than the circuit time constant, the capacitor will not reach the source voltage and most of the energy flowing through the circuit will be wasted in the resistance in the loop. For example if we build an RC circuit with a 1 F supercapacitor and a 1 Ω resistor, and apply a step DC source of 5000 V for a time period of say 100 μs, a simple calculation will show that energy dissipated in the loop resistance will be very much larger than the energy stored in the supercapacitor within the duration of the step DC voltage. A quantification on this will be given later in the chapter.

Based on a unique research project by several authors of this book, presented in Appendix A, this chapter describes the capability of supercapacitors to absorb short duration, high-voltage transient surges. The classical *RC* model of the SC is used to show how SCs with extremely low ESR and very high capacitance can survive short duration, high-voltage transient surges without being overcharged as can other types of (electrolytic and ceramic) capacitors. Based on

Design of Transient Protection Systems. https://doi.org/10.1016/B978-0-12-811664-7.00004-5

experimental observations, an electrical circuit model for the surge behaviour of SCs is also proposed. In the surge model of an SC, the ESR in the classical *RC* model is replaced by a surge-specific ESR called the surge ESR. In Chapter 9, we present the case of designing an SC-based surge protector, based on a patented technique called supercapacitor-assisted surge absorber (SCASA).

4.2 Maximum Current Capability of a Supercapacitor

Due to their low ESR and high power density, SCs can be charged and discharged at much higher currents than other energy storage media. As shown in Table 4.1, SC manufacturers recommend a maximum continuous operating current for each capacitor based on the thermal limit of the internal materials. The theoretical maximum output current, V_{rated}/ESR, is the current which flows when a fully charged capacitor is short circuited by an ideal zero resistance. Table 4.1 illustrates the distinction between these two current ratings.

The manufacturer recommended operating maximum current is a continuous current which can be used repeatedly without damaging the internal electrodes. Overheating of SCs can occur from continuous overcurrent or overvoltage charging. Overheating can lead to increased ESR, gas generation, decreased lifetime, leakage, venting, or rupture. Because SCs are traditionally used for quick power delivery and fast energy storage systems, high currents for time spans greater than few seconds can generate considerable heat through the power dissipation in the ESR. This can cause the maximum operating temperature of 60°C (typical) to be exceeded. The temperature rise, ΔT, of an SC when current, i, flows through the SC is

TABLE 4.1 Comparison of Energy Storage and Delivery Capabilities of 2.7 V Supercapacitors by Maxwell Technologies, Inc.

Capacitance (F)	DC ESR (mΩ)	V_{rated}/ESR (A)	Maximum Rated Current (A)	Energy Storage Capacity $0.5CV^2$ Based (J)
1	700	3.9	0.8	3.6
3.3	290	9.3	1.9	12.0
5	170	15.9	3.2	18.2
10	75	36	7.2	36.4
25	42	64.3	20	91.1
50	20	135	27	182.2
100	15	180	36	364.4
150	14	192	40	546.8

TABLE 4.2 Comparison of Temperature Rise, ΔT, HC Series Supercapacitors by Maxwell Technologies for a 2000 A Peak Current of 8/20 μs Combination Waveform

Capacitance (F)	C_{th} (J/°C)	ESR (mΩ)	ΔT (°C)
1	1	700	76.6
3.3	1.4	290	22.7
5	2	170	9.3
10	3.6	75	2.3
25	6.3	42	0.7
50	13	20	0.2
100	23	15	0.07
150	32	14	0.05

$$\Delta T = \frac{1}{C_{\text{th}}} \int_0^\infty i^2 \; R_{\text{C}} \; dt, \tag{4.1}$$

where R_C is the ESR of the SC and C_{th}(with units J/°C) is the thermal capacitance of the SC. For the 8/20 μs combination current waveform given in Eq. (2.2) the temperature rise will be

$$\Delta T = \frac{R_{\text{C}} \; A_{\text{I}}^2 \; I_{\text{p}}^2}{C_{\text{th}}} \int_0^\infty t^6 e^{-2t/\tau} dt = \frac{45}{8} \frac{1}{C_{\text{th}}} R_{\text{C}} \; A_{\text{I}}^2 \; I_{\text{p}}^2 \; \tau^7. \tag{4.2}$$

According to Table 4.2, for larger SCs, as the ESR gets smaller the resulting temperature rise for a single surge reduces drastically. This temperature rise is calculated assuming perfect insulation. In a real-world application the effect is further reduced as the heat dissipates to the environment. Therefore, based on the above preliminary calculations, it is evident that an SC will not be thermally damaged by a single high-voltage transient.

It will be seen at a later stage of this chapter that the effective ESR shown by an SC to surge is much smaller than the datasheet value. Hence the calculated ΔT rise will be further reduced to almost negligible values.

4.3 Behaviour of an *RC* Circuit Subjected to a High-Voltage Transient

To analyze the SC voltage rise, the stored energy in the SC, and the energy dissipation in the ESR, consider applying a short duration, high-voltage square pulse to an SC. The SC is modelled as a simple series *RC* branch for simplicity.

The surge voltage of an SC [1] is defined in datasheets as the maximum voltage that the SC can withstand for few seconds without irreversible damage to the ESR, life time or cell opening. It has been observed experimentally that high-voltage transient surges with peaks of several thousand volts cannot destroy an SC even with repeated surges (see Appendix A). For a detailed illustration of this phenomenon, consider a situation where a high-voltage rectangular pulse with voltage, $V_{\max}$, is applied to a simple RC network for a duration T, starting from $t = 0$.

The input source, v_{in}, with path resistance, R_p, is connected to capacitor, C, with internal resistance, R, as shown in Fig. 4.1A. The voltage rise and the current decay in an ideal capacitor when connected to an ideal voltage source are described by Eq. (4.3).

$$v_{ch}(t) = E(1 - e^{-t/CR}), \tag{4.3a}$$

$$i_{ch}(t) = \frac{E}{R} e^{-t/CR}. \tag{4.3b}$$

Applying Kirchhoff's voltage law to the circuit in Fig. 4.1A using

$$v_{in} = iR + iR_p + \frac{1}{C}\int_0^t i\,dt, \tag{4.4}$$

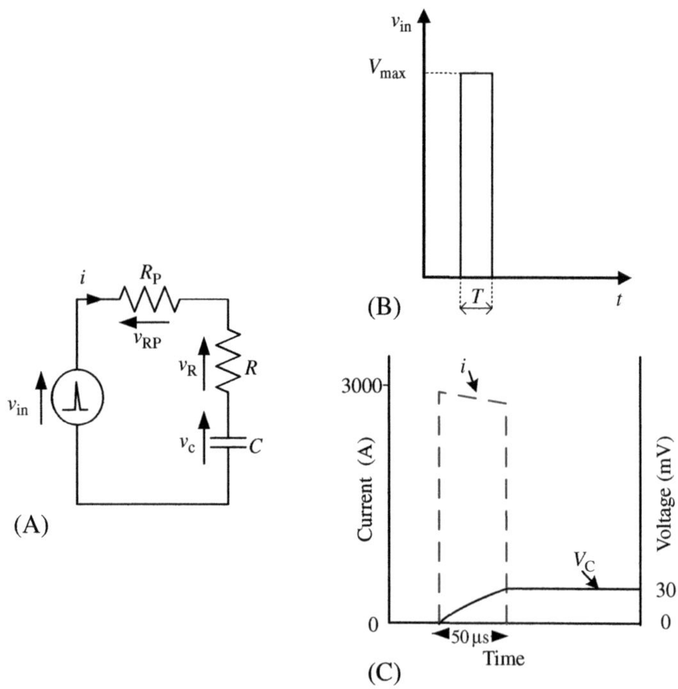

FIG. 4.1 Applying short duration pulse on supercapacitor: (A) step voltage on RC circuit; (B) applied step waveform; (C) resulting current and voltage waveforms for the supercapacitor.

$$i(t:\ 0<t<T) = \frac{V_{\max}}{R_{\mathrm{T}}}\ e^{-t/CR_{\mathrm{T}}}, \tag{4.5}$$

$$v_{\mathrm{c}}(t:\ 0<t<T) = V_{\max}\left(1-e^{-t/CR_{\mathrm{T}}}\right), \tag{4.6}$$

where $R_{\mathrm{T}} = R + R_{\mathrm{p}}$ and $V_{\max}$ is the pulse voltage. We also assume that the capacitor is initially fully discharged.

Energy dissipated in the capacitor ESR, E_{R}, is

$$E_{\mathrm{R}} = \int_0^T i^2 R\, dt = \frac{CR}{2R_{\mathrm{T}}} V_{\max}^2 \left(1-e^{-2T/CR_{\mathrm{T}}}\right). \tag{4.7}$$

At time T, the capacitor has energy, E_{C},

$$E_{\mathrm{C}} = \frac{1}{2} C v_{\mathrm{c}}^2, \tag{4.8}$$

where $v_{\mathrm{c}} = V_{\max}\left(1-e^{-t/CR_{\mathrm{T}}}\right)$. Fig. 4.2 shows the voltage rise in several electrolytic and SCs for a 6 kV, 50 μs pulse. It is seen that only SCs can withstand the high-voltage surges, while the other types will overcharge and be destroyed by the single pulse. For example as per graph in Fig 4.2 with a logarithmic Y axis, a 100 μF capacitor will have to have a DC voltage rating of over 4500 V. By contrast, a 1 F SC could withstand such a surge of 6 kV, 50 μs duration, even if the SC has a DC voltage rating of only 1 V. The research publication in Appendix A gives a more detailed view with experimental results for SC families from three different SC manufacturers.

The reason that SCs do not charge up significantly during a brief high-voltage pulse is illustrated in Fig. 4.1C. In a practical SC charging circuit with a large capacitance, small ESR and path resistance will offer a high RC-time constant that limits the rate of charge accumulation in the SC. Hence most of the energy in the pulse is dissipated in the ESR and path resistance rather than charging the capacitor. This phenomenon can be further elaborated by comparing the ESR loss to the energy stored in the capacitor.

To compare how much energy is dissipated in the capacitor ESR compared to the energy stored in the capacitor, consider the ratio $E_{\mathrm{R}}/E_{\mathrm{C}}$,

$$\frac{E_{\mathrm{R}}}{E_{\mathrm{C}}} = \frac{\frac{1}{2}\frac{CR}{R_{\mathrm{T}}} V_{\max}^2 \left(1-e^{-2T/CR_{\mathrm{T}}}\right)}{\frac{1}{2} C V_{\max}^2 \left(1-e^{-T/CR_{\mathrm{T}}}\right)^2} \tag{4.9}$$

$$= \frac{R}{R_{\mathrm{T}}} \frac{\left(1+e^{-T/CR_{\mathrm{T}}}\right)}{\left(1-e^{-T/CR_{\mathrm{T}}}\right)}. \tag{4.10}$$

Consider a practical example of an uncharged 5 F SC with 170 mΩ ESR, connected to a 6 kV step voltage source for 50 μs. The path resistance is 2 Ω. Hence the final capacitor terminal voltage, v_{c} will be,

$$v_{\mathrm{c}} = 6\mathrm{kV}\left[1-\exp\frac{-50\ \mu s}{(0.170+2)\ \Omega\ \ 5\mathrm{F}}\right] = 27.6\mathrm{mV}.$$

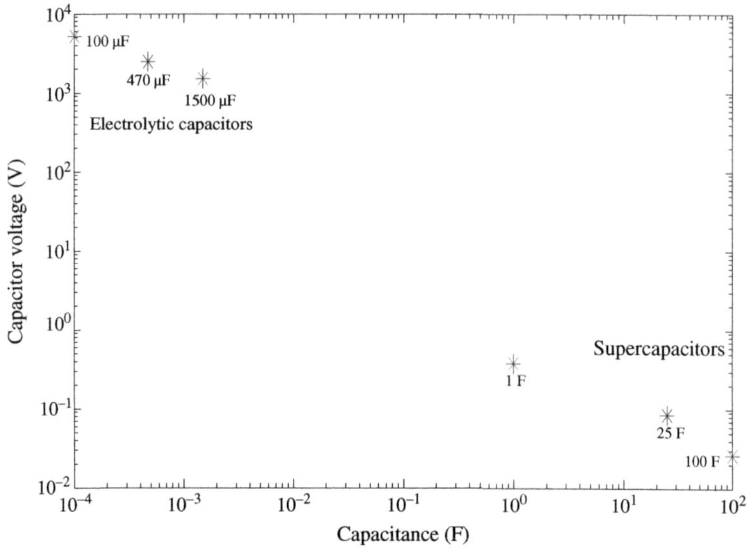

FIG. 4.2 The voltage build up on different capacitors for a 6 kV, 50 μs single pulse. Capacitors chosen are 100, 470, and 1500 μF electrolytic capacitors, and 1, 25, and 100 F SCs with 15, 100, 150, 700, 42, and 15 mΩ ESR, respectively.

During this short duration of 50 μ*s*, the energy stored in the SC is will be

$$E_C = \frac{1}{2}C\{v_c^2\} = \frac{1}{2} \times 5\text{F} \times 27.5\text{mV}^2 = 1.9\text{mJ}.$$

The energy lost in the capacitor ESR

$$E_R = \frac{170\text{m}\Omega \times 5\text{F}}{2 \times 2.170\Omega} \times 6\text{kV}^2 \left[1 - \exp\frac{-2 \times 50\mu\text{s}}{2.170\Omega \times 5\text{F}}\right] = 64.9\text{J}.$$

Hence

$$\frac{E_R}{E_C} = \frac{64.9\text{ J}}{1.9\text{ mJ}} = 3.4 \times 10^4.$$

This shows that a very small voltage is building up on the capacitor while storing small amounts of energy during the short duration of the high-voltage rectangular pulse input.

Fig. 4.3 shows a comparison of E_R/E_c ratio for several electrolytic and SCs. It can be observed that an SC develops only a minimal voltage rise across its terminals after a short duration transient surge, while dissipating transient energy in the ESR of the device. This is due to the difference in the characteristic time constant, τ_1 (i.e., the product of capacitance and ESR). The characteristic time constant is governed by the electrochemical properties and the construction of the capacitor. Hence an energy ratio of 10^3–10^4 variation can

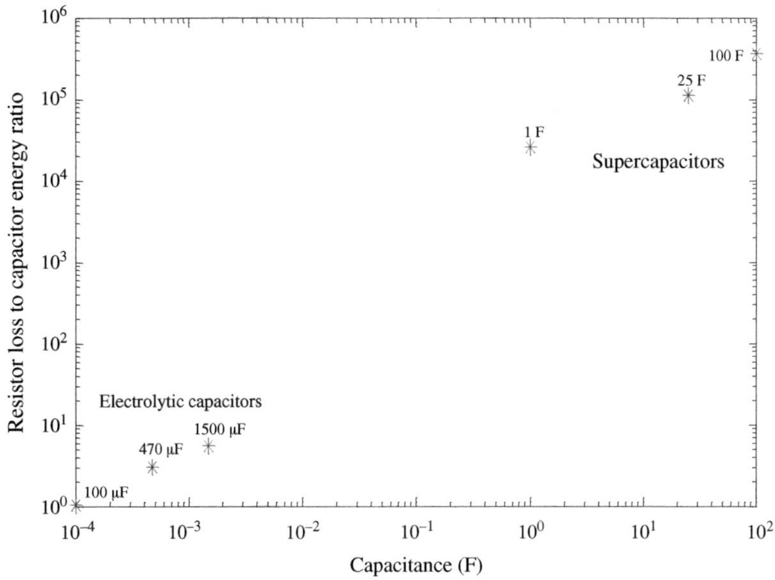

FIG. 4.3 Comparison of energy lost and stored for electrolytic capacitors and SCs subjected to a 6 kV, 50 μs pulse. Capacitors chosen are 100, 470, and 1500 μF electrolytic capacitors, and 1, 25, and 100 F SCs with 15, 100, 150, 700, 42, and 15 mΩ ESR, respectively.

be observed. This observation further affirms the previous conclusion of the inability of non-SCs to withstand high-voltage surges.

As a result of an SC charging up to a negligible voltage when a high voltage is applied for a short duration, the effect of capacitor voltage on the circuit current can be neglected. Further, because the ESR of an SC is very small, the short circuit current waveform (8/20 μs combination current waveform) discussed in Chapter 2, can be assumed for the current when a surge is directly applied to an SC.

4.4 Applying a High-Voltage Surge to a Supercapacitor

Now since the ability of SCs to withstand high voltages for short duration is theoretically indicated, the following discussion provides analytical and experimental proof for the case of applying consecutive lightning surges to an SC using a lightning surge simulator. For all experiments, high-voltage surges were generated from a NoiseKen lightning simulator; model LSS-6230. The test setup is shown in Fig. 4.4, which includes the lightning surge simulator and an oscilloscope. The SC is connected to the LSS through a pair of high-voltage test leads. The LSS has inbuilt 1000:1 attenuated outputs for surge voltage and current. When applying a surge to an SC, the SC voltage is measured using a Tektronix P5122 100:1 attenuated voltage probe. The oscilloscope a Tektronix

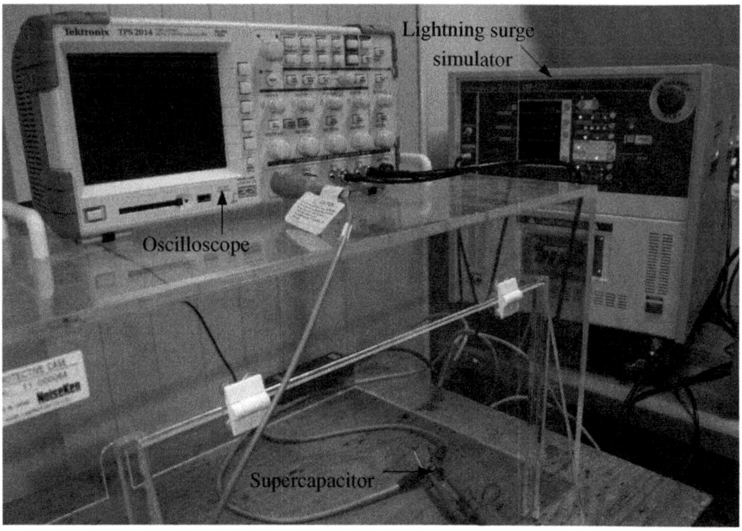

FIG. 4.4 Instruments and the test setup for supercapacitor surge testing.

TPS2014, is used for all waveform capturing; it has isolated channels to separately monitor LSS voltage, current, and SC voltage.

The lightning surge simulator can generate the standard combination waveform and ring waveform as described in Chapter 2. The peak output voltage of these waveforms can be varied from 100 V to 6.6 kV. The LSS is capable of generating up to 999 repetitive surges with a minimum 10-s interval. Appendix B provides technical details of the LSS indicated in Fig. 4.4 and some useful data for a few selected commercial supercapacitors usable in surge protectors which indicates their ESR values and other essential data.

For cross-validation of numerical simulation and actual test, the parameters of the mathematical equation specified in Eq. (2.3) were fine tuned to match the short-circuit current waveform generated by a 6 kV 8/20 μs surge. Because the positive peak of the surge waveform is of most importance to the equivalent circuit analysis, simulated and experimental waveform positive peaks were matched. The matched simulated and experimental waveforms are shown in Fig. 4.5.

4.4.1 Applying a Surge to an *RC* Circuit

When a current, i, passes through a capacitor, C, the voltage, v_c, built up across the capacitor is

$$v_c = \frac{1}{C}\int_0^t i\,dt. \tag{4.11}$$

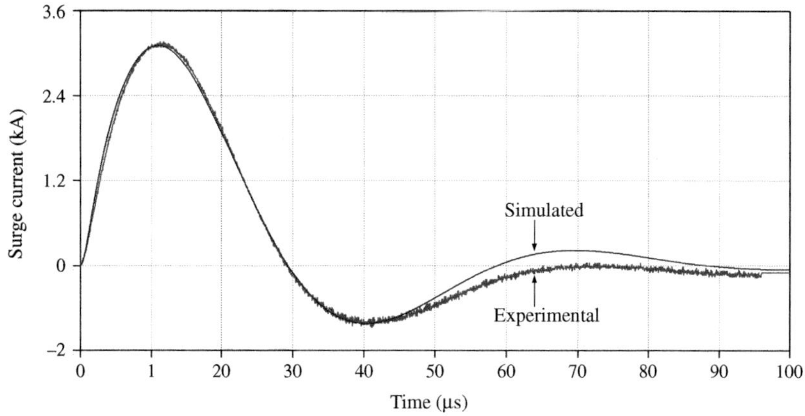

FIG. 4.5 Comparison of practical and simulated surge current waveforms based on Eq. (2.3).

This assumes the capacitor is initially fully discharged. As seen in Fig. 4.5, because the surge current relaxes to zero when $t \to \infty$, the total energy absorbed by the SC for a single surge can be expressed as

$$E_{\rm c} = \frac{1}{2} C V_{\rm c}^2, \tag{4.12}$$

where $V_{\rm c} = v_{\rm c}(t \to \infty)$. Substituting Eq. (2.3) in Eq. (4.11), gives

$$V_{\rm c} = \frac{A I_{\rm p} \omega \tau_1^2}{C} \left[\frac{1}{\omega^2 \tau_1 + 1} - \frac{\tau_1^2}{\omega^2 (\tau_1 + 1)^2 + \tau_1^2} \right]. \tag{4.13}$$

The energy dissipated in the ESR, R, is

$$E_{\rm R} = A^2 I_{\rm p}^2 R \int_0^\infty \sin^2(\omega t) e^{-2t/\tau_1} (1 - e^{-t})^2 dt. \tag{4.14}$$

Hence the ratio between the energy dissipated in the resistor to the energy stored in a capacitor for a 3000 A simulated lightning surge can be found as follows:

$$E_{\rm R} = 4.53 \times 10^6 A^2 I_{\rm p}^2 \; R, \tag{4.15}$$

$$E_{\rm C} = \frac{1}{2} C \left(8.19 \times \frac{A T_{\rm p}}{C} \right)^2 = 3.35 \times 10^7 \times \frac{A^2 I_{\rm p}^2}{C} \tag{4.16}$$

$$\therefore \quad \frac{E_{\rm R}}{E_{\rm C}} = 1.4 \times 10^5 \times RC. \tag{4.17}$$

Now consider the same 5 F SC with 170 mΩ ESR, used for the example in the previous section. For this case when $A = 1.4$, $I_{\rm P} = 3000$ A, $R = 170$ mΩ, and $C = 5$ F, and thus

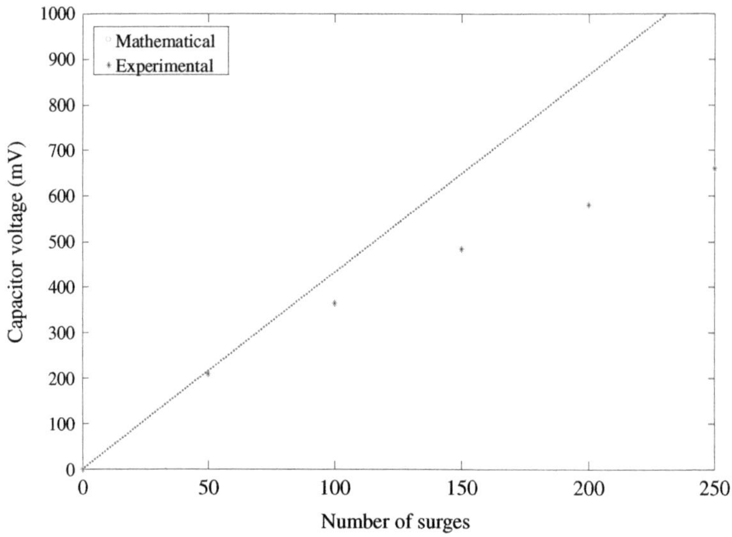

FIG. 4.6 Applying consecutive surges to a 1 F supercapacitor; experimental vs. simulated data.

$$\frac{E_R}{E_C} = 1.4 \times 10^5 \times 0.17 \times 5 = 11.9 \times 10^4. \quad (4.18)$$

Comparing this with the ratio obtained in the previous section for a square wave, 50 μs pulse as 3.4×10^4 for the same capacitor and peak surge current, it is 3.5 times higher for the lightning surge current waveform. This can be explained by the ringing wave shape in the experimental current waveform. During the positive cycle the capacitor will be charged and it will be discharged during the negative cycle. But in both cycles the resistor loss will be accumulated. Hence E_R is higher than for case of a unipolar square pulse.

Fig. 4.6 illustrates the effect of applying consecutive 6 kV 8/20 μs surges to a 1 F capacitor. Simulation data are based on Eq. (4.13). As the number of surges increases beyond 50, the mathematical estimate deviates from the experimental value. This discrepancy between simulated and experimental values is due to the internal construction of the double-layer electrode [2]. The simple *RC* model used for this analysis cannot reflect the actual electrical behaviour of SCs. This is because an activated carbon electrode has various pores. Ion movement is easy and quick into shallow pores, and slow into deep pores due to physical resistance. Therefore, multiple parallel R–C branches are required to interpret the different pore sizes.

Fig. 4.7 gives further details on applying different voltage surges to different capacitors. It can easily be seen that even after hundreds of consecutive LSS generated surges, the capacitor voltage rise is negligible in comparison to the surge voltage. Table 4.3 lists the surge absorption capability of SCs from different manufacturers and different sizes. The precharge voltage is set at several different settings to emulate different operating conditions.

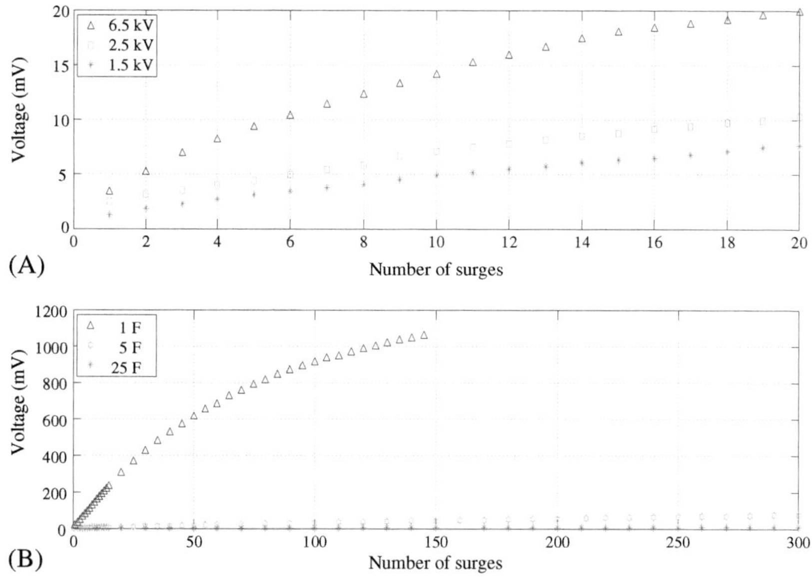

FIG. 4.7 Applying consecutive surges to supercapacitor: (A) different surge voltages to a 5 F capacitor; (B) 6 kV peak surges to different capacitors.

Up to this section, all previous discussions on the surge behaviour assumed the SC to be a simple resistor and a capacitor. Fig. 4.8A shows the experimental voltage and current waveforms across a capacitor when a lightning surge is applied to a 1-F SC. A clear shift between the voltage and the current waveforms can be observed in the figure. This can be simulated by using a series inductor in the equivalent capacitor *RC* model, as shown by the voltage and current waveforms in Fig. 4.8B. The simulation with an *RC* model has both the current and voltage peaks aligned, but the *RLC* model demonstrates a peak shift, Δt. 1 F capacitor with 50 mΩ resistance and 50 nH inductance in series is used for the *RLC* circuit and *RC* circuit used the same resistor and capacitor values for the simulation. The following section of this chapter will quantify the *RLC* equivalent circuit model components using the peak shift between the current and voltage pulses, the voltage and current peak values.

4.4.2 Applying a Surge to an *RLC* Circuit

Consider applying a high voltage, v_{in}, to an *RLC* circuit shown in Fig. 4.9. Applying Kirchoff's voltage law to the loop in Fig. 4.9,

$$v_c = v_R + v_L + v_C = iR + L\frac{di}{dt} + \frac{1}{C}\int_0^t i\,dt.$$

TABLE 4.3 **A Summary of Test Results Applicable to the Three Supercapacitor Families Used in the Surge Capability Testing [5]**

Capacitor Data	LSS Output Peak Voltage (kV)	Precharge Voltage (mV)	Number of Surges to Destroy the Device
	6.6	0	220 to over 250
Cap-xx		250	Over 220
0.18 F, 2.3 V		500	Over 180
		1800	Over 160
	6.6	0	Did not fail after 1000
Maxwell		200	Did not fail after 600; charge accumulation was not observed
230 F,2.5 V		1000	Did not fail after 600; over the period of 700 repeated surges 0.1 V discharge was observed[a]
		2500	Did not fail after 600; over the period of 700 surges a discharge of 0.2 V was observed[a]
	6.6	0	Did not fail after 600
Maxwell		500	Did not fail after 600; no charge accumulation observed
90 F, 2.7 V		1000	Did not fail after 600; slight discharge of 0.15 V observed during 700 surges[a]
		2000	Did not fail after 600; discharge of approx. 350 mV was observed over 700 surges[a]

[a] *This early observation was not further analyzed because the research team decided not to investigate the electrochemistry aspects of packaged SCs. In practical circumstances such number of repetition surges may not occur.*

When the current through the circuit maximizes,

$$\frac{di}{dt} = 0, \tag{4.19}$$

$$\Longrightarrow v_c = iR + \frac{1}{C}\int_0^t i\,dt. \tag{4.20}$$

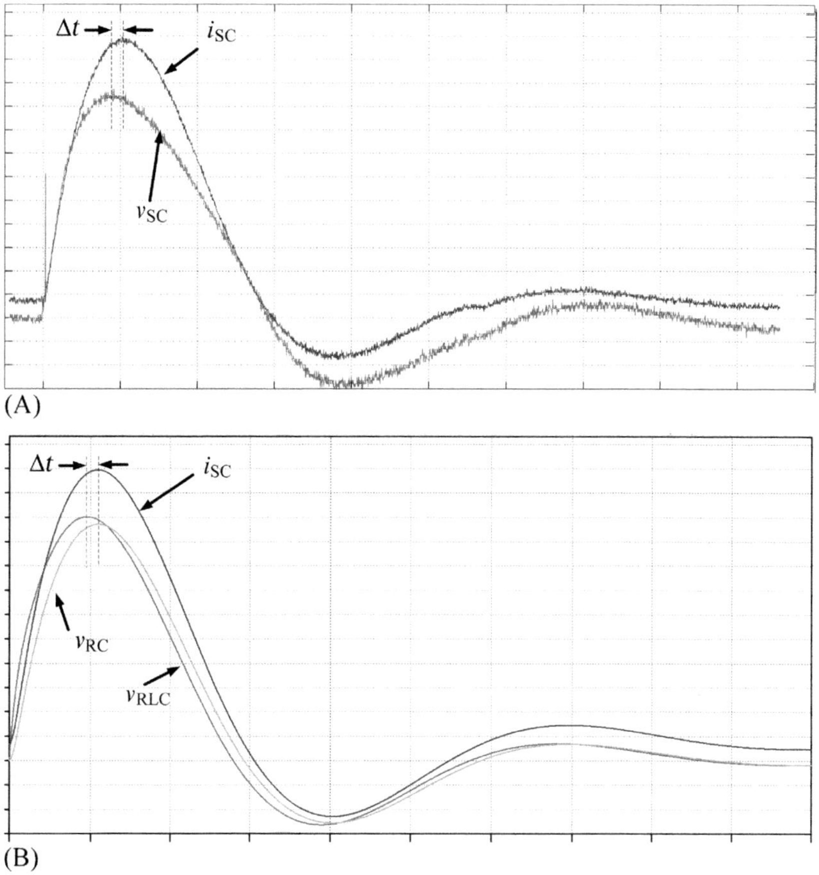

FIG. 4.8 Comparison of *RC* and *RLC* circuit simulation with experimental results for a supercapacitor: (A) experimental results; (B) simulation for *RLC* and *RC* circuit.

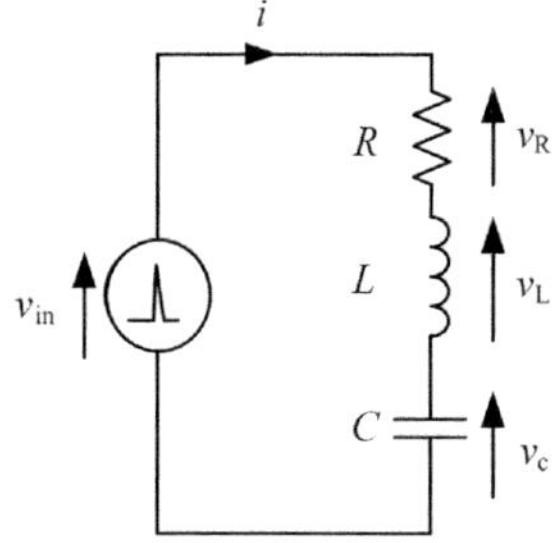

FIG. 4.9 Applying high-voltage surge to an *RLC* series circuit.

From previous observations in comparison to the surge voltage, the capacitor voltage rise (accumulated charge) can be neglected. Therefore, Eq. (4.20) will reduce to

$$v_c = iR, \tag{4.21}$$

$$\Rightarrow R = \left.\frac{v_c}{i}\right|_{\frac{di}{dt}=0} = R_{surge}. \tag{4.22}$$

Therefore, if the peak current and the voltage are known at the time of the peak current, by dividing the voltage by the current, surge ESR, R_{surge}, can be obtained.

It can be observed in Fig. 4.8A that the voltage leads the circuit current in the *RLC* circuit. Now to obtain an analytical value for the time difference between the voltage and the current peak, Δt, consider the time at which $\frac{di}{dt} = 0$ and the time when $\frac{dv_c}{dt} = 0$. Hence the peak shift, Δt, observed in an *RLC* can be found as described in Eq. (4.23).

$$\Delta t = \left|\left(t_1 : \frac{dv_c}{dt} = 0\right) - \left(t_2 : \frac{di}{dt} = 0\right)\right|. \tag{4.23}$$

Approximating the circuit current by the short-circuit current $i = AI_p t^3 e^{-t/\tau}$ described previously, t_2 can be found as

$$\left.\frac{di}{dt}\right|_{t=t_2} = 0, \tag{4.24}$$

$$\Leftrightarrow t_2^2 e^{-t_2/\tau}\left[3 - t_{2/\tau}\right] = 0, \tag{4.25}$$

$$t_2 \neq 0 \text{ and } e^{-t_2/\tau} \rightarrow 0 \text{ when } t_2 \rightarrow \infty \tag{4.26}$$

$$\Leftrightarrow t_2 = 3\tau. \tag{4.27}$$

t_1 can be found by equating $\frac{dv_c}{dt}$ to zero

$$\left[L\frac{di^2}{dt} + R\frac{di}{dt} + \frac{1}{C}i\right]_{t=t_1} = 0,$$

thus,

$$t_1^2 e^{-t_1/\tau}\left[\left(\frac{1}{C} - \frac{R}{\tau} + \frac{L}{\tau^2}\right)t_1^2 + \left(3R - \frac{6L}{\tau}\right)t_1 + 6L\right] = 0.$$

$t_1 \neq 0$ and $e^{-t_1/\tau} \rightarrow 0$ when $t_1 \rightarrow \infty$.

$$t_1 = \left[\frac{-(3R - 6L/\tau) \pm \left\{ (3R - 6L/\tau)^2 - 4\left(1/C - R/\tau + L/\tau^2\right)6L \right\}^{1/2}}{2(1/C - R/\tau + L/\tau^2)} \right]. \quad (4.28)$$

Based on simulation data, the solution for t_1 by choosing the negative sign in Eq. (4.28) is too small to be experimentally measured. Hence only the t_1 answer corresponding to the positive sign is considered for the Δt calculation and further analysis. By substituting t_1 and t_2 in Eq. (4.23), L can be found as

$$L = \frac{\Delta t^2\tau^2 - CR_{\text{surge}}\Delta t^2\tau + 6\Delta t\tau^3 - CR_{\text{surge}}\Delta t\tau^2 + 9\tau^4}{C\Delta t^2 - 3C\tau^2}. \quad (4.29)$$

Eqs (4.22), (4.29) describe the surge model of an SC. The equation system requires the voltage and current peak values and the peak shift to calculate the surge resistance and the equivalent inductance. The following section provides three examples to validate the accuracy of the *RLC* model.

4.5 *RLC* Model Validation

To validate the proposed model, the method followed was to simulate the analytically found R and L values for a chosen capacitor. The inputs to the analytical model are based on experimentally found voltage peak, current peak, and peak shift values. Fig. 4.10 gives the surge current and voltage waveforms for 1, 5, and 25 F SCs. Current and voltage peak values are identified by inspection. Table 4.4 lists the analytical solution for surge ESR and the surge inductance for each case using the experimental data from Fig. 4.10 in Eqs (4.29), (4.23). It can be observed that for every capacitor, the surge ESR is lower than the datasheet provided DC ESR. Similar observation was made for low-voltage electrochemical spectroscopy measurement of SCs in [3, 4]. Now the obtained surge ESR and the surge inductance with the manufacturer specified capacitance can be simulated for a 6 kV input lightning surge using LTSpice. The resulting current and voltage waveforms are given in Fig. 4.11. Based on the simulation results the percentage match with the experimental data are listed in Table 4.5. Percentage errors are calculated with respect to the experiment values and are denoted as E_{Vp}, E_{Ip} and $E_{\Delta t}$ for peak voltage, peak current, and peak shift, respectively.

In Chapter 9, the proposed novel surge protector topology using the above-discussed surge absorption capability of SCs is detailed.

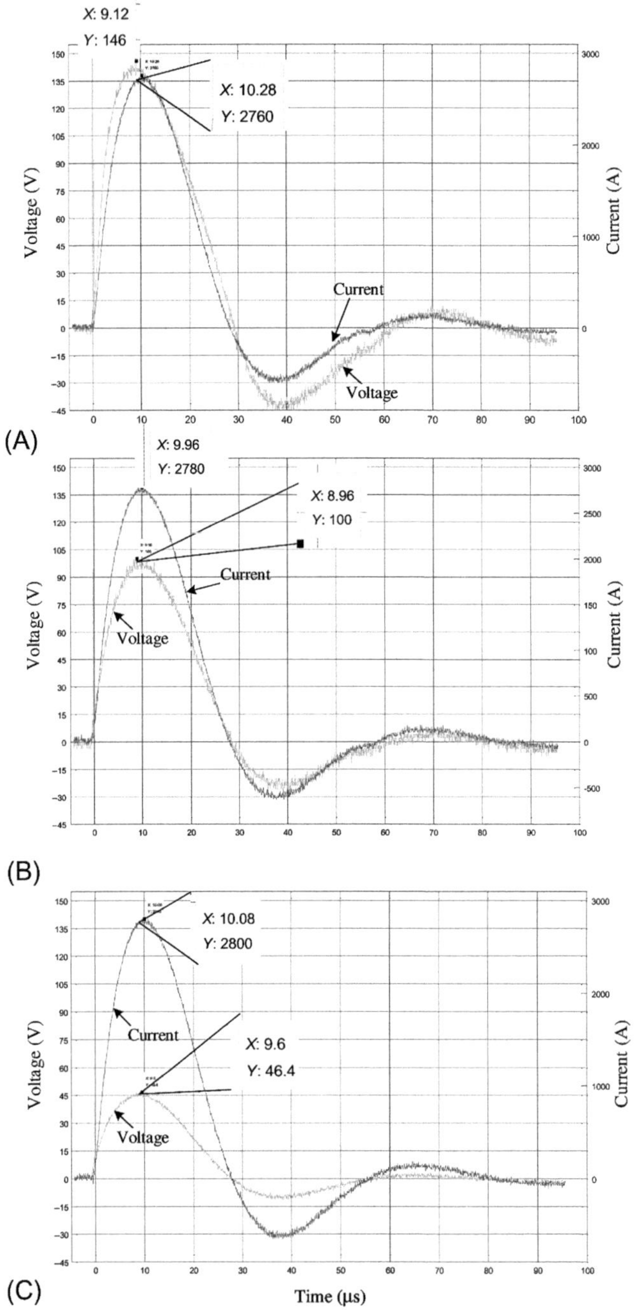

FIG. 4.10 Experimental results for the peak voltage and current shift for different capacitors subjected to 6 kV surge: (A) 1 F; (B) 5 F; (C) 25 F.

TABLE 4.4 Supercapacitor *RLC* Surge Model; Calculated Equivalent Circuit Parameters Based on Experimental Data as per Eqs (4.22), (4.29) and Fig. 4.10

Capacitance (*F*)	DC ESR (*mΩ*)	V_p (*V*)	I_p (*A*)	Δt (μs)	Surge ESR (*mΩ*)	Surge Inductance (*nH*)
1 ± 20%	700	146	2760	10.28 − 9.12 = 1.16	52.8	69.3
5 ± 20%	170	100	2780	9.96 − 8.96 = 1	36	39.9
25 + 20%	42	46.4	2800	10.08 − 9.6 = 0.48	16.6	8.3

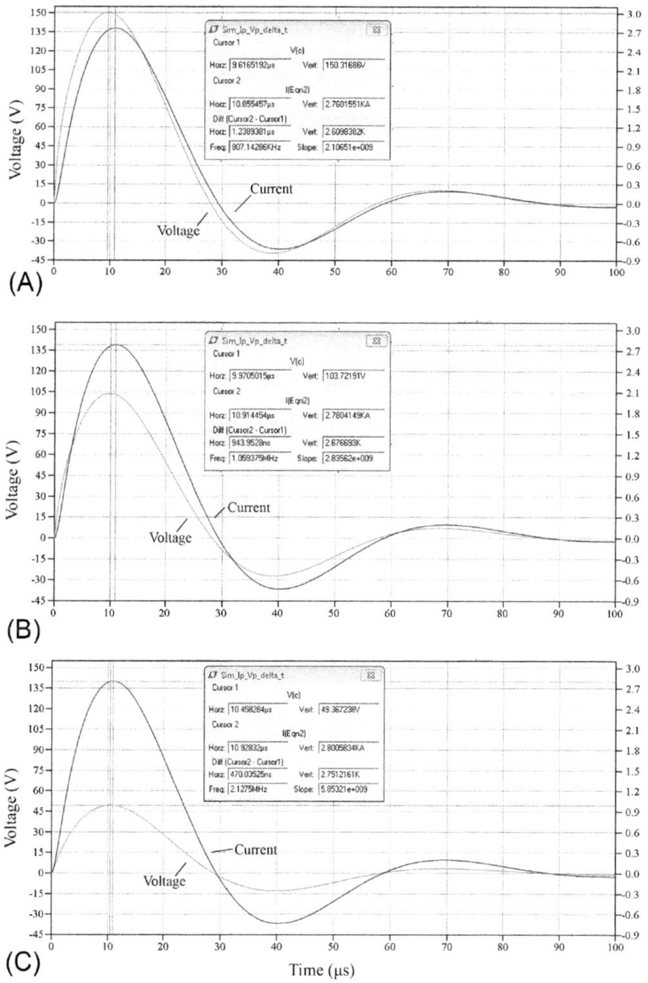

FIG. 4.11 LTSpice simulation results for the peak voltage and current shift for different capacitors subjected to 6 kV surge: (A) 1 F; (B) 5 F; (C) 25 F.

TABLE 4.5 Supercapacitor *RLC* Surge Model Verification: Simulated Data and Percentage Error With Respect to Experimental Data as per Eq. (4.29), Figs 4.10 and 4.11

Capacitance (F)	V_p (V)	$\%E_{V_p}$	I_p (A)	$\%E_{I_p}$	Δt (μs)	$\%E_{\Delta t}$
1 ± 20%	150.3	2.9	2760	0	10.85 − 9.61 = 1.24	6.9
5 ± 20%	103.7	3.7	2780	0	10.91 − 9.97 = 0.94	−6
25 + 20%	49.4	6.3	2751	−1.8	10.93 − 10.46 = 0.47	−2.1

References

[1] Maxwell Technologies, Product guide-Boostcap ultracapacitors, 2009. Maxwell Technologies, Technical Report.

[2] muRata, High performance electrical double-layer capacitors, 2016. Available from: www.murata.com//media/webrenewal/campaign/ads/america/edlc/promo_dmf_productguide1.ashx?la=en-us. Retrieved May 2018.

[3] F. Rafik, H. Gualous, R. Gallay, A. Crausaz, A. Berthon, Frequency, thermal and voltage supercapacitor characterization and modeling, J. Power Sources 165 (2) (2007) 928–934.

[4] R. Ktz, M. Hahn, R. Gallay, Temperature behavior and impedance fundamentals of supercapacitors, J. Power Sources 154 (2) (2006) 550–555.

[5] N. Kularatna, J. Fernando, A. Pandey, S. James, Surge capability testing of supercapacitor families using a lightning surge simulator, IEEE Trans. Ind. Electron. 58 (10) (2011) 4942–4949.

Chapter 5

Characterization of Metal Oxide Varistors and Transient Voltage Suppressor Diodes

5.1 Introduction

Metal oxide varistors (MOV) and transient voltage suppressors (TVS) such as bidirectional breakover diodes (BBDs) are nonlinear devices that are central to the operation of an SPD. These devices are manufactured by many companies, and the datasheets provide essential specifications for an SPD designer.

This chapter provides an overview on characterization of MOVs and TVS devices, going through an analytical approach. This chapter will be useful for test engineers, model developers, and SPD circuit designers in general.

5.2 Nonlinearity of Surge Absorbing Components and Analytical Methods for Characterization: An Overview

Although MOVs and semiconductor components used in SPDs are in general nonlinear, the analysis of transient propagation in linear systems is important for several reasons. First, significant parts of the overall SPD circuit we need to analyze could be linear. Second, the exact solutions available for the behaviour of linear systems through standard techniques, such as the Laplace transform method, allow us to analyze the linear parts easily. Third, if nonlinear differential equations (DEs) are formulated in order to simulate the nonlinear systems, formulation becomes an extension of the linear DEs that are formed for the linear portions of the overall system.

The Laplace transform method, which is a standard technique for solving linear differential and integro-differential equations, is used to analyze linear circuits containing resistors and reactive components such as capacitors and inductors. The Laplace transform method is, in many circumstances, simpler and more convenient to use than the other classical methods of solving ordinary DEs. Applications of this method relevant to the study of this surge propagation are presented in Appendix E of [1].

Design of Transient Protection Systems. https://doi.org/10.1016/B978-0-12-811664-7.00005-7

SPDs consist of nonlinear devices and therefore their analysis gives rise to nonlinear DEs. In general, nonlinear DEs cannot be solved analytically by a method such as Laplace transforms. This limitation of the Laplace transform is highlighted in Section 5.3.

Until the 1960s, practical ways of solving nonlinear problems involved graphical and experimental approaches. Since that time, computer simulation has become a powerful tool in solving nonlinear problems. In order to predict power dissipation and energy absorption in surge protection devices, now we can resort to computer simulation of the circuits that contain them.

Circuit simulation software such as Simulation Program with Integrated Circuit Emphasis (SPICE) and its derivatives are suitable for transient propagation studies under certain conditions, the most important being the availability of suitable models for the nonlinear surge protector devices used.

In the absence of models for the nonlinear devices, we could employ numerical simulation using a high-level language such as MATLAB. A good example of such an exercise can be found in Chapter 7. In order to carry out this method successfully, the nonlinear devices used must be characterized in order to develop suitablc mathematical models for them. This chapter will discuss characterization of two versatile surge protection components, an MOV and a TVS diode.

Section 7.2.2 looks at circuit simulation of an SPD using PSpice software, and compares the results to the simulation of relevant DEs through MATLAB coding.

5.3 Limitation of Laplace Transform Method

As mentioned earlier in Section 3.2, the SPD circuits we need to analyze contain nonlinear devices such as MOVs which can be represented by the power law relationship of Eq. (3.1), which is repeated here,

$$i(v) = kv^{\alpha},$$

where k is a constant dependent on the device geometry and α defines the degree of nonlinearity of the characteristic. The nonlinearity is the deviation from $\alpha = 1$. For MOVs the value of α can be in the range of 15–30.

Although nonlinear systems can be solved by extending the techniques developed, such as Laplace transform, for the solution of linear systems, there are a number of difficulties associated with such an implementation, one being the necessity to handle a very large number of terms [2]. Hence, although a solution would be possible, it will be more straightforward to solve the state equations that describe the system directly using numerical techniques as carried out in Chapters 6 and 7. By appropriate device characterization (see Section 5.5), we can obtain the mathematical models required for such numerical computation.

5.4 Simulation Software for Numerical Computing for Testing SPDs

SPICE is a general-purpose analogue electronic circuit simulator. If the nonlinear devices have SPICE models readily available, circuit performance can be simulated with the help of a SPICE2 or SPICE3-based circuit simulator (such as PSpice) [3]. PSpice is a prominent commercial version of SPICE available from Cadence Design Systems. PSpice models for certain surge protection components are available from manufacturers and it becomes a useful simulation package when matching components are used in SPDs under investigation. Almost identical results can be obtained by such simulation compared with MATLAB-based numerical simulations. These simulations can be done prior to the prototype stage as an additional step of testing to ensure that the circuit works properly before investing in prototype development.

Electromagnetic transients including DC (EMTDC) solves DEs in the time domain. EMTDC is most suitable for simulating the time domain instantaneous responses of electrical systems. The value of EMTDC is enhanced by its state-of-the-art graphical user interface called Power Systems Computer Aided Design (PSCAD). PSCAD allows the user to graphically assemble the circuit, run the simulation, analyze the results, and manage the data in an integrated graphical environment. Although the device library of this software boasts many power electronic devices, there are no representations of some important surge protection components such as low-voltage (under 1000 V) MOVs and TVS diodes. In spite of this shortcoming it has provision for the creation of a model for a nonlinear device with a known characteristic, making it a candidate for simulating surge protection circuitry with nonlinear components.

The advantages of a high-level language-based numerical computation over the software packages discussed earlier are:

- The exposed DEs defining the system and their solution provide more insight into the working of the system.
- Provides a good understanding of how circuit analysis techniques and mathematical techniques of solving a set of nonlinear DEs together make up a good software tool for the simulation of nonlinear power electronic interfaces.
- Suitable models can be quickly found for off-the-shelf surge protection components by experiment, when popular software-based models are not readily available.

Keeping this in mind, we will characterize some important nonlinear devices used in SPDs in the following sections of this chapter. An in-depth study of numerical simulation for the analysis of an SPD is presented in Chapter 7.

5.5 Surge-Absorbent Device Characterization

So far we have discussed four different types of practical devices used in surge protection circuits: MOVs, gas discharge tubes (GDTs), TVS diodes, and TVS thyristors. Except GDTs the others are solid-state devices. These practical devices are used extensively in category-A and -B protection units. In Section 2.3, we briefly looked at surge propagation in an SPD using a category-B protection unit shown in Fig. 2.4, which is a two-wire version of a category-A protection unit with larger protection devices to handle the larger transients anticipated for category-B. This circuit uses an MOV for the first level of protection and a TVS diode (which is in the form of a back-to-back break-over diode or BBD) for the second level of protection. For a comprehensive investigation of surge propagation through this circuit using numerical simulation, it is necessary to characterize the nonlinear devices and obtain suitable mathematical models. The work done in the development of the relevant models for the MOV and the TVS diode is presented next.

5.5.1 MOV Characterization

MOV manufacturers such as Littelfuse offer different types of MOV products to suit particular electrical environments. A sample set of MOVs from their product line is given in Table 5.1 [4].

TABLE 5.1 Selection Criteria for Littelfuse MOV Devices

Voltage (V)	Energy (J)	Packaging and Other Considerations	Preferred Series
AC applications			
130–1000	11–360	Through-hole mounting	LA
		Low/medium AC power line	'C' III
			UltraMOV
130–750	270–1050	High-energy applications	DA
		Shock/vibration environment	HA, HB
			NA
			DB
DC applications			
4–460	0.1–35	Through-hole mounting	ZA
		Automotive and low-voltage applications	

For our characterization experiments here, we have selected the devices 275L40C, 20V275, and V271HA32 from the 'C' III, UltraMOV, and HA series, respectively. Each of these devices is rated to operate at a maximum RMS voltage of 275 V.

5.5.1.1 Measuring Set-Up

The measuring circuit diagram for obtaining the *V–I* characteristics of the MOV is shown in Fig. 5.1. The NoiseKen LSS-6110 lightning surge simulator (LSS) is used as an impulse source.

The LSS-6110 simulator, like the LSS-6230 mentioned in Section 4.4, is a combination wave generator or hybrid generator that can provide a 1.2/50 μs impulse voltage in an open circuit (1.2 μs corresponds to the front time and 50 μs corresponds to the time to half-value as depicted Fig. 5.2) and an 8/20 μs

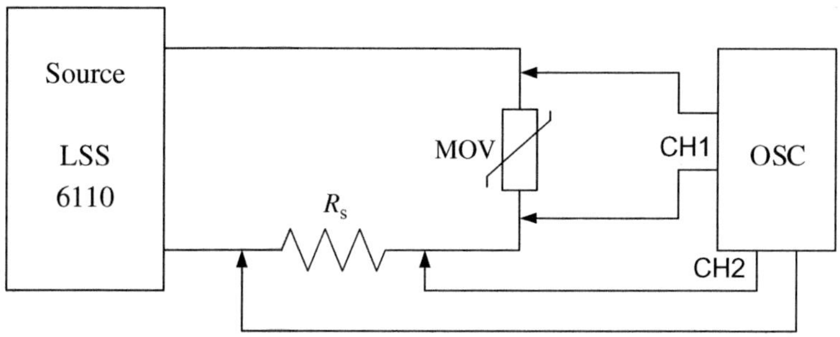

FIG. 5.1 Circuit diagram for the measurement of the MOV voltage and current; impulse source is an LSS. A two-channel oscilloscope is used for measurements.

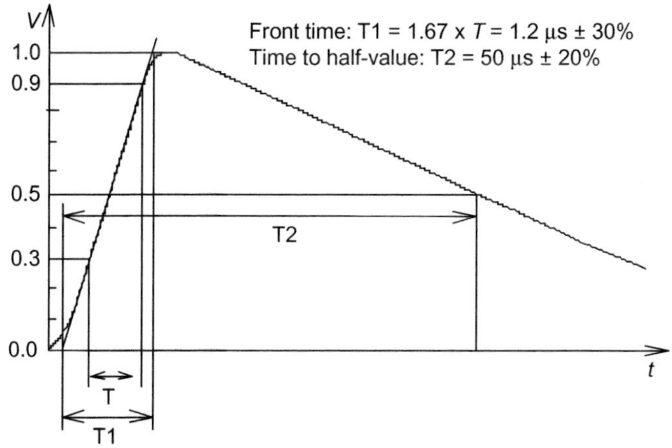

FIG. 5.2 Waveform of 1.2/50 μs open-circuit voltage (waveform definition according to IEC 60060-1).

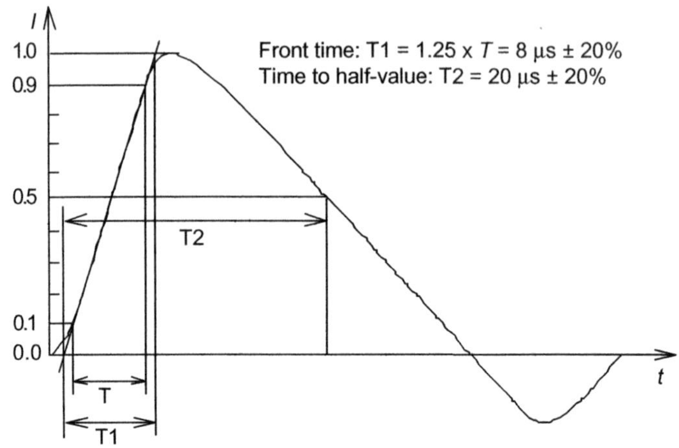

FIG. 5.3 Waveform of 8/20 μs short-circuit current (waveform definition according to IEC 60060-1).

impulse current in a short circuit (8 μs front time and 20 μs time to half-value; see Fig. 5.3). These waveforms conform to IEC publications 61000-4-5 [5] and 60060-1 [6] and the IEEE publication C62.41.2 [7].

The charging voltage of the impulse generator can be adjusted in 100 V steps from 0.1 to 6.6 kV.

The voltages are recorded with a Tektronix TPS2024 oscilloscope at a sampling rate of 2 GS/s. Tektronix P5120 1 kV 20× (bandwidth 200 MHz) and P6015A 40 kV 1000× (bandwidth 75 MHz) high-voltage probes are used for voltage measurements.

Of the MOVs characterized, types 275L40C and 20V275 have diameters of 20 mm. The third-type V271HA32 has a diameter of 32 mm. All three types are rated for a steady-state maximum RMS voltage of 275 V. A series of pulses with different voltage amplitudes is applied to each one of the varistors.

5.5.1.2 Waveforms Obtained and Their Analysis

Measurements described in the previous section fall into two distinct categories: leakage region and the normal varistor operation region as depicted in Fig. 5.4.

Leakage Region Analysis

All of the tested varistors operate in the leakage region for low impulse voltages ranging from 100 to 500 V, and turn on beyond 500 V. A typical varistor current waveform obtained in the leakage region of operation is shown in Fig. 5.5.

In order to find an explanation for the resonance in the above current waveform we consider the MOV equivalent circuit in the leakage region shown in Fig. 5.6A as well as the equivalent circuit of the impulse generator LSS-6110 shown in Fig. 5.6B.

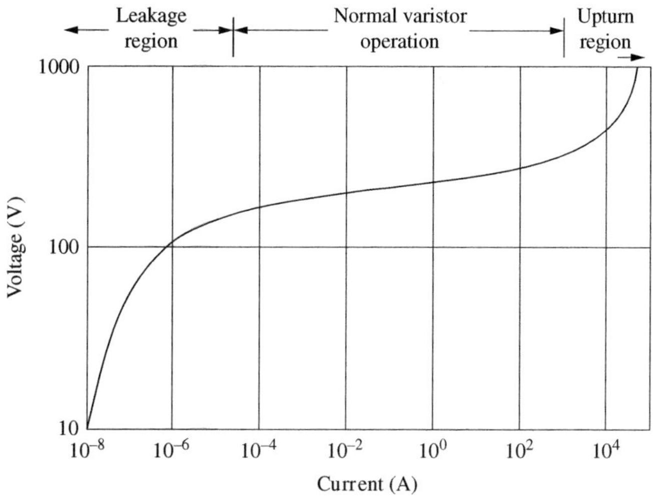

FIG. 5.4 Typical varistor characteristic showing different regions of operation.

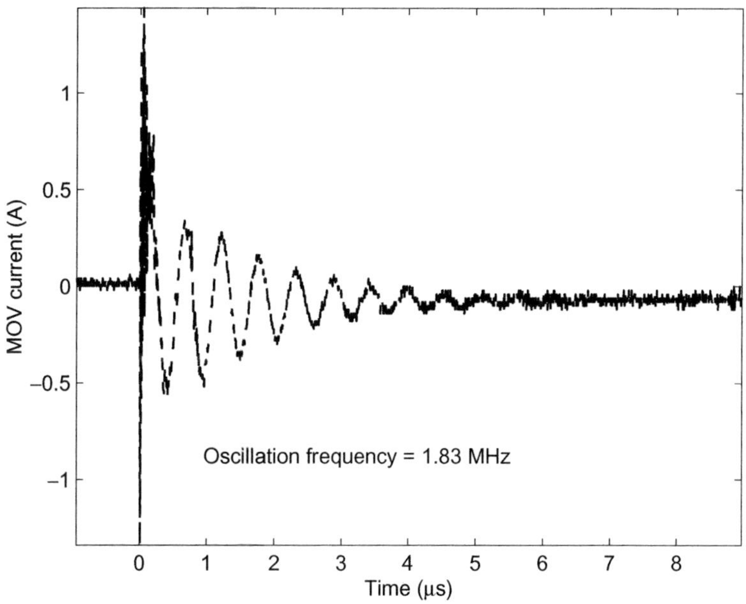

FIG. 5.5 Leakage region varistor current produced by a low-voltage impulse.

The combined equivalent circuit of Fig. 5.7 drawn for the measuring circuit of Fig. 5.1 disregards R_{OFF} as it approaches 10^9 Ω in the leakage region. The following paragraphs discuss the values of C_m and L_m used for the analysis of this circuit.

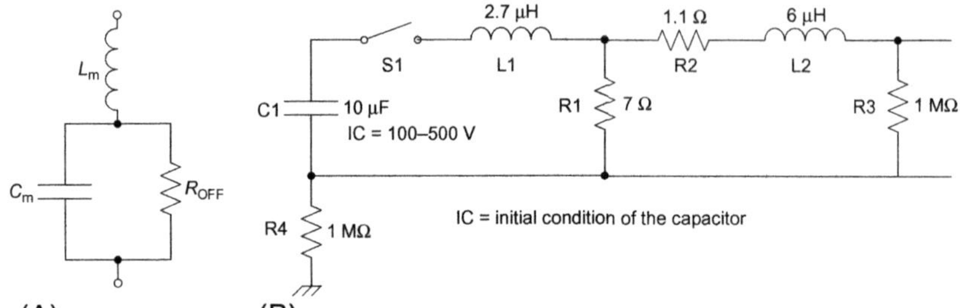

FIG. 5.6 The equivalent circuits required to analyze the working of the circuit of Fig. 5.1 for low-voltage impulses. (A) MOV leakage region. (B) Equivalent circuit of the LSS-6110 impulse generator (Noise Laboratory Co., Ltd., Japan).

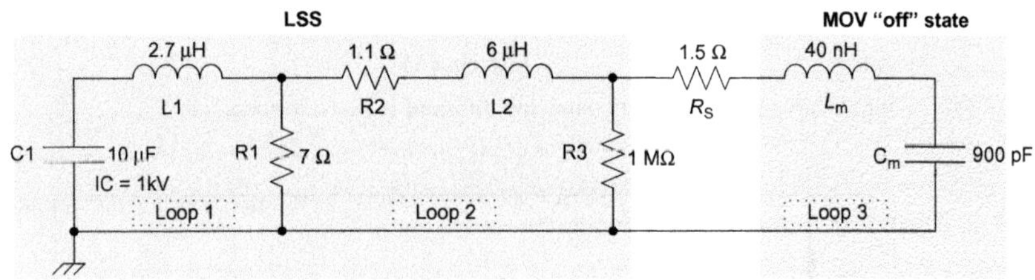

FIG. 5.7 Combined equivalent of the measuring circuit shown in Fig. 5.2 for the leakage region of the varistor.

The typical capacitance value C_m for a 20V275 UltraMOV series varistor is available from the datasheet as 900 pF [8]. Because the bulk region of the varistor acts as a dielectric, the device capacitance depends on its area and varies inversely with its thickness. Hence the capacitance of the varistor is a function of its voltage and energy ratings. The voltage rating is determined by device thickness, and the energy rating is directly proportional to volume [9].

L_m the lead inductance of the MOV, which is not explicitly given in the datasheet, can be calculated using the mechanical dimensions given in Fig. 5.8 which have been taken from the respective datasheets.

The lead inductance, the self-inductance of a pair of parallel wires, is given by [11],

$$L_m = \frac{\mu_0 h}{\pi} \ln \frac{d-a}{a}, \tag{5.1}$$

where h, d, and a are illustrated in Fig. 5.8. Multiplication by the symbol μ_0, which denotes the relative permeability of free space ($4\pi \times 10^{-7}$ H/m), gives the result in henries. The lead inductances calculated, using the average values of d and a, gives 32.2 nH for the 20V275 varistor and 29.9 nH for the 275L40C varistor.

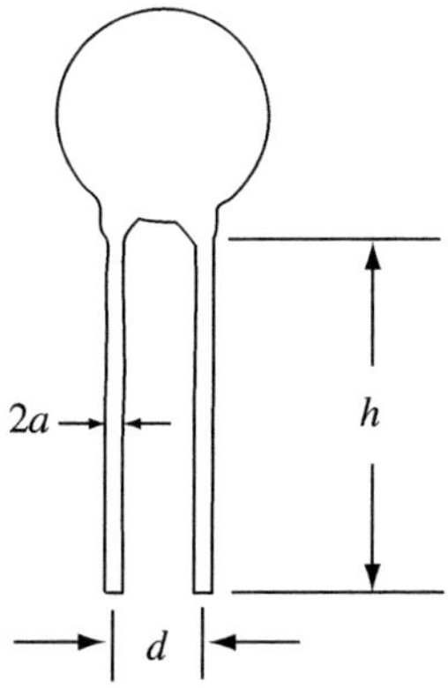

Varistor model	275L40C	20V275
h	25.4	25.4
$2a$	0.76 (min) 0.86(max)	0.76 (min) 0.86(max)
d	6.5 (min) 8.5 (max)	9 (min) 11(max)

Note: Measurements in mm

FIG. 5.8 Mechanical dimensions of radial lead varistors used [8, 10].

Analysis of the measuring circuit tells us that the loop-3 current is almost equal to loop-2 current due to the high resistance of R_3 and therefore the ringing seen in the loop-3 current is mainly due to the combined inductance of L_2 and L_{m} resonating with C_{m}. This resonant frequency, which is sensitive mainly to L_2, L_{m}, and C_{m}, is calculated as 2.15 MHz using Eq. (5.2)

$$f_{r3} = \frac{1}{\sqrt{(L_2 + L_{\mathrm{m}})C_{\mathrm{m}}}}. \tag{5.2}$$

The observed value of approximately 1.83 MHz is fairly close to the calculated figure of 2.15 MHz. It must be noted that very small inductance and capacitance values associated with the cables and setup have been ignored in this calculation. It is important to recognize that the interconnection cables are of the length between 150 cm to over 250 cm, and the corresponding additional inductance introduced into the loops will affect the measurement.

As expected the rate of decay in the ringing waveform is sensitive to both R_2 and R_S. This was verified by simulating the circuit given in Fig. 5.7 using a SPICE simulator.

Normal Varistor Operation Region Observations

When the impulse voltage exceeds 600 V, the tested varistors tend to move from the leakage region to the normal operation region. Typical varistor current and voltage waveforms obtained in this region of operation are shown in Fig. 5.9.

In Fig. 5.9 measured voltage and current waveshapes for the varistor 275L40C are shown. The LSS was set to output a 700 V surge. It is observed that the voltage peak occurs before the current peak. This tells us that the varistor during conduction is not purely resistive.

Waveshapes similar to the ones shown in Fig. 5.9 were obtained for the other two types of tested varistors as well. A time delay between the voltage and

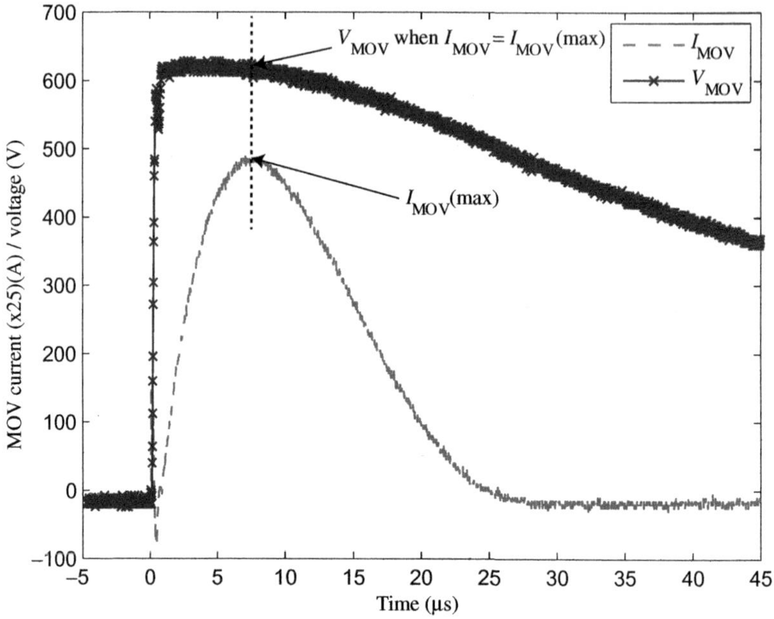

FIG. 5.9 Measured voltage across and current through a varistor in the normal operation region. The LSS was set to deliver a surge of 700 V.

current peaks was observed in all measurements taken in the normal operation region of the varistors. The difference in this delay for different samples of the same type of varistor was found to be negligible. Fig. 5.10 shows the associated dynamic *V–I* curve for traces shown in Fig. 5.9. If we are to simulate the varistor in its normal region of operation we will have to develop a varistor model capable of reproducing the major features displayed by the curves in Figs 5.9 and 5.10.

Figs 5.9 and 5.10 clearly show that when the voltage exceeds a certain threshold (around 620 V in this case), the current increases extremely fast. Although the current starts to decrease after reaching a maximum, this decrease follows a different more gradual path with the disappearing surge voltage. This phenomenon gives rise to a hysteresis loop in the *V–I* characteristic of a varistor. If a very accurate model of MOV dynamics is needed, then this phenomenon will have to be taken into account [12–15].

Most application notes and international standards which discuss varistor characteristics ignore this hysteresis phenomenon and present an idealized *I–V* characteristic which is a slight variation of the one shown in Fig. 5.4, where the upturn region is usually omitted and the axes interchanged as shown in Fig. 5.11 [16, 17]. Some of the important definitions indicated in Fig 5.11 and Table 5.2 will be discussed in the following sections where we will characterize several individual varistors.

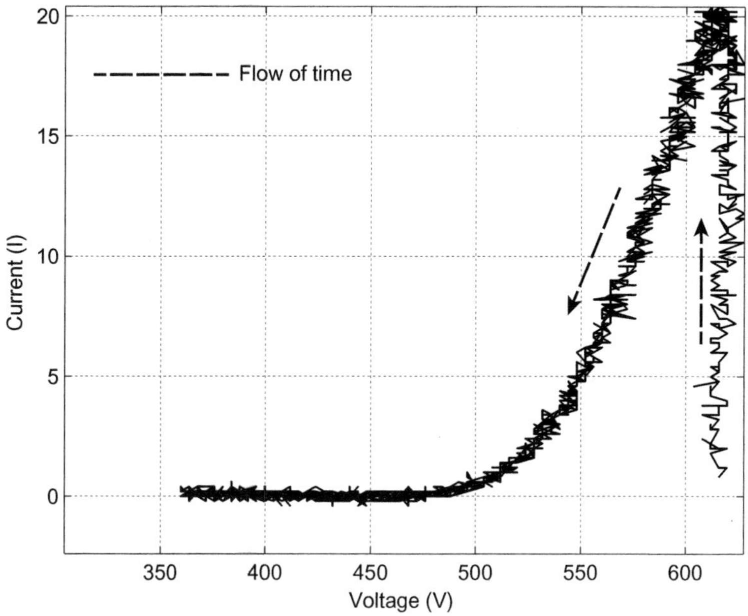

FIG. 5.10 Associated dynamic I–V curve for traces shown in Fig. 5.9.

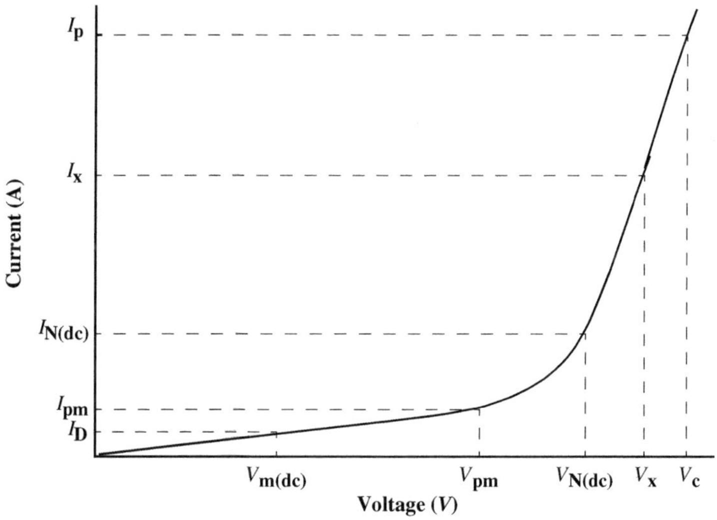

FIG. 5.11 I–V graph of an MOV illustrating symbols and definitions (see Table 5.2).

The most important property of a varistor is its nonlinear I–V characteristic in the normal operation region. As indicated in references [16, 18, 19], this property can be expressed by the power law relationship seen earlier $i(v) = kv^{\alpha}$, where $\alpha \gg 1$ is the coefficient of nonlinearity. The following argument tells

TABLE 5.2 Description of Terms and Letter Symbols Used in Defining a Varistor

Terms and Descriptions	Symbol
Clamping voltage. Peak voltage across the varistor measured under conditions of a specified peak pulse current and specified waveform	V_C
Rated DC voltage (varistor). Maximum continuous DC voltage which may be applied	$V_{m(dc)}$
DC standby current (varistor). Varistor current measured at rated voltage, $V_{M(DC)}$	I_D
Nominal varistor voltage. Voltage across the varistor measured at a specified pulsed DC current, $I_{N(DC)}$, of specific duration. IN(DC) is specified by the varistor manufacturer	$V_{N(dc)}$
Rated recurrent peak voltage (varistor). Maximum recurrent peak voltage which may be applied for a specified duty cycle and waveform	V_{pm}
Varistor voltage. Voltage across the varistor measured at a given current, I_X	V_X

us that α can be determined graphically if the I–V characteristic of a varistor is drawn as a log–log plot. Assume that MOV current follows a power-law function of MOV voltage:

$$i(v) = kv^{\alpha}. \tag{5.3}$$

Taking the logarithm of both sides of Eq. (5.3)

$$\log i = \log k + \alpha \log v. \tag{5.4}$$

Because Eq. (5.4) is the equation of a straight line, $y = mx + c$, the normal operation region of the varistor should appear as a straight line in a log–log plot of its I–V characteristic. The coefficient of nonlinearity α is given by the slope of this line.

5.5.1.3 Approximation of the Coefficient of Nonlinearity

Table 5.3 shows the voltage and current values obtained for the characterization of the above varistor by using the measuring set-up described in Section 5.5.1.1 and the wave shapes described in Section 5.5.1.2. In order to obtain these values, surge voltages from 100 V to 2 kV (in 100 V steps) were applied to the test circuit. The varistor current was calculated by measuring the voltage across a high-wattage 1 Ω resistor connected in series with the varistor. In the leakage region of operation (from 0 to 500 V) the current is recorded as 0 V due to the negligibly small values obtained.

Fig. 5.12 shows the linear and log–log I–V plots done using the values recorded in Table 5.3.

TABLE 5.3 Voltage and Current Readings Obtained to Characterize a 275L40C Varistor

V_{MOV} (V)	I_{MOV} (A)	V_{MOV} (V)	I_{MOV} (A)	V_{MOV} (V)	I_{MOV} (A)	V_{MOV} (V)	I_{MOV} (A)
98	0.0	584	1.4	664	60.8	704	136.0
197	0.0	620	10.2	668	76.8	708	152.0
300	0.0	632	22.2	692	91.2	712	166.0
400	0.0	636	34.8	688	105.6	716	184.0
500	0.0	656	48.8	696	120.8	724	196.0

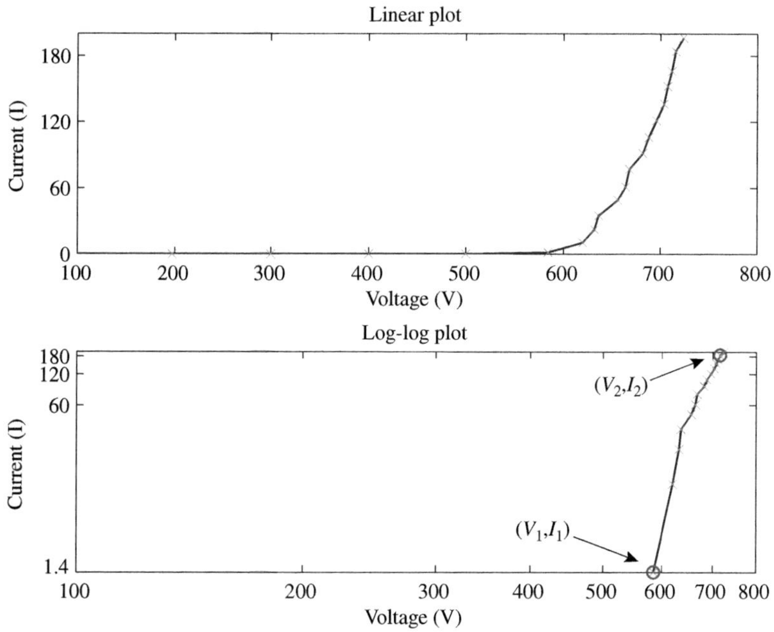

FIG. 5.12 Linear and log–log *I–V* plots for the varistor 275L40C.

It follows from Eq. (5.4) that the coefficient of nonlinearity can be determined from the log–log *I–V* plot by using the following equation:

$$\alpha = \frac{\log I_2 - \log I_1}{\log V_2 - \log V_1} = \frac{\log (I_2/I_1)}{\log (V_2/V_1)}, \tag{5.5}$$

where (V_2, I_2) and (V_1, I_1) are two suitably chosen points on the log–log *V–I* plot. The chosen points are shown in Fig. 5.12 for the log–log plot of varistor 275L40C. The calculated value of α works out to be 23.94. Depending on

TABLE 5.4 Calculated Values of α With Percentage Uncertainties

Device	Type	Exponent α (± %)
MOV	275L40C	23.94 ± 10
MOV	20V275	20.18 ± 8
MOV	V271HA32	16.9 ± 10

the points chosen for the above calculation of α, we see a deviation of ± 10%. The values of α obtained for all the tested varistor types are summarized in Table 5.4.

5.5.1.4 Accurate Measurement of Leakage Region Currents

As we know now the measuring set-up shown in Fig. 5.1 is not able to provide accurate measurements of the small currents (<1 mA) resulting from voltages applicable for the leakage region of the varistor. This range of voltages can be approximately up to the $V_{N(DC)}$ value, which is specified as 473 V for the 275L40C varistor [10].

The measuring set-up shown in Fig. 5.13 can be used to characterize the leakage region of the varistor. Here the impulse generator has been replaced by a Glassman HV regulated DC power supply EW5R120 as the source. This supply is capable of 0–5 kV continuous output voltage and 0–120 mA output current. A Fluke 19 digital voltmeter was used for measuring the voltages. As the varistors have low average power dissipations, R_s must be chosen to limit the current to a safe value. For example, for a Littelfuse C-III 20 mm varistor, the average power dissipation should not exceed 1 W.

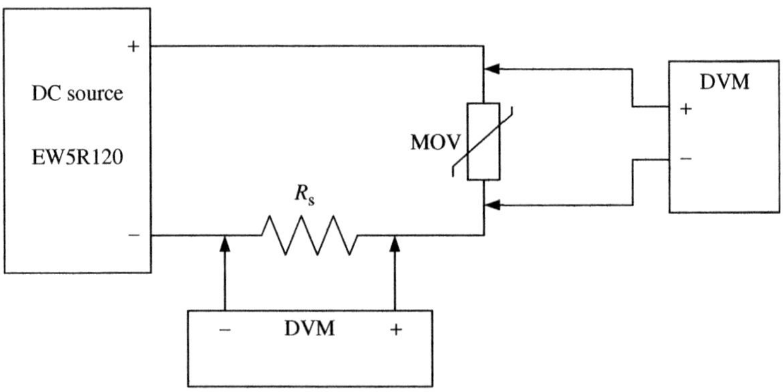

FIG. 5.13 Circuit for measuring MOV voltage and current in the leakage region of operation.

5.5.1.5 A Varistor I–V *Characteristic for Leakage and Normal Regions of Operation*

The varistor voltage and current values in Table 5.5 were obtained by using the measuring set ups shown in Figs 5.1 and 5.13 separately in order obtain a complete *I–V* characteristic to cover both the leakage as well as the normal region of operation. The device under test (DUT) was a 275L40C varistor.

The varistor voltage at 1 mA DC test current is specified as variable from 389 to 473 V in the datasheet [10]. The data in Table 5.5 confirm this. Fig. 5.14 shows the linear and log *I–V* plots using the values recorded in Table 5.5.

The log plot in Fig. 5.14 is approximately piecewise linear and could be modelled by using a power law expression, such as the one given in Eq. (5.3), for each of the linear portions. Let the leakage region indicated as the 'off' model in the figure be represented by $K_1 v^{\alpha 1}$ and the normal region indicated as the 'on' model in the figure be represented by $K_2 v^{\alpha 2}$.

5.5.1.6 Development of a Mathematical Model for the Varistor

In order to get an estimation of the parameters for the 'off' and 'on' models we will use the MATLAB command `polyfit(x,y,n)` to return the coefficients of our single degree polynomial. The single degree polynomial in our case is given by Eq. (5.4). Hence polyfit would return the values for α (slope) and $\log K$ (intercept) for the straight line given by Eq. (5.4).

In order to illustrate the development of a combined model for the leakage and normal regions of operation for a varistor, the data points of Table 5.5 are used again in the log–log plot of Fig. 5.15. The positioning of the straight line segments in Fig. 5.15, represented by $i = K_1 v^{\alpha 1}$ and $i = K_2 v^{\alpha 2}$, was done with the

TABLE 5.5 Voltage and Current Readings Obtained to Characterize the Leakage and Normal Regions of Operation of a 275L40C Varistor

Source	V_{MOV}	I_{MOV}	Source	V_{MOV}	I_{MOV}
DC	100.3	1.5E-07	DC	461	4.16E-03
DC	199.4	1.1E-06	DC	462	4.7E-03
DC	300.3	8.8E-06	DC	464	5.3E-03
DC	397	4.1E-05	Impulse	600	2
DC	446	7.2E-04	Impulse	640	21
DC	452	1.9E-03	Impulse	660	46
DC	457	3.0E-03	Impulse	680	72
DC	458	3.5E-03	Impulse	700	104

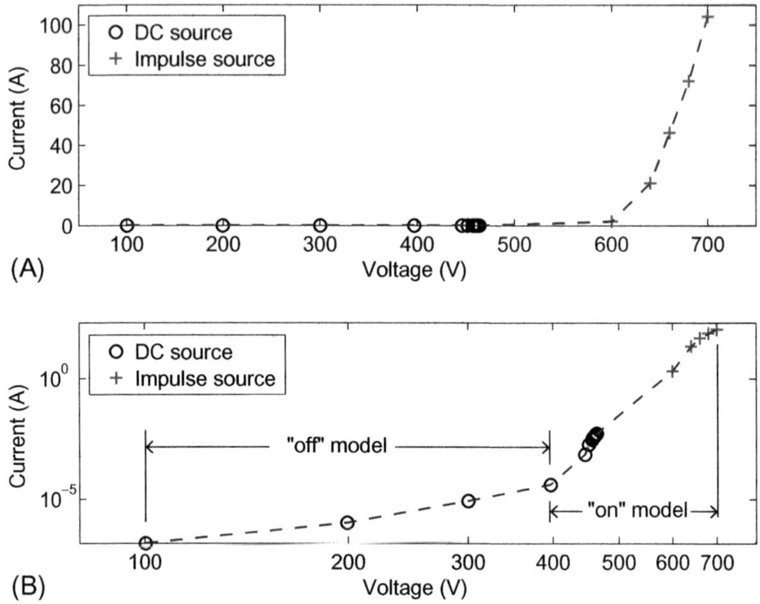

FIG. 5.14 *I–V* plots that show the leakage and normal regions of operation for a 275L40C varistor: (A) linear plot; (B) log–log plot.

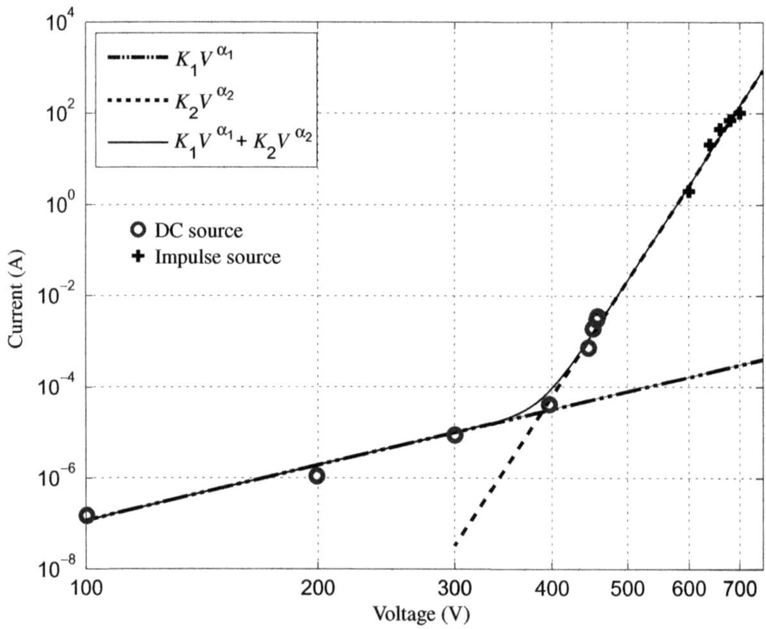

FIG. 5.15 Development of a combined model for the leakage and normal regions of operations for a varistor.

following values obtained from appropriate curve fitting using `polyfit` for the tested 275L40C varistor.

$K_1 = 9.3231 \times 10^{-16}$, $\alpha_1 = 4.0413$ (*off* model parameters)
$K_2 = 3.5040 \times 10^{-73}$, $\alpha_2 = 26.2396$ (*on* model parameters)

The curve for the combined model $i = K_1 v^{\alpha_1} + K_2 v^{\alpha_2}$ is shown in Fig. 5.15 along with its two straight line component segments. The following section verifies the validity of this model.

5.5.1.7 Verification of the Varistor Model

The circuit in Fig. 5.16 was chosen to verify the mathematical model developed for the varistor in the previous section. In this circuit a set of DC voltages ranging from 20 to 1200 V is applied to the input and the resulting varistor current and voltage are simulated along with the load current. An experimental validation is subsequently performed by taking maximum values resulting from surging the circuit using an LSS.

Now an equation for the input current I_i is formulated; we can solve for I_i by finding the roots of this equation. Once I_i is known, the remaining unknowns can be found by the equations listed below, for each value of V_i, within the range of interest.

By applying Kirchoff's voltage law, we have

$$V_L = I_L R_L = (I_i - I_m) R_L = V_i - I_i R_S. \tag{5.6}$$

The varistor current I_m represented by the model developed in the previous section is given by:

$$i_m = K_1 v^{\alpha_1} + K_2 v^{\alpha_2}. \tag{5.7}$$

Hence I_m can be defined as a function of V_L and therefore

$$i_m = f(V_L) = f(V_i - I_i R_S). \tag{5.8}$$

Substitution of Eq. (5.8) in Eq. (5.6) results in,

$$[I_i - f(V_i - I_i R_S)] R_L = V_i - I_i R_S. \tag{5.9}$$

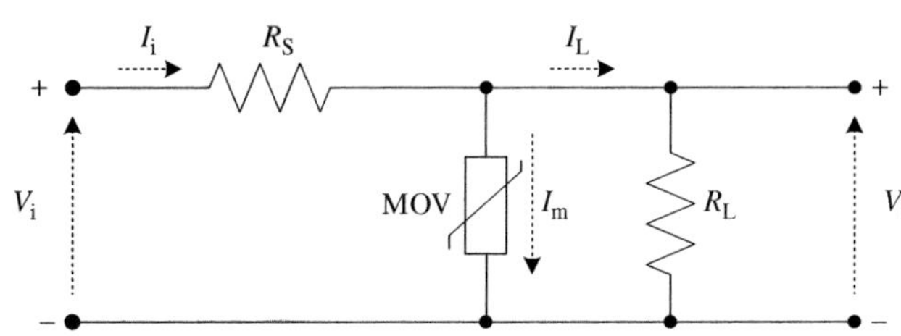

FIG. 5.16 The circuit used for an initial verification of the model developed for the varistor.

Rearranging Eq. (5.8) gives

$$[I_i(R_L + R_S) - V_i]/R_L - f(V_i - I_i R_S) = 0. \tag{5.10}$$

We see that the only unknown in Eq. (5.10) is I_i. For each input voltage value, this equation is solved to find the corresponding value of I_i. This is done by finding the range of values the equation would take for a given span of values for I_i. Because the required value of I_i is given by the roots of the equation, it can be found by locating the value of I_i at the zero crossing point of the equation curve. This is illustrated in Fig. 5.17 for a sample input voltage of 800 V. MATLAB command `fzero` that uses the bisection iteration method was used in the calculation of I_i.

The simulations done with the help of the above solution are shown in Fig. 5.18, where the circuit resistances were set to $R_S = 1\ \Omega$ and $R_L = 1\ k\Omega$. As expected, when the input voltage exceeds the MOV's threshold voltage almost all of the input current is carried by the MOV. In order to validate these simulations and check the suitability of the developed varistor model, an experimental validation was performed by surging the circuit with the LSS. The data collected for this validation are given in Table 5.6 (the italicized cell values were calculated) and the points are plotted in Fig. 5.18. It can be clearly seen that the simulation and the experimental results compare well. We are now convinced that we can experiment with this model to investigate transient propagation through surge protection circuitry.

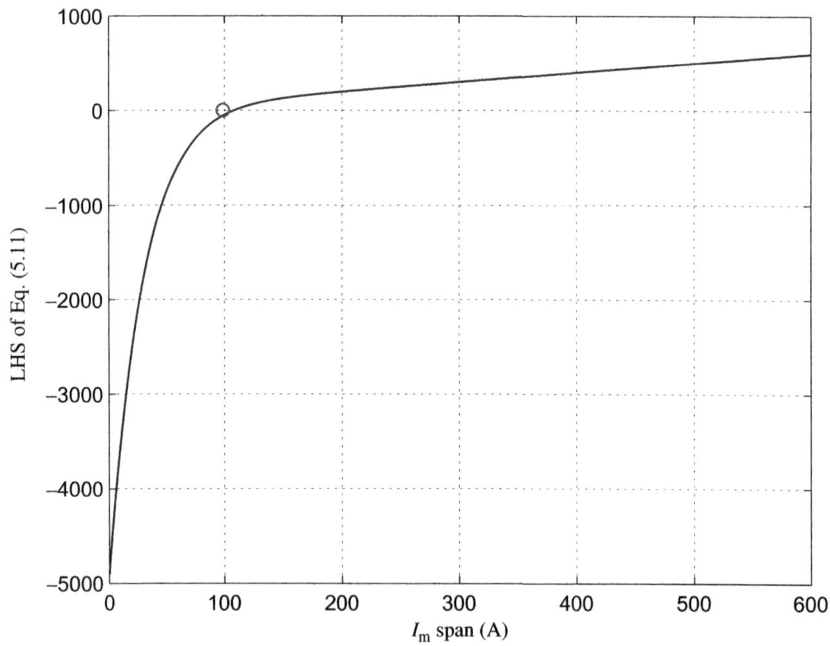

FIG. 5.17 Illustration to show the solving of Eq. (5.10) to find the value of I_m for $V_i = 800$ V.

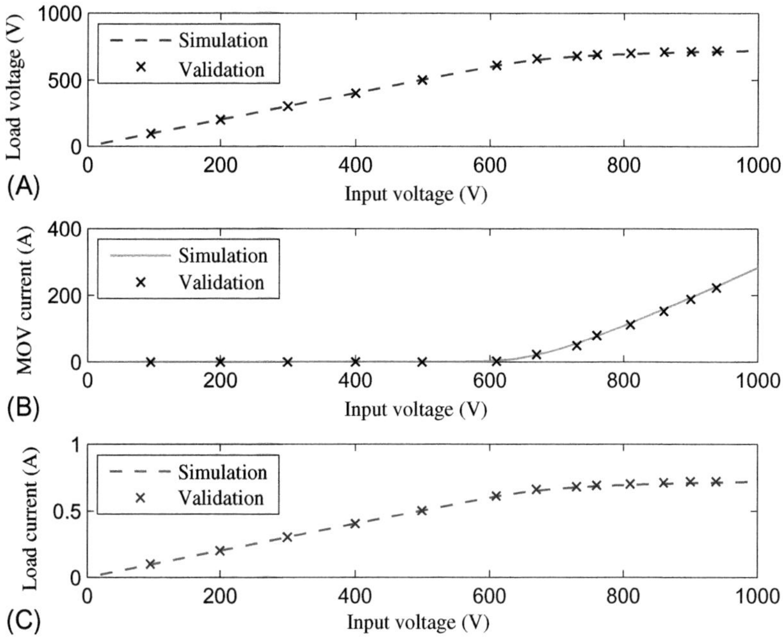

FIG. 5.18 Simulation results for the circuit of Fig. 5.17 along with experimental validation using an LSS: (A) V_i vs. V_L; (B) V_i vs. I_m; (C) V_i vs. I_L.

TABLE 5.6 Experimental Data Obtained From Testing the Circuit of Fig. 5.16

V_i (V)	V_{R_S} (V)[a]	$I_{R_S} = V_{R_S}/R_S$ (A)	V_L (V)	$I_L = V_L/R_L$ (mA)	$I_m = V_{R_S} - I_L$ (A)
95	Very small		95		
200	Very small		200		
300	Very small		300		
400	Very small		400		
500	Very small		500		
610	2	*2*	610	*610*	*1.5*
670	22	*22*	660	*660*	*21.5*
730	49	*49*	680	*680*	*48.5*
760	79	*79*	690	*690*	*78.5*
810	111	*111*	700	*700*	*110.5*
860	152	*152*	710	*710*	*151.5*
900	188	*188*	713	*713*	*187.5*
938	222	*222*	718	*718*	*221.5*

[a]Very small values indicated accordingly are not accurately measurable. Italicized values were calculated.

5.5.1.8 Study of Varistor Conductance, Power, and Energy During Surge Propagation

Once the varistor voltage $v(t)$ and current $i(t)$ variations with respect to time are known, it is possible to study the conductance, power, and energy variations using the following relationships.

$$\text{Conductance,}\quad G(t)=I(t)/V(t), \tag{5.11}$$

$$\text{Power,}\quad P(t)=V(t)I(t), \tag{5.12}$$

$$\text{Energy,}\quad E=\int P(t)dt=\int V(t)I(t)dt. \tag{5.13}$$

In Section 5.5.1.2, we saw the varistor voltage and current waveshapes during the propagation of a typical transient. These variations can be used to study the varistor conductance, power, and energy variations as illustrated in Fig. 5.19.

Detailed observations of the conductance plot (Fig. 5.19) indicate that the conductance follows the pattern of the MOV current, due to the small drop in the MOV voltage. When the conduction is at its peak, the resistance of the MOV drops to a very low value which is close to 7 Ω.

Fig. 5.19 also shows the instantaneous power variations for the varistor and this curve too tends to follow the shape of the varistor current, again because the varistor voltages stay almost constant during conduction. The last plot in Fig. 5.19 gives an idea as to the total energy absorbed by the varistor at the end of the transient. The study of energy absorption by simulation is important in predicting device failure.

5.5.1.9 Study of an MOV's Industrial Characterization That Uses a Logarithmic-Term Model

As mentioned in Section 3.2.2, the SPICE-based industrial model of an MOV is based on a logarithmic-term model for the nonlinear resistance, which is an extension of the model given by Eq. (3.2). In order to understand the industrial characterization, we will study the characterization of the MOV in this section using the logarithmic-term model

$$\log(v)=B_1+B_2\log_{10}(i)+B_3e^{-\log_{10}(i)}+B_4e^{\log_{10}(i)}, \tag{5.14}$$

where i is the current through the varistor and v is the voltage across the varistor.

We will now use the interpolation parameters given in Table 5.7 for the S20K250 varistor [12] to study the implication of each of the terms in Eq. (5.14) in shaping the V–I characteristic of the MOV.

First, let us first consider the effect of the coefficient B_1 by setting $B_2=B_3=B_4=0$. Then, from Eq. (5.14), we have

$$\log(v)=B_1=2.6830619. \tag{5.15}$$

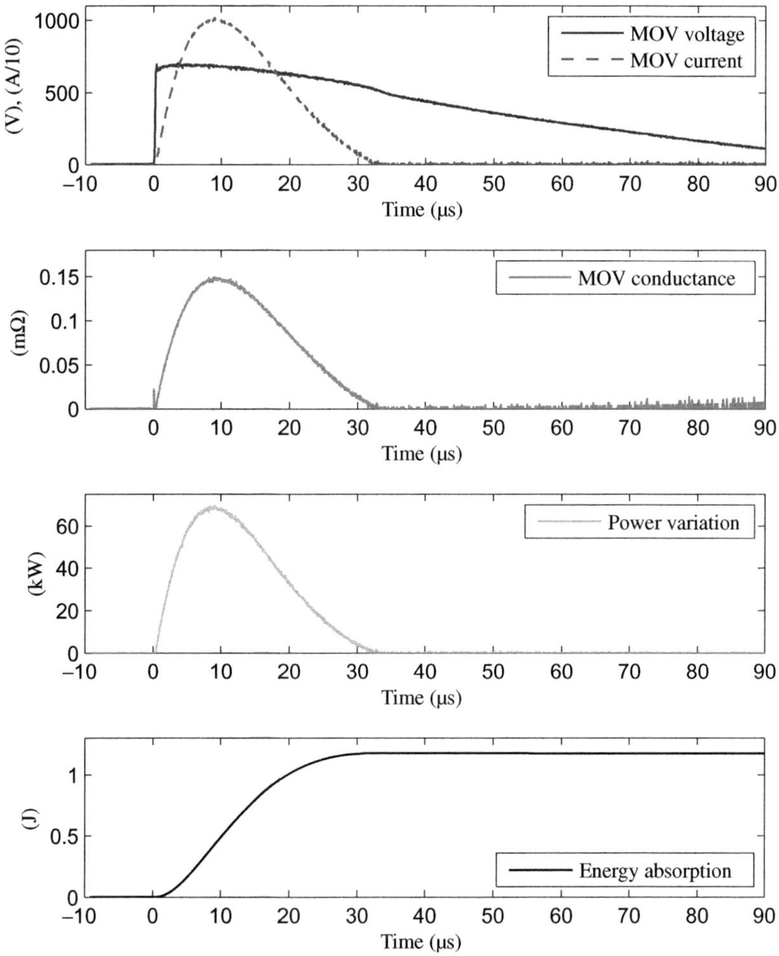

FIG. 5.19 Characterization of the varistor conductance, power dissipation, and energy absorption during the propagation of a transient.

TABLE 5.7 Parameters Used for the Basic Varistor Model

Coefficient	Varistor S20K250
B1	2.6830619
B2	0.0261918
B3	−0.0006173
B4	0.0045183

Taking antilogs, we get

$$v = 10^{2.6830619} = 482.016. \tag{5.16}$$

We see that the contribution of this term is a constant voltage offset independent of current. If we change B_1, first to 2.5830619 and then to 2.7830619, the respective values for v will be 482 and 606 V. Careful examination of the plots around the cut-in points tells us that these voltage values correspond to the voltage across the device when the current is approximately 1 A. The effect of varying B_1 on the I–V characteristic of the MOV is shown in Fig. 5.20.

Next, let us first consider the effect of the coefficient B_2 on the characteristic. If we set $B_1 = B_3 = B_4 = 0$, then, from Eq. (5.14), we have

$$\log(v) = B_2 \log_{10}(i). \tag{5.17}$$

Taking antilogs, we get

$$v = i^{B_2} \text{ or } i = v^{1/B_2} = v^{1/0.0261918} = v^{38.18}. \tag{5.18}$$

Here we have a power-law relationship between the MOV current and the voltage, similar to the relationship given by Eq. (5.3). Our coefficient of nonlinearity α in Eq. (5.3) is similar to the reciprocal of B_2.

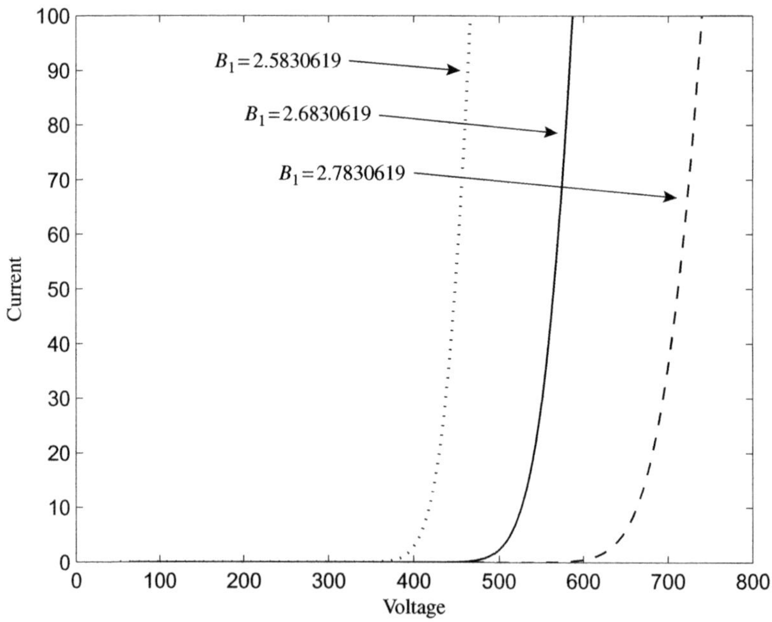

FIG. 5.20 The effect of changing the coefficient B_1, on the I–V characteristic of the MOV.

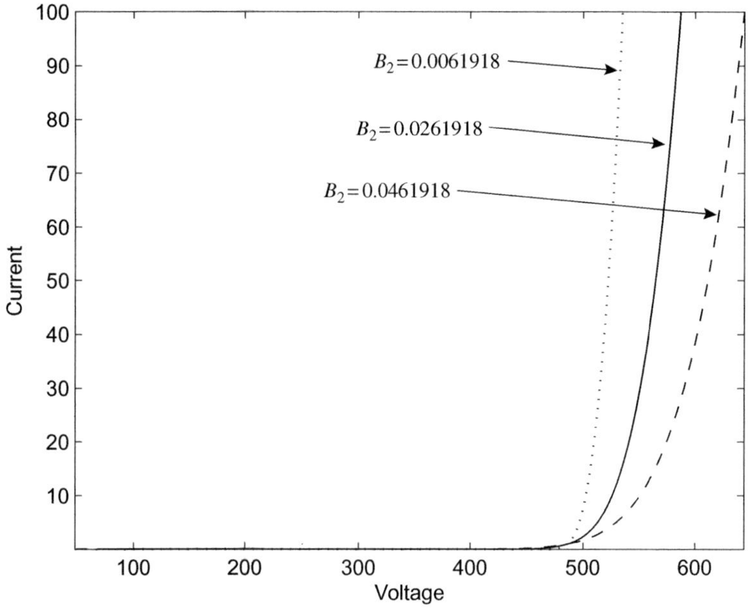

FIG. 5.21 The effect of changing the coefficient B_2, on the I–V characteristic of the MOV.

The effect of varying B_2 only on the I–V characteristic of the MOV is shown in Fig. 5.21.

The coefficient B_3 describes the V–I characteristic of the 'off' model, but has no effect on the shape of the 'on' model characteristic of the varistor. Because we are mostly interested in dissipated energies in individual components during the 'on' time of the varistor, this coefficient can be safely ignored.

Now let us consider the effect of the last coefficient B_4 on the characteristic. If we let $B_1 = B_2 = B_3 = 0$, then, from Eq. (5.14), we have

$$\log(v) = B_4 e^{\log_{10}(i)}. \tag{5.19}$$

Taking antilog and using the log identity $\log_a x / \log_a y = \log_y x$, we can show that

$$v = 10^{B_4 i^{0.434}}. \tag{5.20}$$

The B_4 term contributes to the degree of nonlinearity of the I–V characteristic, in a way similar to the B_2 term, as shown in Fig. 5.22. The changes in the degree of nonlinearity can be finer in the case of B_4.

This study of the varistor model given by Eq. (5.14) shows us that the coefficients B_1, B_2, and B_4 together can achieve a good curve fit for the I–V characteristic of the varistor. We have already seen that a similar curve fit can be obtained by the coefficient k and the exponent α, if we use the alternate model given by Eq. (5.3). In the numerical simulation work that has been done to

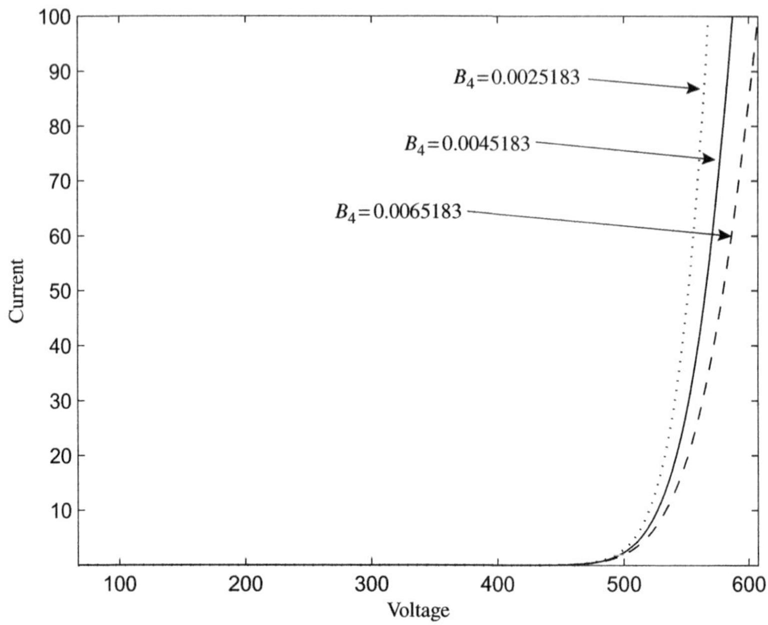

FIG. 5.22 The effect of changing the coefficient B_4, on the I–V characteristic of the MOV.

investigate surge propagation in Chapter 7, the simpler model given by Eq. (5.3) is chosen, as our interest here is focused mainly on the energy distribution within the power electronics interface.

We note that, despite its complexity, the industrial characterization studied here does not account for the hysteresis effect discussed in Section 5.5.1.2. A parallel-branch model where each branch contains a nonlinear resistance based on the model studied here is proposed in [12], to account for the hysteresis effect.

5.5.2 TVS Diode Characterization

Manufacturers such as Littelfuse offer a broad range of TVS diodes, including high peak pulse current and peak pulse power options up to 10 kA and 30 kW, respectively. A sample set of TVS diodes from their product line is given in Table 5.8. They are available in both unidirectional (unipolar) or bidirectional (bipolar) diode circuit configurations. It is quite important to note that the peak pulse power values in the Table 5.8 are occurring only for very short time periods and hence they do not creating any destructive effect on the devices.

For our experiments of characterization and model development, we selected the devices 1.5KE170CA and 1.5KE400CA from the 1.5KE series, which are bidirectional. This series includes matching unidirectional devices as well. Fig. 5.23 illustrates a typical I–V characteristic for a bidirectional

TABLE 5.8 Some Characteristics of Samples From a TVS Diode Product Selection Table [20]

Series Name	Reverse Standoff Voltage (V_R)	Peak Pulse Power Range (P_{PP})	Peak Pulse Current (I_{PP}8 × 20μs)
Surface mount: standard applications (400–5000 W)			
SMAJ	5.0–440 V	400 W	NA
SMCJ	5.0–440 V	1500 W	NA
Axial leaded: standard applications (400–5000 W)			
1.5KE	5.8–495 V	1500 W	NA
5KP	5.0–250 V	5000 W	NA
Axial leaded: high power			
30KPA	28.0–288 V	30,000 W	NA
AK10	58–430 V	NA	10,000 A
Automotive applications			
SLD	10–24 V	2200 W based on 1 μs/150 ms pulse	NA

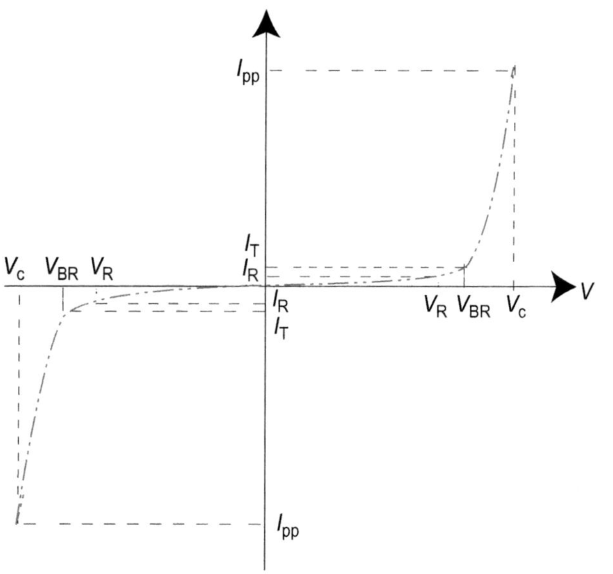

FIG. 5.23 Typical *I–V* curve characteristics for a bidirectional TVS diode [21].

TVS diode. Because the peak pulse power range is constant for all devices in the 1.5KE series, the devices with higher breakdown voltages have lower peak pulse current (I_{PP}) at the clamping voltage (V_C).

Terms and descriptions for the important symbols shown in Fig. 5.23 follow:

V_R (Stand-off voltage)—Maximum voltage that can be applied to the TVS without significant conduction.
V_{BR} (Breakdown voltage)—Measured at a specified DC test current typically 1 mA.
I_{PP} (Peak pulse current)—Identifies the maximum current the TVS diode can withstand without damage.
V_C (Clamping voltage)—This is the peak voltage that will appear across the TVS diode when subjected to a specified I_{PP} (peak impulse current).
I_R (Reverse leakage current)—Current measured at V_R.

The measuring circuit diagram used for obtaining the *I–V* characteristics of the TVS diode is identical to the one shown in Fig. 5.1. The Noiseken LSS-6110 LSS is used as an impulse source. A series of pulses with different voltage amplitudes are applied to each of the TVS diodes tested.

5.5.2.1 Waveforms Obtained and Their Analysis

The 1.5KE400CA TVS diode moves from a high impedance state to normal operation when the impulse voltage exceeds 500 V because its breakdown voltage lies above 420 V. Typical current and voltage waveforms obtained for this device in the region of normal operation are shown in Fig. 5.24.

The waveforms shown in Fig. 5.24 were obtained by setting the LSS to output an impulse of 900 V. It is clearly seen that the speed of response of conduction in this TVS diode is far superior to that of the varistor tested earlier (Fig. 5.9). In the case of the TVS diode, the current maximizes in approximately 2.5 μs, compared to 7.5 μs for the varistor. In spite of this difference, both devices are fast enough to respond to real-world transient events [22].

The dynamic *I–V* curve associated with the waveforms of Fig. 5.24 is shown in Fig. 5.25. Unlike in the case of the varistors investigated, we do not observe a significant hysteresis effect.

5.5.2.2 A TVS Diode (1.5KE170CA) I–V Characteristic for Leakage and Normal Regions of Operation

The voltage and current values in Table 5.9 were obtained by testing a TVS diode using the measuring set ups shown in Figs 5.1 and 5.13. A 1.5KE170CA TVS diode was used as the DUT.

We note that the leakage current $I_{RM} = 5\mu A$ @ $V_{RM} = 145$ V specified in the datasheet [23] is very close to corresponding datapoints recorded in Table 5.9.

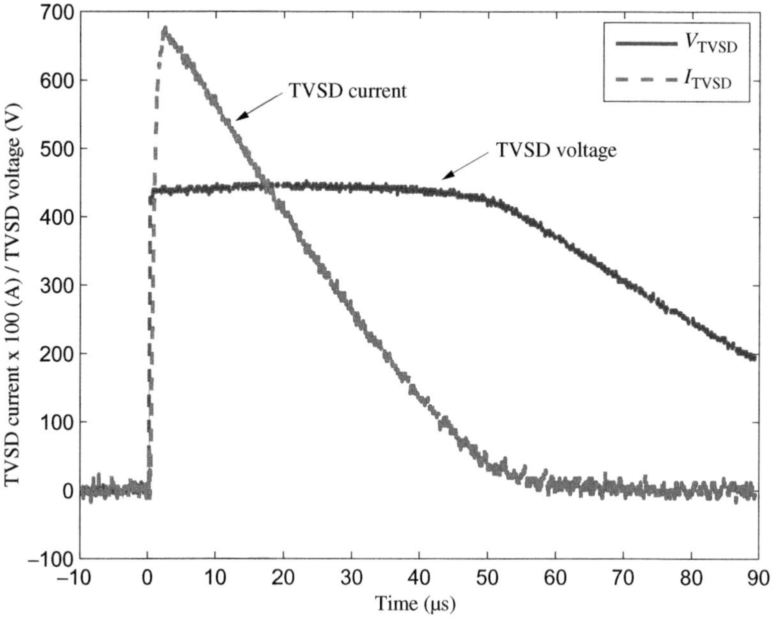

FIG. 5.24 Voltage across and the current through a TVS diode triggered by an impulse input.

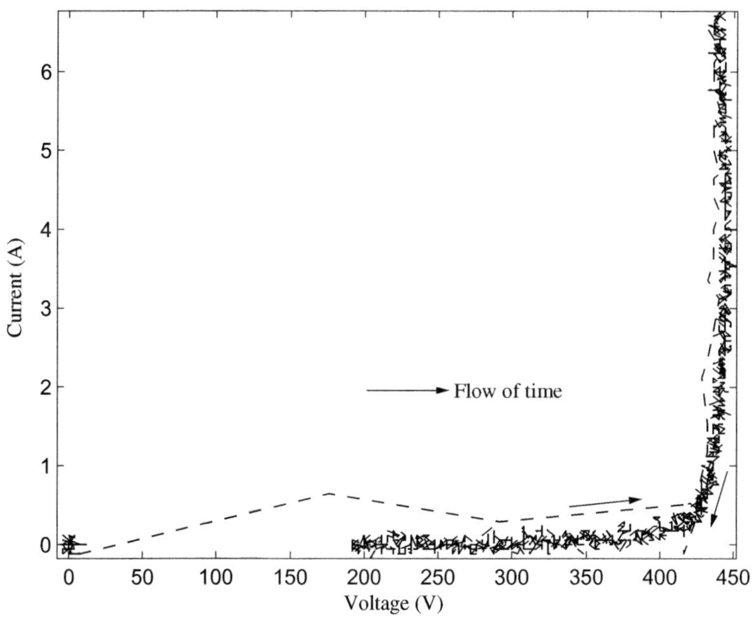

FIG. 5.25 Associated dynamic I–V curve for traces shown in Fig. 5.24 for the TVS diode 1.5KE400CA.

TABLE 5.9 Voltage and Current Readings Obtained to Characterize the Leakage and Normal Regions of Operation of a 1.5KE170CA TVS Diode

Source	V_{TVSD}	I_{TVSD}	Source	V_{TVSD}	I_{TVSD}
DC	100.3	0.92E-06	DC	169.6	35.71E-06
DC	106.2	2.12E-06	DC	169.6	44.16E-06
DC	110.6	2.94E-06	DC	169.7	73.34E-06
DC	138.5	8.67E-06	DC	170	84.05E-06
DC	169.4	17.24E-06	Impulse (200 V)	194	2.00
DC	169.5	26.93E-06	Impulse (300 V)	208	23.10

Fig. 5.26 shows the linear and log–log I–V plots done using the values recorded in Table 5.9. The nearly straight-line segments of the log–log plot suggest piecewise modelling similar to the varistor modelling done in Sections 5.5.1.5 and 5.5.1.6. Again, the *off* model and the *on* model shown in the figure may be represented by $K_1 v^{\alpha 1}$ and $K_2 v^{\alpha 2}$, respectively. Calculation of the constants K_1, K_2, α_1, and α_2 is illustrated in the following section.

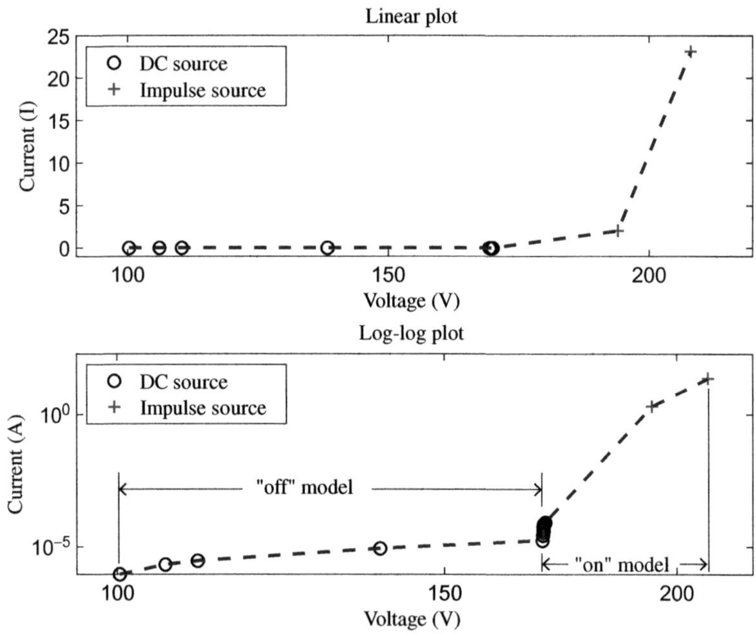

FIG. 5.26 Linear and log–log V–I plots that show the leakage and normal regions of operation for a 1.5KE170CA TVS diode.

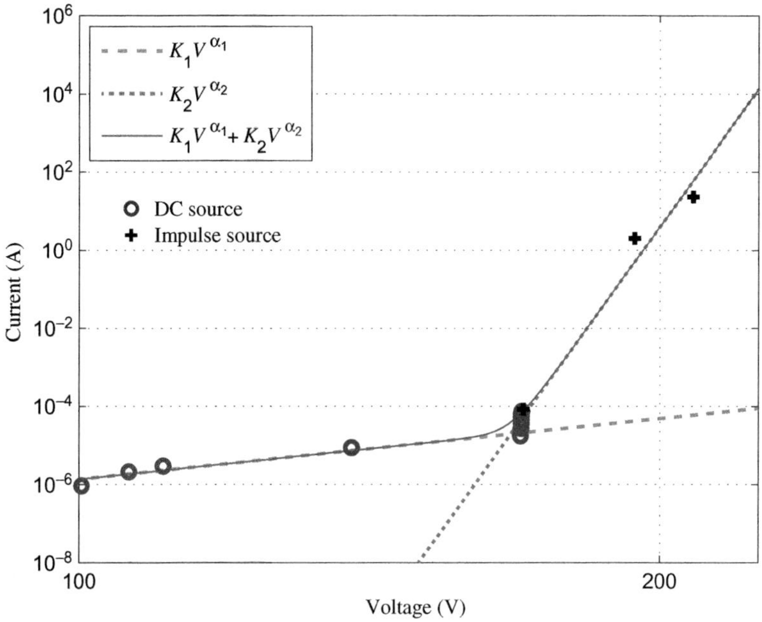

FIG. 5.27 Development of a combined model for the *off* and *on* regions of a TVS diode.

5.5.2.3 *Complete Mathematical Model for the TVS Diode*

In order to develop a combined model for the *off* and *on* regions of the TVS diode, the data points of Table 5.9 are used again in the plot of Fig. 5.27. The curve fitting exercise carried out here is very similar to the one done for the varistor, where the slope of the linear segments was found with the use of MATLAB's `polyfit` command.

The positioning of the straight line segments in Fig. 5.27, represented by $i = K_1 v^{\alpha_1}$ and $K_2 v^{\alpha_2}$, was done with the following values obtained from appropriate curve fitting using `polyfit` for the tested 1.5KE170CA TVS diode.

$K_1 = 5.5217 \times 10^{-17}$, $\alpha_1 = 5.1561$ (*off* model parameters)
$K_2 = 9.8813 \times 10^{-159}$, $\alpha_2 = 68.8466$ (*on* model parameters)

The curve for the combined model $i = K_1 v^{\alpha_1} + K_2 v^{\alpha_2}$ was also developed earlier for the varistor and was shown in Fig. 5.15.

5.5.3 Comparison of Transient Suppressor Models

The models that we developed for the 275L40C varistor and the 1.5KE170CA TVS diode are displayed side by side in Fig. 5.28 for comparison. Two important parameters, the degree of nonlinearity α, and the leakage current can be easily compared using these plots.

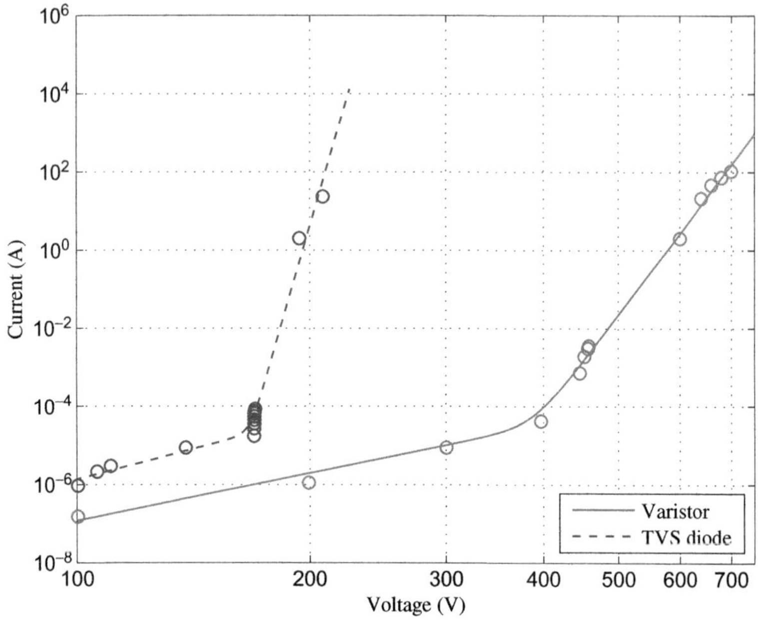

FIG. 5.28 Varistor and TVS diode *I–V* characteristics drawn on log–log graph for comparison of the degrees of nonlinearity.

It can be seen that the *on* model alpha factor is much higher for the TVS diode than that of the varistor. For higher *on* model alpha factors, the voltage–current slope of the curve becomes very steep and approaches an almost constant voltage. High *on* model alphas are desirable for clamping applications that require operation over a wide range of currents [22].

Leakage current can be an area of misconception when comparing a varistor and a TVS diode. For example, we see from Fig. 5.28, that the TVS diode leakage current is about 10 times higher at 100 V than the varistor. The leakage current of a TVS diode can be reduced by specifying a higher voltage device [22].

References

[1] S. James, Investigation of Surge Propagation in Transient Voltage Suppressors and Experimental Verification (Ph.D. thesis), University of Waikato, 2014.

[2] S.B. Karmakar, Laplace transform solution of nonlinear differential equations, Indian J. Pure Appl. Math. 11 (4) (1980) 407–412.

[3] Littelfuse, Varistor spice models, 2018. www.littelfuse.com/technical-resources_old/spice-models/varistor-spice-models.aspx. Retrieved May 2018.

[4] Littelfuse, The ABCs of MOVs: application note AN9311.6, 1999. www.secomtel.com/UpFiles/Attach/0/2007/5/17/141245.pdf. Retrieved November 2012.

[5] IEC 61000-4-5: 2005, Electromagnetic Compatibility (EMC)—Part 4–5: Testing and Measurement Techniques—Surge Immunity Test, International Electrotechnical Commission, Geneva, 2005.

[6] IEC 60060-1: 2010, High Voltage Test Techniques—Part 1: General Definitions and Test Requirements, International Electrotechnical Commission, Geneva, 2010.

[7] IEEE C62.41.2-2002, IEEE Recommended Practice on Characterization of Surges in Low-Voltage (1000 V and Less) AC Power Circuits, 2003, p. 17.

[8] Littelfuse, Littelfuse—UltraMOV varistor series, 2012. www.littelfuse.com/data/en/Data_sheets/Littelfuse_MOV_UltraMOV.pdf. Retrieved December 2012.

[9] Littelfuse, Varistor testing: application note AN9773, 1998. www.littelfuse.com/data/en/Application_Notes/an9773.pdf. Retrieved November 2012.

[10] Littelfuse, Littelfuse: C-III varistor series, 2012. www.littelfuse.com//media/electronics/datasheets/varistors/littelfuse_varistor_ciii_datasheet.pdf. Retrieved December 2012.

[11] D.C. Tayal, Electricity and Magnetism, Global media, 2009.

[12] B. Zitnik, M. Babuder, M. Muhr, R. Thottappillil, Numerical modelling of metal oxide varistors, in: Proceedings of the XIVth International Symposium on High Voltage Engineering, Tsinghua University, Beijing, China, 2005, pp. 1–6.

[13] I. Kim, T. Funabashi, H. Sasaki, T. Hagiwara, M. Kobayashi, Study of ZnO arrester model for steep front wave, IEEE Trans. Power Delivery 11 (1996) 834–841.

[14] R.A. Jones, P.R. Clifton, G. Grotz, M. Lat, F. Lembo, D.J. Melvold, D. Nigol, J.P. Skivtas, A. Sweetana, D.F. Goodwin, J.L. Koepfinger, D.W. Lenk, Y. Latour, R.T. Leskovich, Y. Musa, J. Adinzh, K. Stump, E.R. Taylor, Modeling of metal oxide surge arresters, IEEE Trans. Power Delivery 7 (1992) 302–309.

[15] C. Chrysanthou, J. Boksiner, Analysis of coordination between primary and secondary protectors, IEEE Trans. Power Delivery 12 (1997) 1501–1507.

[16] Littelfuse varistors—basic properties, terminology and theory: application note AN9767.1, 1998. www.digikey.co.nz/Web%20Export/Supplier%20Content/Littelfuse_18/PDF/LF_Varistors.pdf. Retrieved November 2012.

[17] IEEE C62.33-1982, IEEE Standard Test Specifications for Varistor Surge-Protective Devices, 1982.

[18] L.M. Levinson, H.R. Philipp, The physics of metal oxide varistors, J. Appl. Phys. 46 (1975) 1332–1341.

[19] G.Z. Zang, J.F. Wang, H.C. Chen, W.B. Su, C.M. Wang, P. Qi, Nonlinear electrical behaviour of the WO3-based system, J. Mater. Sci. 39 (2004) 4373–4374.

[20] Littelfuse, Transient voltage suppression (TVS) diode products, 2008. www.newark.com/pdfs/techarticles/littelfuse/TVS_Diodes.pdf. Retrieved November 2012.

[21] Littelfuse, Littelfuse: 1.5KE transient voltage suppression diode series, 2012. www.mouser.com/ds/2/240/Littelfuse_TVS-Diode_1.5KE-41903.pdf. Retrieved September 2013.

[22] Littelfuse, Transient suppression devices and principles: application note AN9768, 1998. www.digikey.co.nz/Web%20Export/Supplier%20Content/Littelfuse_18/PDF/LF_TransientSuppressionDevices.pdf. Retrieved November 2012.

[23] Fairchild, Transient voltage suppressors 1V5KE6V8(C)A–1V5KE440(C)A, 2002. projects.uniprecision.com/sc_upload/images/1.5KE36AFAIRCHILD%20%20%20%20KE07575.pdf. Retrieved September 2013.

Chapter 6

Investigation of Surge Propagation in Linear Systems

As mentioned in the beginning of Chapter 5 significant parts of the surge protection device (SPD) to be analysed are linear. In order to perform experimental validations for the investigations of transient propagation, a lightning surge simulator (LSS) will be used as the source of the transient. As seen in Section 5.4.4.1, the LSS equivalent circuit is a linear circuit with R, L, and C elements. In addition, important surge protection devices such as varistors and transient voltage suppressor (TVS) diodes also have linear circuit elements in their mathematical models. As a result, investigation of transient propagation through related linear circuitry is the way forward in preparing for the inclusion of the nonlinear devices in Chapter 7 to complete the study transient propagation through an SPD.

Among standard methods available for the analysis of transient response in linear circuits, Laplace transform stands out since it gives systematic and routine solutions for differential equations. Here we apply Laplace methods to study surge propagation through the linear circuits of interest to this project. The solutions obtained for the output responses of the LSS equivalent circuit will be verified with experimentally obtained waveforms. Numerical simulation will also be used for validation.

In Sections 6.2 and 6.3 numerical simulations are used to study the behaviour of a metal oxide varistor (MOV) driven by the LSS with the help of limited linear models for the MOV. These simulations will be validated using LTspice which is a SPICE (Simulation Program with Integrated Circuit Emphasis) based circuit simulation application.

6.1 Analysis of LSS Responses Using Laplace Methods

Let us consider the Noiseken LSS 6110 equivalent circuit shown in Fig. 6.1: We will be using this LSS for experimental validation of our investigations throughout this research.

Here we need to solve for the open-circuit voltage and the short-circuit current. In producing these waveforms the LSS behaves as a combination wave

Design of Transient Protection Systems. https://doi.org/10.1016/B978-0-12-811664-7.00006-9

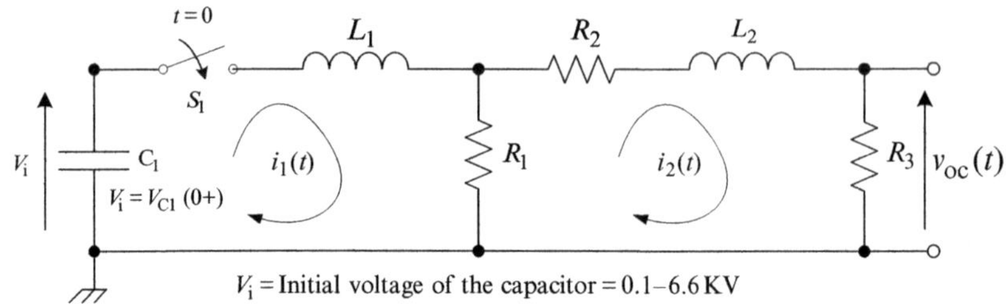

FIG. 6.1 The LSS equivalent circuit for finding the 1.2/50 μs open-circuit voltage (Noise Laboratory Co., Ltd., Japan).

(hybrid) generator. The open-circuit voltage should match the 1.2/50 μs waveform and the short-circuit current should match the 8/20 μs waveform, both specified by the IEC 61000-4-5 standard.

6.1.1 Laplace Solution for 1.2/50 μs Open-Circuit Voltage and its Validations

The capacitor C_1 of the LSS (Fig. 6.1) can be charged to any value from 100 V to 6.6 kV in 100 V steps. When switch S_1, a mercury relay discharge switch, is closed, the capacitor C_1 discharges through the wave-shaping circuit to produce the required open-circuit voltage across R_3. In order to solve this 2-loop LSS circuit for the open-circuit voltage, we will assume that the loop currents are i_1 and i_2. In order to match the given standards, the model parameters given in Table 6.1 are specified by the manufacturer.

The Laplace transformed network representation of Fig. 6.1 is given in Fig. 6.2.

The loop equations for the circuit shown in Fig. 6.2 are given by

$$-\frac{V_i}{s}+\left(sL_1+\frac{1}{sC_1}\right)\tilde{I}_1(s)+R_1\left[\tilde{I}_1(s)-\tilde{I}_2(s)\right]=0 \tag{6.1}$$

TABLE 6.1 Parameters for the Noiseken LSS-6110

Parameter	Value	Parameter	Value
C_1	10 μF	R_1	7 Ω
L_1	2.7 μH	R_2	1.1 Ω
L_2	6 μH	R_3	1 MΩ

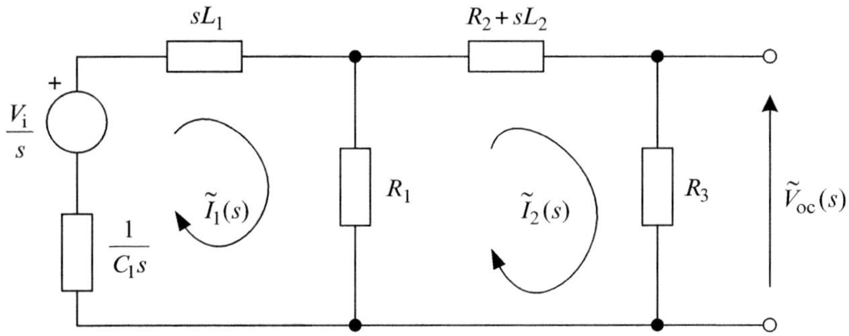

FIG. 6.2 Transformed network for the circuit in Fig. 6.1.

$$R_1\left[\tilde{I}_2(s)-\tilde{I}_1(s)\right]+(R_2+R_3+sL_2)\tilde{I}_2(s)=0 \tag{6.2}$$

Rearranging Eq. (5.1) and Eq. (5.2),

$$\tilde{I}_1(s)\left[R_1+sL_1+\frac{1}{sC_1}\right]+\tilde{I}_2(s)[-R_1]=\frac{V_i}{s} \tag{6.3}$$

$$I_1(s)[-R_1]+I_2(s)[R_1+R_2+R_3+sL_2]=0 \tag{6.4}$$

Using Cramer's rule to solve for $\tilde{I}_2$

$$\tilde{I}_2(s)=\frac{\begin{vmatrix}\left[R_1+sL_1+\frac{1}{sC_1}\right] & \frac{V_i}{s}\\ -R_1 & 0\end{vmatrix}}{\begin{vmatrix}\left[R_1+sL_1+\frac{1}{sC_1}\right] & -R_1\\ -R_1 & [R_1+R_2+R_3+sL_2]\end{vmatrix}} \tag{6.5}$$

The fact that the determinant of a matrix A is denoted by $|A|$ is used in Eq. (6.5).

Simplifying Eq. (6.5),

$$\tilde{I}_2(s)=\frac{R_1\left(\frac{V_i}{s}\right)}{\left(R_1+sL_1+\frac{1}{sC_1}\right)(R_1+R_2+R_3+sL_2)-R_1^2} \tag{6.6}$$

Further simplifying

$$\tilde{I}_2(s)=\frac{C_1R_1V_i}{(sC_1R_1+s^2L_1C_1+1)(R_1+R_2+R_3+sL_2)-sC_1R_1^2} \tag{6.7}$$

After expanding the denominator and rearranging, if we let

$$a_0=C_1R_1V_i,$$

$$b_3 = L_1 L_2 C_1,$$
$$b_2 = C_1 R_1 L_2 + (R_1 + R_2 + R_3) L_1 C_1,$$
$$b_1 = (R_1 + R_2 + R_3) C_1 R_1 + L_2 - C_1 R_1^2, \text{ and}$$
$$b_0 = R_1 + R_2 + R_3$$

then Eq. (5.7) reduces to

$$I_2(s) = \frac{a_0}{b_3 s^3 + b_2 s^2 + b_1 s + b_0} \tag{6.8}$$

When the MATLAB Symbolic Toolbox command ilaplace is used to find the inverse Laplace transform of Eq. (5.8), we get

$$i_2(t) = a_0 \sum_{\alpha=\%1} \frac{e^{\alpha t}}{3b_3\alpha^2 + 2b_2\alpha + b_1} \tag{6.9}$$

where $\%1 =$ roots of $(b_3 s^3 + b_2 s^2 + b_1 s + b_0)$.

The LSS output voltage is given by $v_{oc}(t) = R_3 i_2(t)$. The code written in MATLAB to find v_{oc} is given in Appendix C.1. This code also includes data generated by the SPICE simulation, which is used here to validate the Laplace solution. The resulting plots for an initial capacitor voltage of 1 kV are given in Fig. 6.3. The SPICE implementation of the circuit in Fig. 6.1, used for the validating simulation, is given in Appendix C.2.

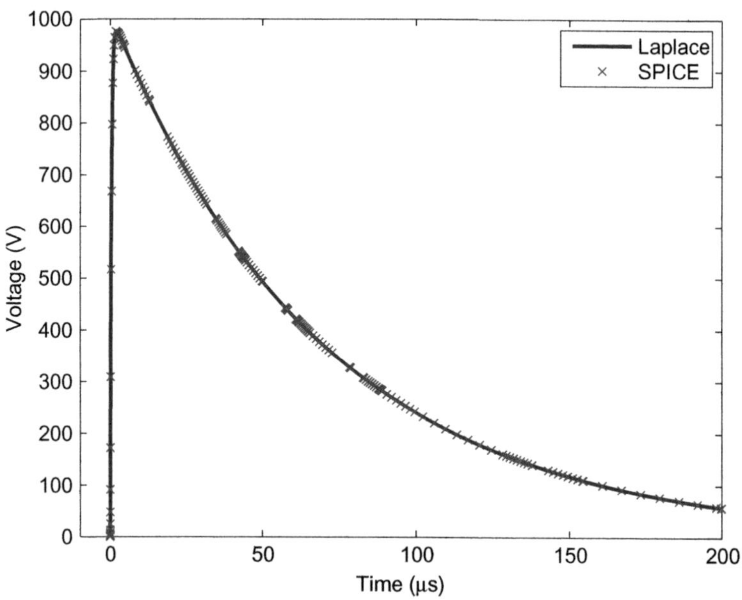

FIG. 6.3 Laplace solution for the open-circuit voltage of the LSS validated with a SPICE simulation.

It is seen that the Laplace solution can be matched very well by SPICE simulation.

In addition, the Laplace solution of the short-circuit current was validated with the experimental result from the LSS. The Tektronix TPS2024 oscilloscope which was used to display the experimental waveforms captures the data by saving each of its input channels in a *comma-separated values* (CSV) file. All CSV files generated by a single display event can be combined into a single CSV file and this file is imported by MATLAB using the command `csvread` to produce the validation curve. Fig. 6.4 shows the results obtained for the Laplace solution and the experimental result and the two curves match well during both rising and decaying parts of the surge.

In studying the earlier waveform generated by the circuit of Fig. 6.1, we found it to be sensitive to the component values C_1, L_1, and R_1 in the first loop, provided R_3 was large. The very short rise-time was sensitive to the value of L_1. In order to understand the decaying part of the open-circuit voltage, we have plotted the curves again in Fig. 6.5, but this time with voltage indicated by a log scale. The straight line seen for the decaying voltage suggests that only a single-time-constant circuit is dominant during this period in spite of the circuit having three reactive components. By changing the values of the reactive components it was found that the dominant time constant for the decaying portion is decided by the first loop component values of C_1 and R_1 (R_1C_1 time constant $= 70\,\mu s$).

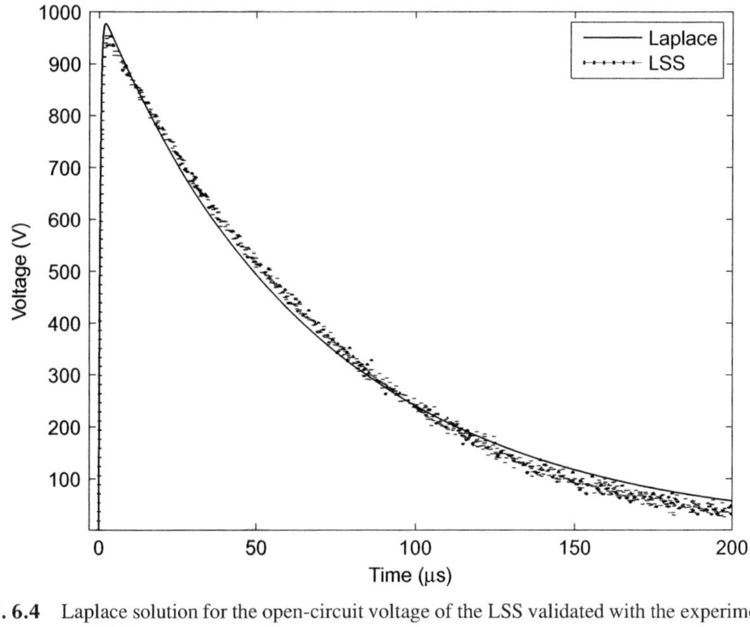

FIG. 6.4 Laplace solution for the open-circuit voltage of the LSS validated with the experimental result from the LSS.

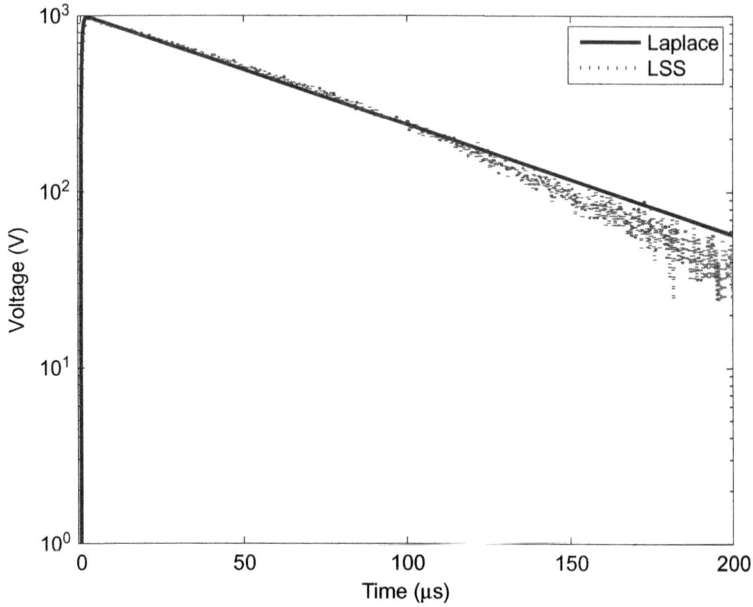

FIG. 6.5 Semilog plots for the curves shown in Fig. 6.4.

Here the experimental curve deviates from the straight line (Laplace solution) after about 100 μs. This deviation is not sensitive to any of the components in the LSS equivalent circuit and is thought to be due to stray reactive components coming to play in the coupling circuit of the LSS.

6.1.2 Comparison of Laplace Solutions With Numerical Simulation

For reasons indicated in Section 5.3, it is important to develop robust numerical simulation techniques to investigate transient propagation through nonlinear circuit elements.

For the LSS circuit shown in Figs 6.1 and 6.2, the following network equations can be formulated as a system of coupled first-order equations which govern the dynamic behaviour of the network.

$$\frac{di_1}{dt} = j \tag{6.10}$$

$$\frac{dj}{dt} = -aj - (b - ac)i_1 - adi_2 + \frac{V_{C1}(0+)}{L}\delta(t) \tag{6.11}$$

$$\frac{di_2}{dt} = ci_1 - di_2 \tag{6.12}$$

where $a=R_1/L_1$, $b=1/L_1C$, $c=R_1/L_2$, and $d=R_1+R_2+R_3/L_2$.

Again the impulse response that results from the delta function input in Eq. (6.11) can be taken care of by using the initial condition that arises from the impulse input, and making the input itself equal to zero. Hence the initial conditions would be $i_1(0)=0$, $j(0)=V_{C1}(0+)/L$, and $i_2(0)=0$.

State equations (6.10)–(6.12) were included in a MATLAB program to be used by the Euler method and by the more accurate rk4fixed and ode15s DE solvers. The MATLAB code LSS_OCV_Euler_RK4.m used to simulate the open-circuit voltage is made available in Appendix C.3. Once i_2 has been obtained, the open-circuit voltage can be determined by finding the voltage drop across resistor R_3. A comparison of the RK4-based simulation with the Laplace solution of the open-circuit voltage is given in Fig. 6.6A.

The MATLAB-based simulations were run on a Windows-based PC which had 3 GB of RAM. For the simulation of the extremely fast rising open-circuit voltage, rk4fixed simulation took as much as 17 min to simulate a 100 μs waveform. The simpler less accurate Euler method carried out the same task in 20 s. Both methods worked satisfactorily for a step size (Δt) of 10 ps. The variable step MATLAB ODE solver ode15s which is suitable to handle "stiff" problems produced satisfactory results in 30 s.

The LSS short-circuit current was also simulated using rk4fixed solver. When we short circuit the output of the LSS in order to determine the short-circuit current we will be short circuiting the resistor R_3. Hence we could find i_2 provided we substitute $R_3=0$ at the beginning of all calculations. The same MATLAB code can be used simply by setting the value of R_3 to zero. The simulation worked satisfactorily with step sizes ranging from 100 ns to 10 ps. A comparison of the rk4fixed-based simulation with the Laplace solution of the short-circuit current is given in Fig. 6.6B.

6.1.3 Laplace Solution for 8/20 μs Short-Circuit Current Validated With Experimental Data From the LSS

Here we will validate the Laplace solution for the short-circuit current of the LSS with the experimental result from the LSS. The resulting short-circuit current for an initial capacitor voltage of 1 kV is shown in Fig. 6.7A along with its validation obtained experimentally from the LSS.

Fig. 6.7B shows the absolute error between the Laplace solution and the experimental validation. We found the area of maximum error to be sensitive in particular to the values of C_1 and R_1 and the short-circuit current curves of Fig. 6.8 were obtained with the changed values of $C_1=9.0$ μF and $L_1=2.0$ μH. We see that the maximum error value has been reduced by about 40% in this case.

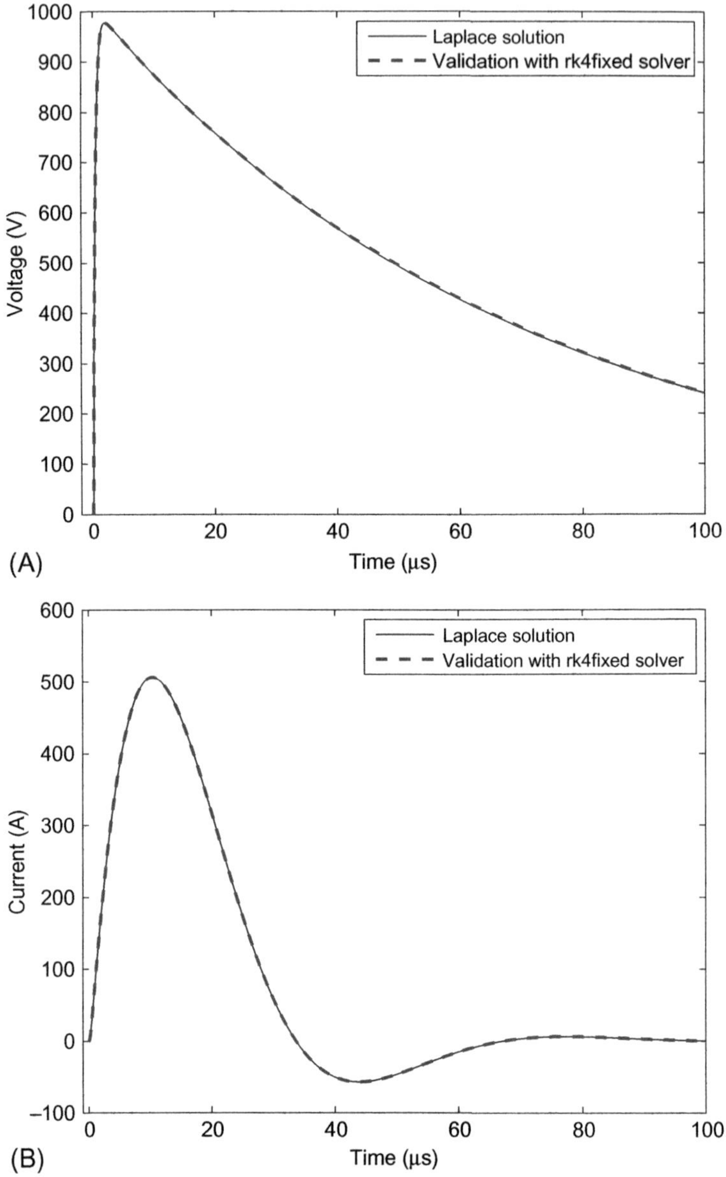

FIG. 6.6 rk4fixed solver-based validations of Laplace solutions with $\Delta t = 10\,\text{ps}$: (A) LSS open-circuit voltage and (B) short-circuit current.

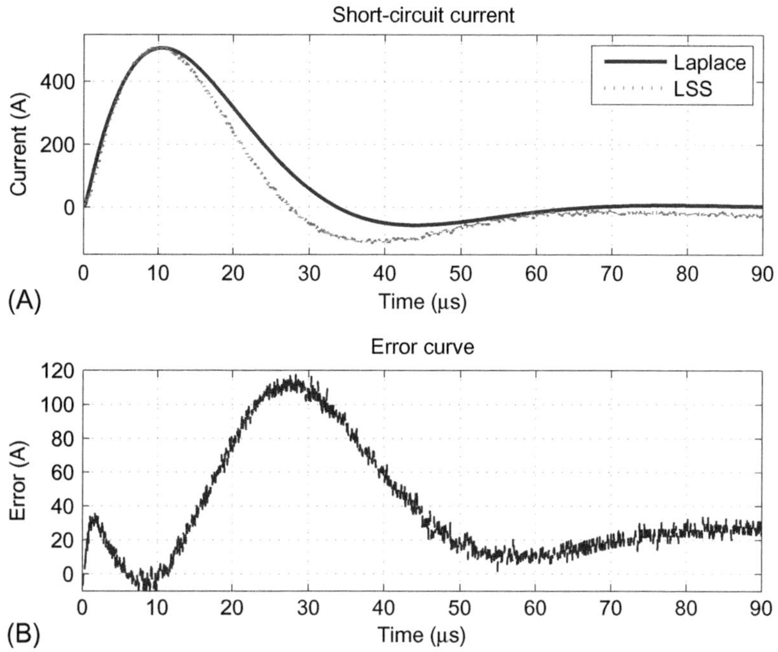

FIG. 6.7 (A) Laplace solution for the short-circuit current of the LSS validated with the experimental result from the LSS. (B) Discrepancy between the Laplace solution and its validation.

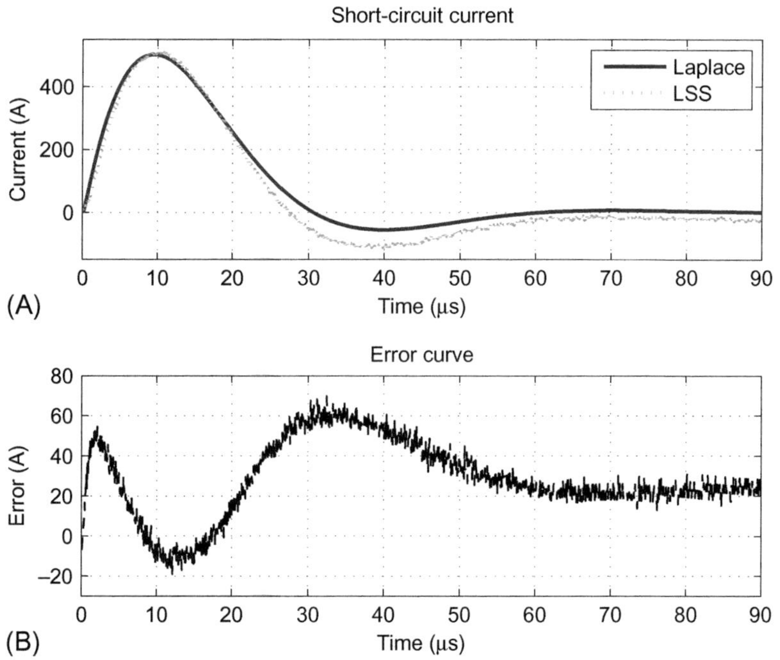

FIG. 6.8 (A) Laplace solution and its validation of Fig. 6.7A changed for better comparison by minor changes to LSS circuit values. (B) Maximum error has reduced by about 40% compared to Fig. 6.7B.

6.2 Investigation of Surge Propagation Through an MOV in the Off State

Chapter 7 investigates surge propagation through a 2-wire version of a category B protection unit shown in Fig. 2.4. Such a unit will have larger protection devices to handle the larger transients anticipated for category B. At the front end of this circuit is an MOV to provide the first level of protection [1]. We have already seen in Chapter 5 that an MOV in the "off" state can be modelled by a purely linear circuit.

In order to build the complete equivalent circuit for this investigation, we need to consider the MOV equivalent circuit for its leakage region which was presented in Fig. 5.6A as well as the equivalent circuit of the Noiseken LSS impulse generator given in Fig. 5.6B. The combined circuit shown in Fig. 6.9 disregards R_{OFF} of the MOV as it approaches $10^9\,\Omega$ in the leakage region.

In order to analyse i_3, the current through the MOV in its "off" state, we consider the loop currents of Fig. 6.9. The state equations for the previous circuit can be formulated as following by using five state variables: three currents and the time derivatives of the currents i_1 and i_3, which are j_1 and j_3, respectively. A complete working to arrive at Eqs (6.13)–(6.17) is shown in Appendix C.4.

$$\frac{di_1}{dt}=j_1 \tag{6.13}$$

$$\frac{dj_1}{dt}=-aj_1-(b-\mathrm{ac})i_1-adi_2+aei_3+\frac{V_{C1}(0+)}{L_1} \tag{6.14}$$

$$\frac{di_2}{dt}=ci_1-di_2+ei_3 \tag{6.15}$$

$$\frac{di_3}{dt}=j_3 \tag{6.16}$$

$$\frac{dj_3}{dt}=-fj_3+cgi_1-gdi_2+(eg-h)i_3 \tag{6.17}$$

where $a=R_1/L_1$, $b=1/L_1C_1$, $c=R_1/L_2$, $d=R_1+R_2+R_3/L_2$, $e=R_3/L_2$, $f=(R_3+R_S)/L_m$, $g=R_3/L_m$, and $h=1/L_mC_m$.

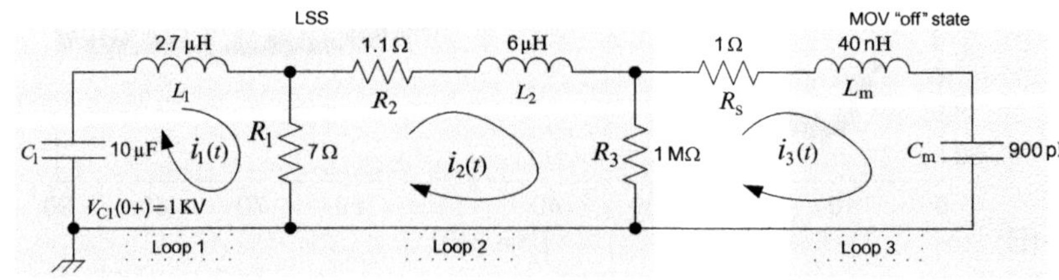

FIG. 6.9 MOV "off" state driven by the LSS (combined equivalent circuit).

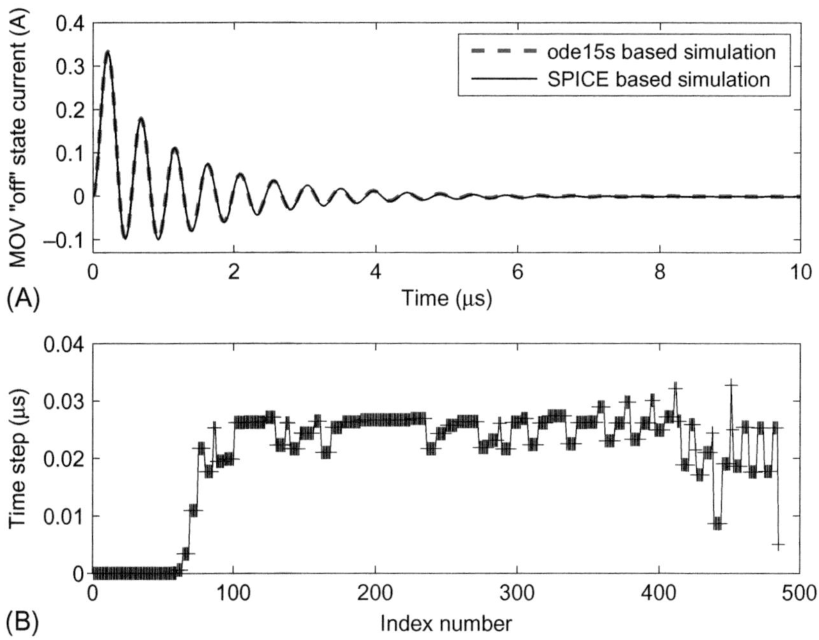

FIG. 6.10 (A) Numerical simulation of the "off" state current in a MOV validated with a SPICE-based simulation; (B) variation in the time step during the simulation; $V_{C1}(0+)=100$ V.

The initial conditions for solving the previous circuit are $i_1(0)=0$, $j_1(0)=V_{C1}(0+)/L$, $i_2(0)=i_3(0)=j_3(0)=0$, where $V_{C1}(0+)$ is the initial voltage on the capacitor C_1.

The simulation plot for i_3, the MOV current in the "off" state is shown in Fig. 6.10A. It shows a high frequency oscillation at the start which decays rapidly. This simulation, for the component values shown in Fig. 6.9, presented a "stiff" problem to the MATLAB's standard ODE solver ode45, as the previous state equations represent several processes of which one has a very small time constant. When ode45 is used the MATLAB application goes into a busy mode and fails to produce the required results within reasonable time. The difficulty was overcome by using ode15s, which is a stiff ODE solver provided by MATLAB. When ode15s is used, the Windows PC mentioned earlier completes the simulation within 1 s. The code using LSS_MOV_three_loops_ode15s.m is given in Appendix C.5.

As mentioned in Section 5.5.1.2 the ringing is due to the MOV capacitance resonating with the MOV lead inductance and the inductance L_2 of the LSS equivalent circuit. A quantitative analysis tells us that most of this current flows through the external current-measuring resistance R_S and the LSS resistances R_1 and R_2; hence these resistances are primarily responsible for the decay in the resonance of the MOV current.

6.3 Investigation of Surge Propagation in Assumed Fixed States of the MOV

In Section 5.5.1.2, we closely looked at the two main regions of operation of an MOV viz. the leakage region and the normal operation region. Here we will study the propagation of a surge through an MOV assuming that it operates at a fixed point in its performance characteristic. Although it is unlikely to operate in such a situation, it would help us understand the behaviour of an MOV under stress using a linear model. The three operating points are shown in the *I*–*V* characteristic of Fig. 6.11 which we developed for the varistor 275L40C in Section 5.5.1.5. They correspond to points in the *leakage region* (34 MΩ), the *knee* (300 Ω), and the *normal operation region* (7 Ω) of the characteristic. Since the MOV's nonlinear resistance is represented by a fixed resistance R_m at each of these points, the overall circuit analysed, shown in Fig. 6.12, remains a linear circuit at each one of these operating points.

In order to numerically simulate the previous circuit and to study the current through the fixed states of the MOV, consider the four loop currents as shown in Fig. 6.12. The state equations for the previous circuit which are given as follows use six state variables; two of the loop equations being second-order DEs were converted to normal form (a system with two coupled first-order DEs). The four loop currents and the time derivatives of the currents i_1 and i_3 are the state

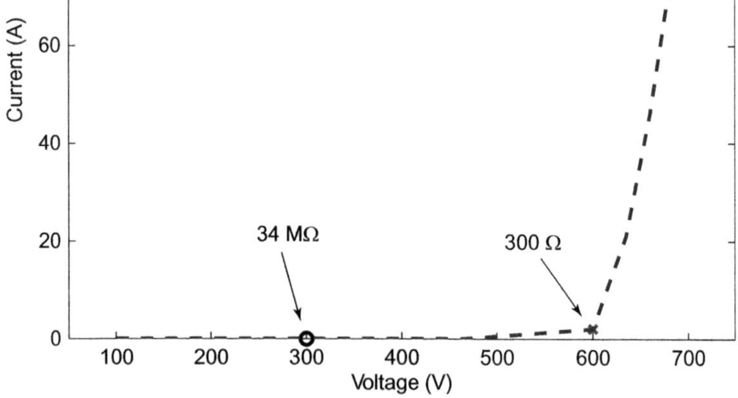

FIG. 6.11 The *I*–*V* characteristic for the 275L40C varistor with a selection of operating points.

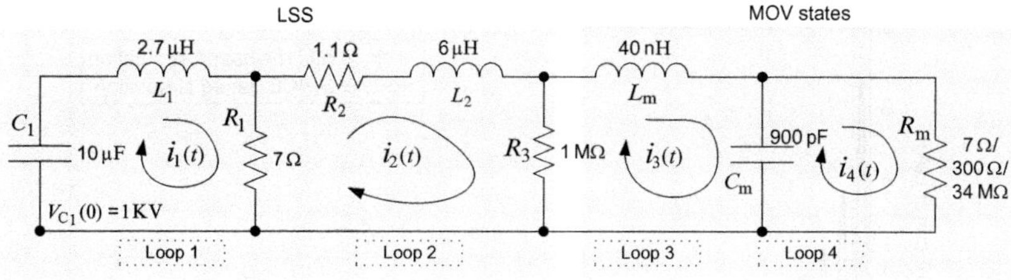

FIG. 6.12 Fixed MOV states driven by transients from the LSS.

variables. The complete derivation of the state equations (Eqs 6.18–6.23) is included in Appendix C.6.

$$\frac{di_1}{dt} = j_1 \tag{6.18}$$

$$\frac{dj_1}{dt} = -aj_1 - (b - ac)i_1 - adi_2 + aei_3 + \frac{V_{c1}(0+)}{L_1} \cdot \delta(t) \tag{6.19}$$

$$\frac{di_2}{dt} = ci_1 - di_2 + ei_3 \tag{6.20}$$

$$\frac{di_3}{dt} = j_3 \tag{6.21}$$

$$\frac{dj_3}{dt} = -fj_3 + cgi_1 - gdi_2 + (eg - h)i_3 + hi_4 \tag{6.22}$$

$$\frac{di_4}{dt} = ki_3 - ki_4 \tag{6.23}$$

where $a = R_1/L_1$, $b = 1/L_1C$, $c = R_1/L_2$, $d = R_1 + R_2 + R_3/L_2$, $e = R_3/L_2$, $f = g = R_3/L_m$, $h = 1/L_mC_m$, and $k = 1/C_mR_m$.

The initial conditions for the simulation are $i_1(0) = 0$, $j_1(0) = V_{C1}(0+)/L$, $i_2(0) = i_3(0) = j_3(0) = i_4(0) = 0$, where $V_{C1}(0+)$ is the initial voltage of the LSS capacitor or the amplitude of the surge voltage generated by the LSS.

In general, ODE solvers have to solve the set of differential equations governing the problem for each time step during the simulation. If the Jacobian matrix of the differential equations is not provided, the efficiency and the accuracy of the simulation suffers. For the three cases studied, the accuracy of the simulations was restricted to the first 7 to 36 μs, since the ODE solver had to calculate the Jacobian matrix of the differential equations for each iteration of the simulation. When the Jacobian matrix was provided, long duration simulations were possible. A complete listing of the code used to simulate the $R_m = 7$ Ω instance, LSS_MOV_four_loops_Rm_7ohms.m is given in Appendix C.7.

Fig. 6.13 shows the simulations for the three cases mentioned. The simulation result for the leakage region ($R_m = 34$ MΩ) is familiar from the discussion

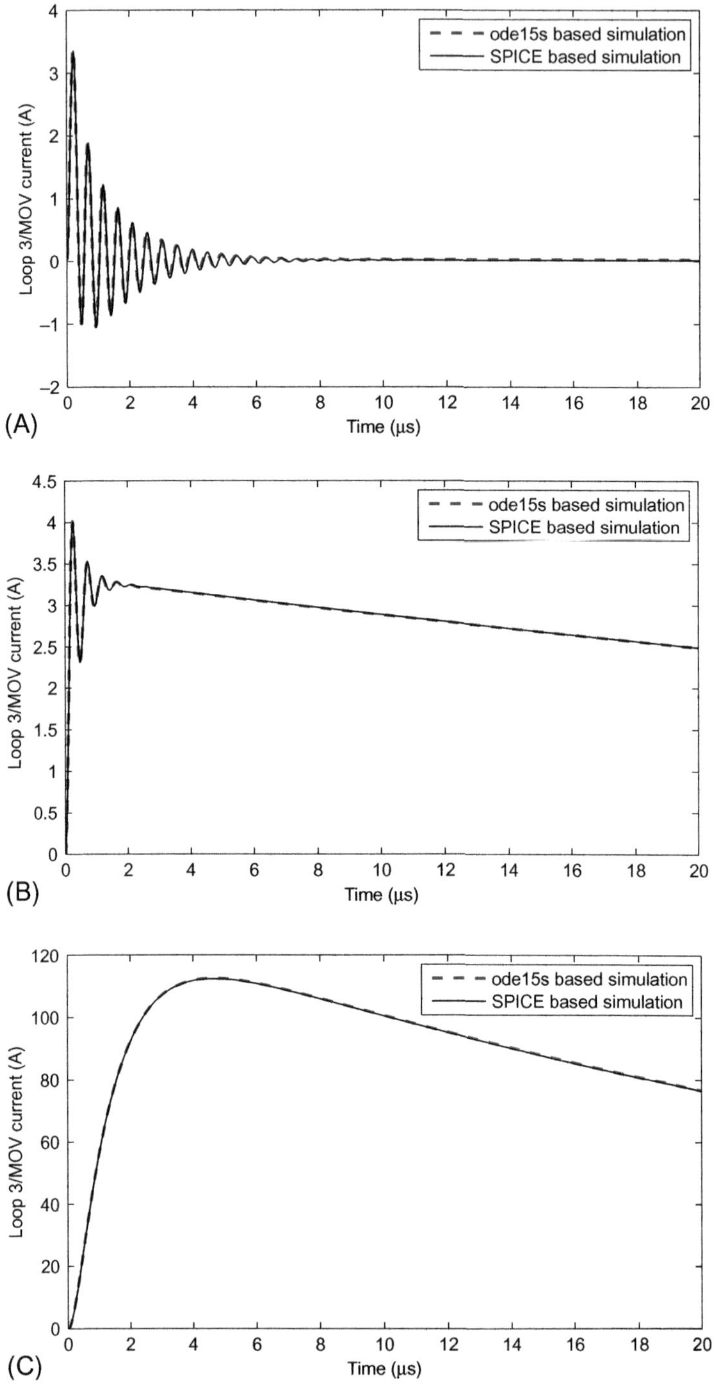

FIG. 6.13 Numerical simulation of assumed fixed states of the MOV validated with SPICE-based simulations (A) $R_m = 34\ \text{M}\Omega$, (B) $R_m = 300\ \Omega$, (C) $R_m = 7\ \Omega$.

in Section 5.5.1.2. The ringing is due to a resonating effect of L_2, L_m, and C_m, and it dies down due the resonating current flowing through R_1 and R_2.

At the knee of the I–V characteristic ($R_m = 300\ \Omega$), significant current flows through the lowered MOV resistance: smaller ringing current flows for a shorter period of time. Fig. 6.13C shows the simulation for the point ($R_m = 7\ \Omega$), well into the normal operation region of the *I–V* curve of the MOV. No ringing is seen in this case. Hence the role played by the lead inductance and the capacitance of the MOV is negligible when the device is operating in its normal region of operation. Therefore it is possible to omit one or both of these components from the MOV equivalent circuit during an investigation of surge propagation as done in Chapter 7.

All numerically simulated plots of Fig. 6.13 were validated with SPICE-based simulations using LTspice. We see that the numerical simulations match very well with the validations.

6.4 Chapter Summary

This chapter has investigated transient propagation through several purely linear circuits which are closely related to the nonlinear surge protection circuit that we intend investigating in this research. We started these investigations by analysing the LSS output waveforms using Laplace transforms and validating them using experimental data as well as numerical and SPICE simulations. Although the Laplace method is very useful in providing accurate results for these linear circuits we will not pursue this method further due to its limitations in handling circuits containing nonlinear device models such as the power-law-based model of an MOV.

The numerical simulations tried and tested here will be put to good use in the investigation of nonlinear circuits presented in Chapter 7.

The following points highlight the ways in which we were able to improve the MATLAB coding of our circuits for fast and accurate simulations:

- The MATLAB ODE solvers ode15s and ode23t enabled us to simulate circuits such as the one given in Section 6.2, where an off state model of an MOV was driven by the LSS. Due to the "stiff" nature of the problem the standard MATLAB ODE solver ode 45 could not handle this problem.
- By supplying the Jacobian matrix to the MATLAB ODE solver we were able to simulate the circuit given in Section 6.3, where a given fixed state of the MOV was driven by the LSS. When the Jacobian was not included in the coding, only short duration simulations were possible.

With these improvements to MATLAB-based numerical simulations we can proceed to simulations of Chapter 7, where the study focuses on the surge propagation through nonlinear surge protection circuitry. Here we also used a SPICE simulator for simulating our electric circuits. These SPICE simulations are useful in validating the numerical simulations. While continuing with the numerical simulations to study the behaviour of nonlinear protection circuitry exposed to transients, an effort is made to study similar circuits using available SPICE models from the surge protection device manufacturers such as *Littelfuse*.

Reference

[1] K.H. Billings, Switchmode Power Supply Handbook, McGraw-Hill, New York, 1999.

Chapter 7

Investigation of Surge Propagation Through a TVSS

7.1 Introduction

In Chapter 5 suitable mathematical models were developed for the metal oxide varistor (MOV) and the transient voltage suppression (TVS) diode. The developed MOV model was validated successfully by using a variable DC source. The required numerical simulation techniques were tested on related linear circuits in Chapter 6 and the results were validated using SPICE simulations. These simulation techniques will be used in this chapter to study surge propagation through a class A/B protection unit with two different nonlinear surge protection components. The study will be limited to class-B operation.

In Chapter 2, we looked at both category A and category B protection units, which are essentially surge protection devices (SPDs) or transient voltage suppressor systems (TVSSs). They usually make up the front end of modern power supplies. As discussed in Chapter 3, four different types of nonlinear devices are used in these circuits: Gas discharge tubes (GDTs) and three different types of solid-state devices: MOVs, TVS diodes, and TVS thyristors. The mathematical models developed for the MOV and the TVS diode in Chapter 5 are used for developing the simulations that are performed in this chapter, as these components make up the 2-level protection in a class A/B protection unit.

The complete TVSS circuit that we will be investigating here is shown in Fig. 2.4 and the same is presented here as Fig. 7.1 for convenience. This circuit uses an MOV for the first level of protection and a TVS diode for the second level of protection. In order to investigate the propagation of a lightning surge through this circuit, it will be driven by the lightning surge simulator (LSS) Noiseken LSS 6110, whose equivalent circuit was presented and examined in Section 6.1. The MOV, the line filter, and the TVS diode will be added in steps to make up the whole class B protection unit during our surge propagation investigations. The resulting circuit at each step is numerically simulated and validated, highlighting the power dissipation and energy absorption of each component. We will start the investigation in the next section by connecting an MOV, which usually provide the first level of protection in a class A/B protector.

Design of Transient Protection Systems. https://doi.org/10.1016/B978-0-12-811664-7.00007-0

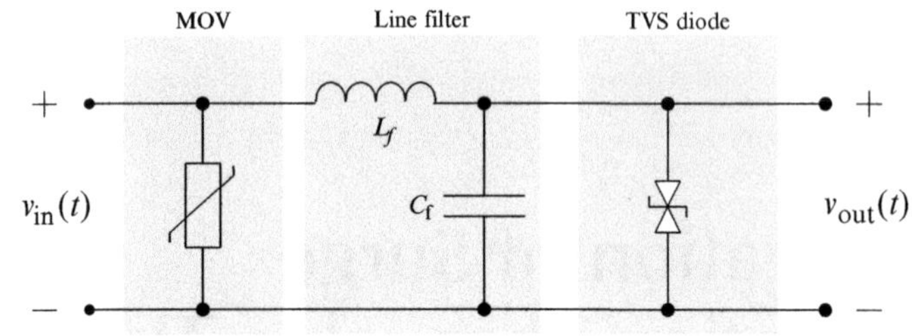

FIG. 7.1 A 2-wire Category B protection unit with two levels of protection.

7.2 Investigation of Surge Propagation Through an MOV

The circuit under test in this case, shown in Fig. 7.2 is simply a stand-alone MOV connected to the lightning surge simulator (LSS) output through a very low-valued resistor; the resistor R_s is required to determine the current through the MOV.

7.2.1 Simulation Using the Power-Law Model of an MOV

In order to develop the set of differential equations required to numerically simulate the previous circuit, we need to consider the equivalent circuit. Replacing the LSS and the MOV with their equivalent circuits gives us the overall circuit shown in Fig. 7.3.

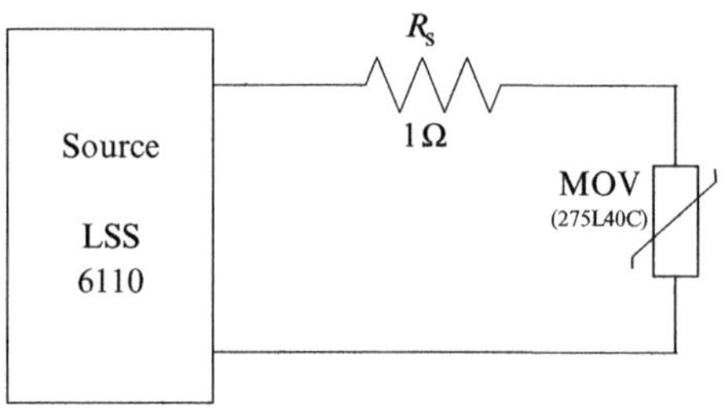

FIG. 7.2 LSS driving the first stage of a category B protection unit.

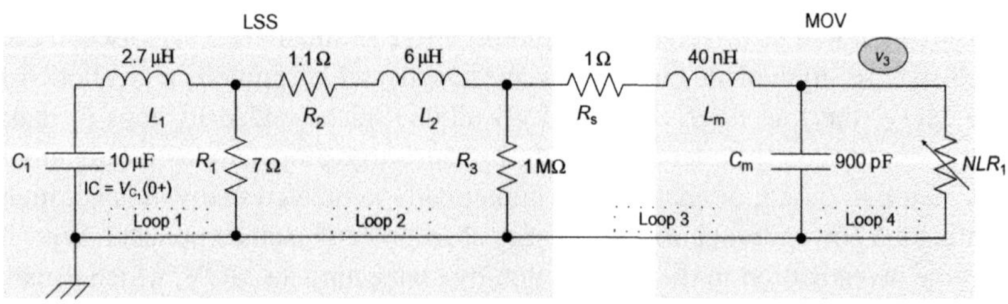

FIG. 7.3 Overall equivalent circuit for the set-up shown in Fig. 6.2.

We will consider four loop currents i_1, i_2, i_3, and i_4 for the loops shown in Fig. 7.3. In addition, we will have to consider the nodal voltage v_3 across the nonlinear resistance NLR_1, as we will be using the following mathematical model for the MOV's nonlinear resistance, which was developed in Chapter 6.

$$i_4 = K_1 {v_3}^{\alpha_1} + K_2 {v_3}^{\alpha_2} \tag{7.1}$$

The set of state equations formulated for the previous circuit is given next. The loop currents i_1, i_2, and i_3; the nodal voltage v_3; and the derivatives of the loop currents $i_1$and i_3 were selected to be the state variables. The complete derivation of the state equations (Eqs 7.2–7.7) is available in Appendix D.1.

$$\frac{di_1}{dt} = j_1 \tag{7.2}$$

$$\frac{dj_1}{dt} = (-a)j_1 - (b - ac)i_1 - (ad)i_2 + (ae)i_3 \tag{7.3}$$

$$\frac{di_2}{dt} = (c)i_1 - (d)i_2 + (e)i_3 \tag{7.4}$$

$$\frac{di_3}{dt} = j_3 \tag{7.5}$$

$$\frac{dj_3}{dt} = -(f)j_3 + (cg)i_1 - (dg)i_2 + (eg - h)i_3 + h\left(k_1 v_3^{\alpha_1} + k_2 v_3^{\alpha_2}\right) \tag{7.6}$$

$$\frac{dv_3}{dt} = k\left(i_3 - k_1 v_3^{\alpha_1} - k_2 v_3^{\alpha_2}\right) \tag{7.7}$$

where $a = R_1/L_1$, $b = 1/L_1C$, $c = R_1/L_2$, $d = (R_1 + R_2 + R_3)/L_2$, $e = R_3/L_2$, $f = (R_3 + R_S)/L_m$, $g = R_3/L_m$, $h = 1/L_mC_m$, and $k = 1/C_m$.

The initial conditions for the simulation are $i_1(0) = 0$, $j_1(0) = V_C(0)/L$, $i_2(0) = i_3(0) = j_3(0) = v_3(0) = 0$. As before, the value given for $j_1(0)$ results due to the initial voltage $V_{C_1}(0)$ of the capacitor C_1. A complete listing of the code used for the simulation LSS_MOV_with_nonlinearR.m is given in Appendix D.2. The values of $K_1 = 9.3231 \times 10^{-16}$, $\alpha_1 = 4.0413$, $K_2 = 3.5040 \times 10^{-73}$, and $\alpha_2 = 26.2396$ for the model of the *nonlinear resistance of the MOV* were obtained from Section 5.5.1.6 where a 274L40C MOV was considered.

In order to study the propagation of the surge through the MOV, we have presented the simulated current through the MOV along with the voltage across the device in Fig. 7.4A. We see that the simulations compare well with the validations with respect to both shape and magnitude in spite of the simulation current peaking slightly before the validation current. In Fig. 7.4B and C we have shown the power variation and energy absorption of the MOV during the period of surge propagation. Simulation of the energy absorption, in particular, is very important as it allows us to predict the incipient failure of the device once the specified threshold is exceeded. Again we see that the curves compare well although we see a difference of 13% in Fig. 7.4C between the simulation

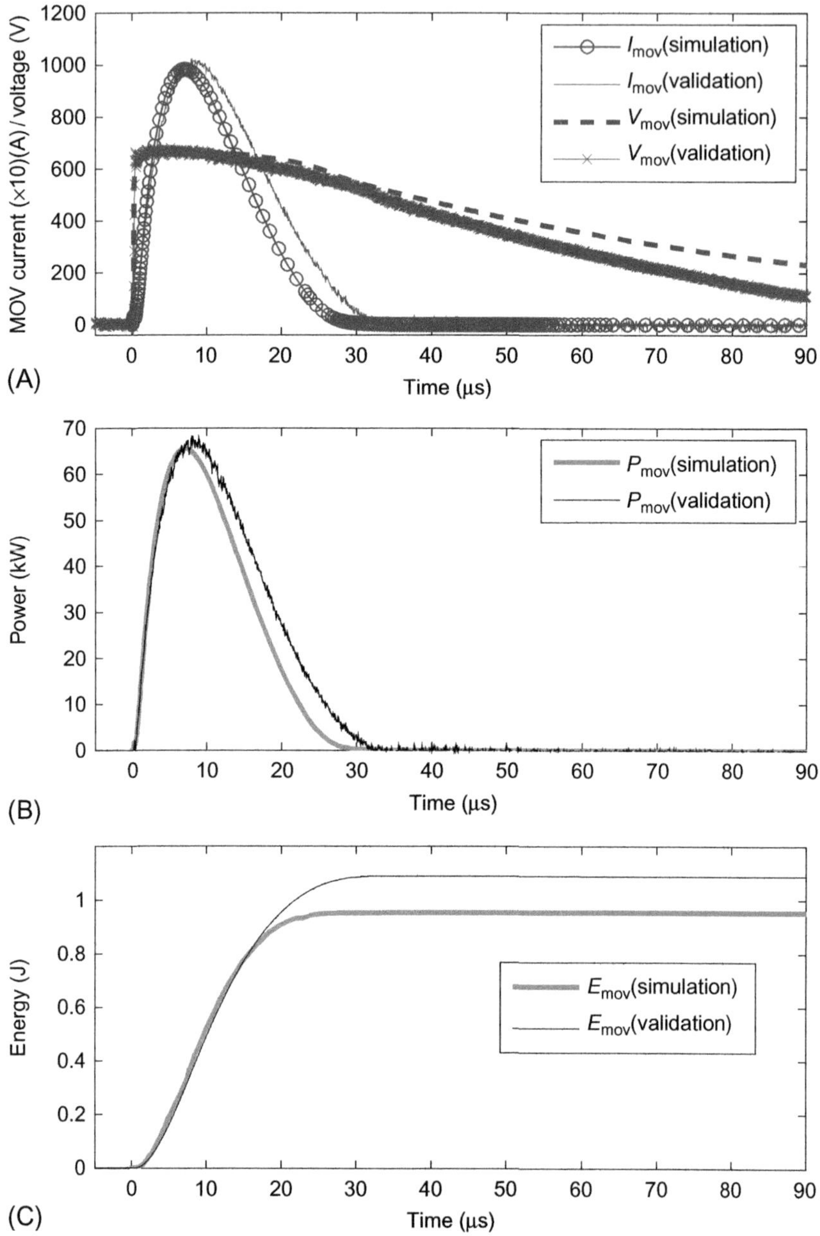

FIG. 7.4 Numerical simulation and validations for the first stage (MOV) of a TVSS (A) MOV current and voltage, (B) power variation for the MOV, and (C) energy absorption in the MOV.

and the validation of the total energy absorbed by the MOV. It was also noted that setting the term $K_1 v_3^{\alpha_1} = 0$ did not have any impact on the curves of Fig. 7.4. This is not surprising as this term is responsible only for modelling of effects in the leakage region of the MOV. Ignoring this term in the model is justified, since the primary goal here is to estimate the total energy absorption of each component of the protector unit, during the propagation of high-voltage transients such as lightning.

In order to understand the discrepancies between simulation and validation, let us look at the phase plots (MOV current vs MOV voltage) for the simulation and validation curves of Fig. 6.4A. These are presented in Fig. 7.5 and it can be seen that the current decrease in the validation curve follows a different more gradual path than the current increase path as per the discussion in Section 5.5.1.2. This phenomenon gives rise to a hysteresis loop in the phase plot of a varistor. The power-law model that we have adopted for the nonlinear resistance of the MOV does not take into account the hysteresis effects and hence, as anticipated, the phase plot for the simulation curve has coincident upward and downward paths for the current within the hysteresis loop of the experimental plot. In spite of this discrepancy in the phase plots, we were able to reduce the error in the total energy absorption to 9% by tuning the parameters values of K_2 and α_2 to $K_2 = 2.4496 \times 10^{-50.15}$ and $\alpha_2 = 18.3338$. Figs 7.6 and 7.7 highlight this improvement to the simulation. The simulation phase plot is now well centred within the hysteresis loop of the validating phase plot.

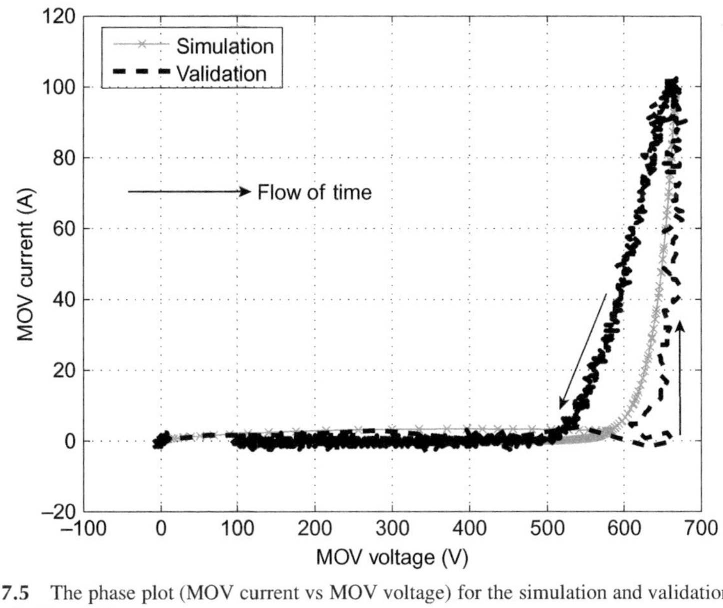

FIG. 7.5 The phase plot (MOV current vs MOV voltage) for the simulation and validation plots shown in Fig. 6.4A.

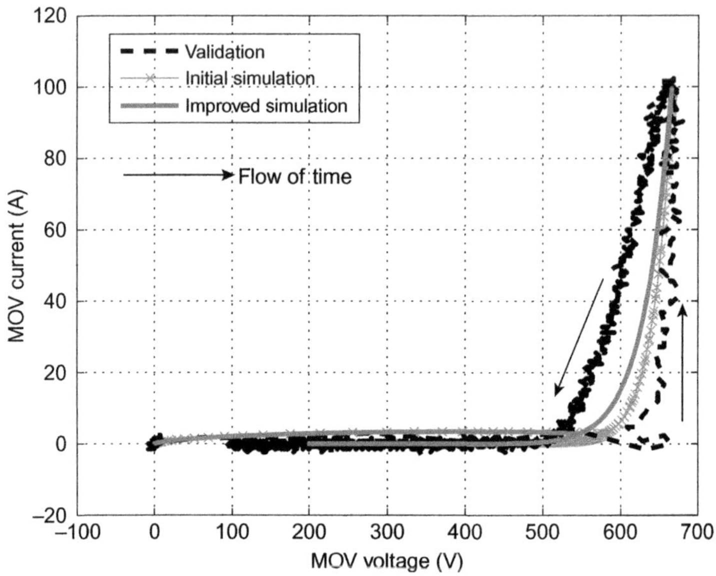

FIG. 7.6 The phase plot of the improved simulation compared with initial simulation and validation plots shown earlier in Fig. 7.5.

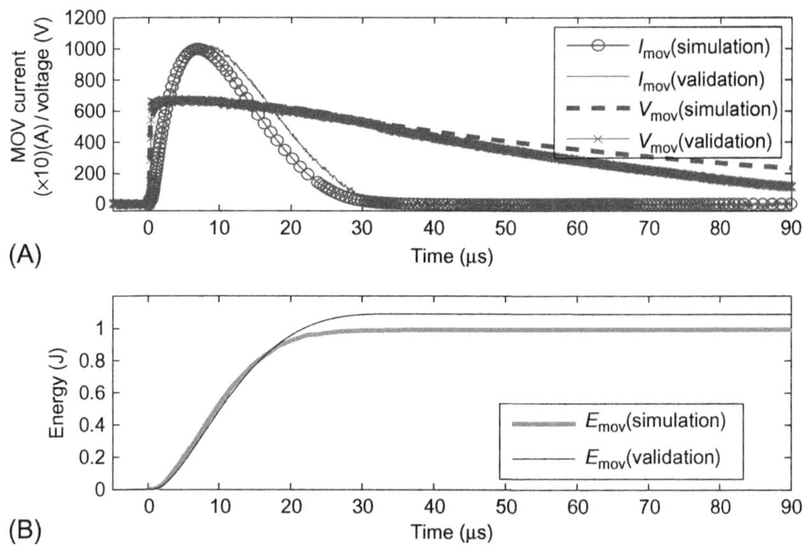

FIG. 7.7 Improved numerical simulations and validations for the (A) MOV current and voltage and (B) energy absorption in the MOV.

7.2.2 SPICE-Based Simulation Using a Manufacturer Supplied Model of an MOV

As mentioned in Chapter 3, the varistor manufacturers such as *Littelfuse* provide SPICE-based models for some of their varistor families. Although we were unable to find any SPICE-based model for the 275L40C varistor from the C-III varistor series, *Littelfuse* does provide a SPICE-based model for the 20V275 varistor which belong to the *UltraMOV* series [1]. This model is used in a PSpice simulation of the circuit shown in Fig. 7.2 with the MOV model 20V275 in place of the 275L40C.

Let us consider the schematic diagram given in Fig. 7.8, which is a representation of Fig. 7.2. The PSpice source file for the simulation is created using this schematic diagram and is given in Appendix D.3. The voltage source $V_x = 0$ V is introduced solely for the purpose of simulating the current through it, which is also the current flowing through the MOV.

The MOV voltage and the current obtained from the PSpice simulation were compared with MATLAB-based numerical simulation and also with experimental results. The MATLAB-based numerical simulation was performed using the power-law-based model for the varistor 20V275. An initial estimate of power-law model parameters resulted in the following: $K_2 = 1.0794 \times 10^{-55}$ and $\alpha_2 = 20.5306$ and the same were used in the simulation that generated the plots of Fig. 7.9.

By marginally changing α_2 from 20.5306 to 20.4006, the discrepancy between the PSpice and MATLAB simulations can be almost eliminated. The resulting plots are shown in Fig. 7.10.

It must be noted that the PSpice model for the MOV given by the manufacturer (seen in the PSpice source file in Appendix D.3) is much more complex mathematically than the power-law model that is used for the MATLAB simulation. In spite of this complexity, it too fails to account for the hysteresis effect shown by MOVs, which was discussed in Section 5.5.1.2.

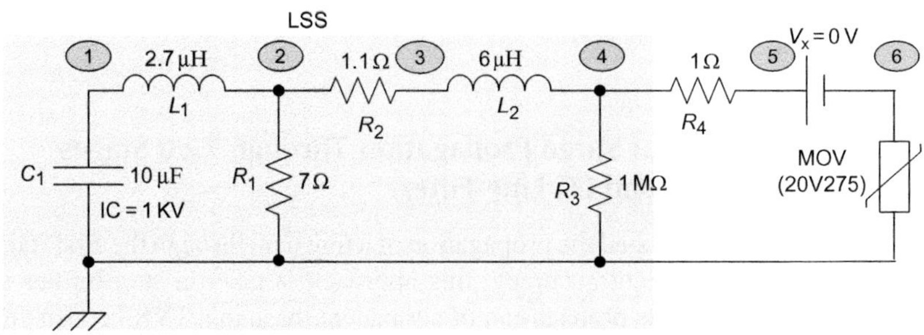

FIG. 7.8 Schematic representation of Fig. 7.2 used to develop the SPICE source file.

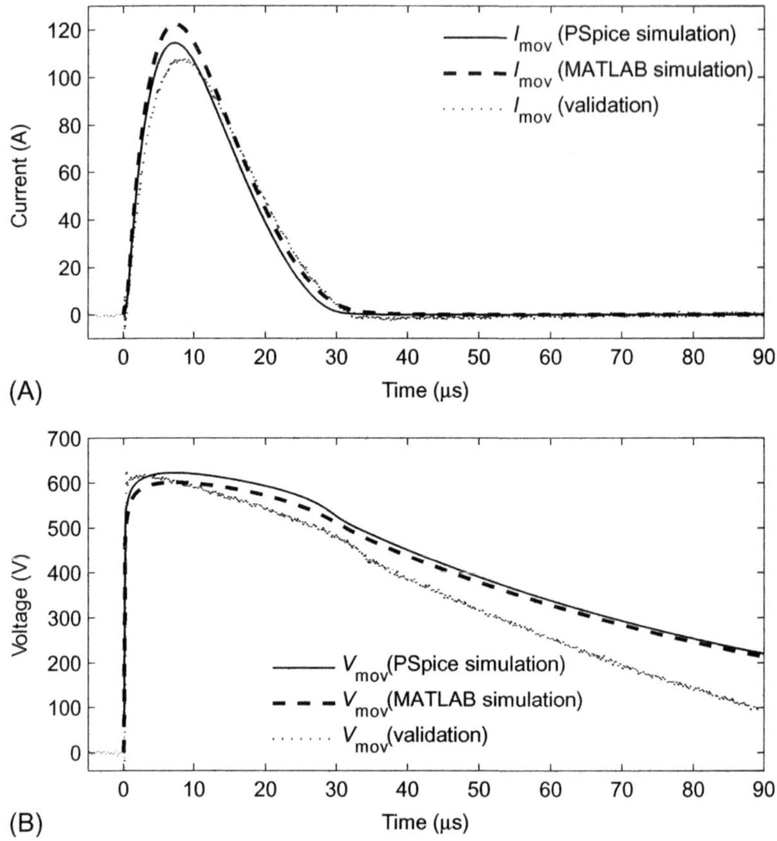

FIG. 7.9 SPICE and MATLAB simulations compared with the validations for the first stage (MOV) of a TVSS (A) MOV current and (B) MOV voltage.

All simulations in the following sections will be done using MATLAB ODE solvers and suitable power-law models for the two nonlinear devices used, viz. the MOV and the TVS diode.

7.3 Investigation of Surge Propagation Through Two Stages of the TVSS (MOV and LC Line-Filter)

Now that we have simulated the propagation of a transient through the first stage of a TVSS with sufficient accuracy, this approach is taken a step further to numerically simulate the propagation of a transient through a TVSS consisting of two stages: an MOV and a line filter. This set-up is shown in Fig. 7.11, where L_f and C_f make up the line filter and the small resistance R_f is included for the purpose of current measurement.

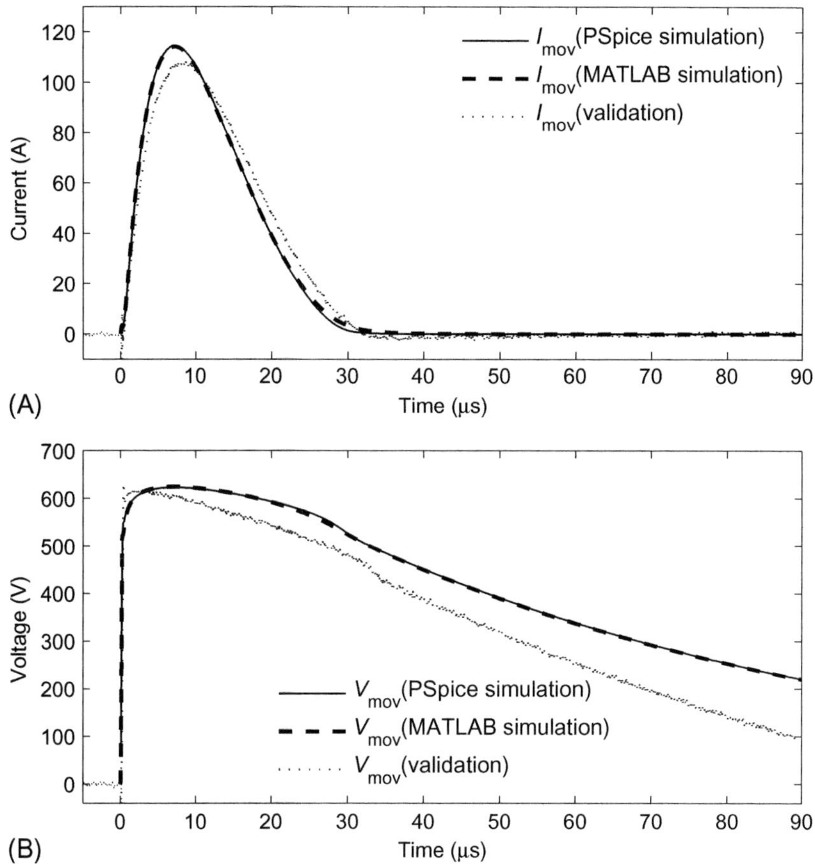

FIG. 7.10 SPICE and MATLAB simulations of Fig. 7.9 made to coincide by changing α_2 from 20.5306 to 20.4006 (A) MOV current and (B) MOV voltage.

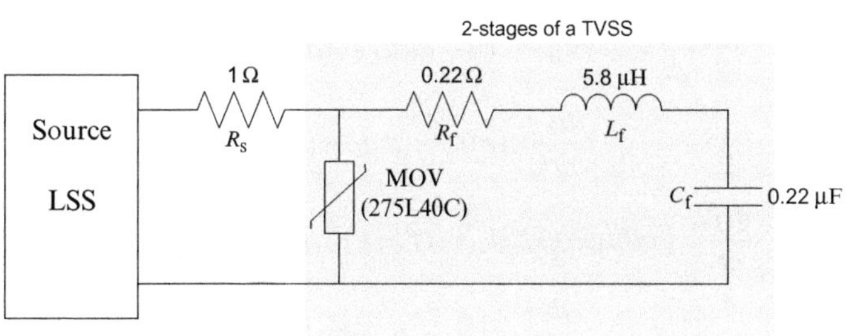

FIG. 7.11 LSS driving the first two stages of a category B protection unit.

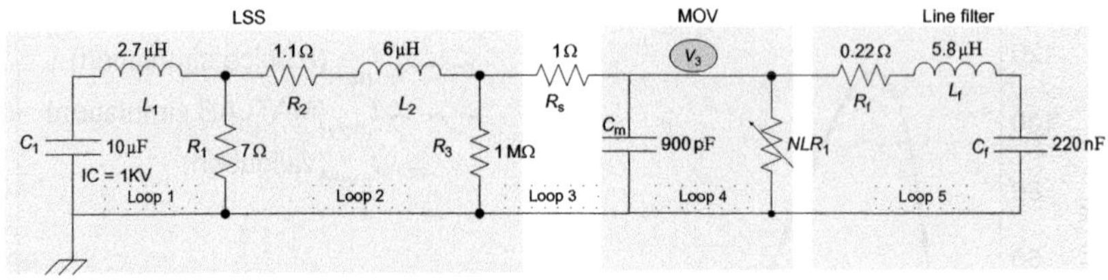

FIG. 7.12 Overall equivalent circuit for the set-up shown in Fig. 7.11.

Consider the complete equivalent circuit for the set-up shown in Fig. 7.11, in order to develop the set of differential equations required to numerically simulate the propagation of a HV transient through it. The equivalent circuit resulting from replacing the LSS and the MOV by their equivalent circuits is shown in Fig. 7.12. The small lead inductance of the MOV, which plays an insignificant role in the normal operation of the MOV, has been omitted from this simulation.

Consider five loop currents i_1, i_2, i_3, i_4, and i_5 for the loops shown in Fig. 7.12. In addition, we need to consider the nodal voltage v_3 across the nonlinear resistance NLR_1, as we will continue to use the same power-law-based mathematical model for the MOV nonlinear resistance which we had used in the previous section. For the current and voltage variables shown in Fig. 7.12, this mathematical model can be represented by

$$i_4 - i_5 = K_2 {v_3}^{\alpha_2} \tag{7.8}$$

The first power-law term $K_1 {v_3}^{\alpha_1}$ has been ignored here due to the reasons mentioned in the previous section. The set of state equations formulated for the simulation of the previous circuit are shown as follows. The loop currents i_1, i_2, i_3, and i_5; the nodal voltage v_3; and the derivatives of the loop currents i_1 and i_5 were selected to be the state variables. The complete derivation of the state equations (Eqs 7.9–7.15) is included in Appendix D.4.

$$\frac{di_1}{dt} = j_1 \tag{7.9}$$

$$\frac{dj_1}{dt} = -(a)j_1 - (b - ac)i_1 - (ad)i_2 + (ae)i_3 \tag{7.10}$$

$$\frac{di_2}{dt} = (c)i_1 - (d)i_2 + (e)i_3 \tag{7.11}$$

$$\frac{di_3}{dt} = (cf)i_1 - (df)i_2 + (ef - g)i_3 + (g)i_5 + (gk_2)v_3^{\alpha_2} \tag{7.12}$$

$$\frac{dv_3}{dt} = l\left(i_3 - i_5 - k_2 v_3^{\alpha_2}\right) \tag{7.13}$$

$$\frac{di_5}{dt} = j_5 \tag{7.14}$$

$$\frac{dj_5}{dt} = (h)i_3 - (h + k)i_5 - (hk_2)v_3^{\alpha_2} - (m)j_5 \tag{7.15}$$

where $a = R_1/L_1$, $b = 1/L_1C_1$, $c = R_1/L_2$, $d = (R_1 + R_2 + R_3)/L_2$, $e = R_3/L_2$, $f = R_3/(R_3 + R_S)$, $g = 1/C_m(R_3 + R_S)$, $h = 1/L_fC_m$, $k = 1/L_fC_f$, $l = 1/C_m$, and $m = R_f/L_f$.

The initial conditions for the simulation are $i_1(0) = 0$, $j_1(0) = V_{C1}(0+)/L$, $i_2(0) = i_3(0) = v_3(0) = i_5(0) = j_5(0) = 0$. As before, the value given for $j_1(0)$ results due to the initial voltage $V_{C1}(0+)$ of the capacitor C_1. A complete listing of the code used for the simulation LSS_MOV_LineFilter.m is given in Appendix D.5. The values of $K_2 = 2.4496 \times 10^{-50.15}$ and $\alpha_2 = 18.3338$ for the model of the nonlinear resistance of the 275L40C MOV are taken from the improved model discussed in Section 7.2.

We will use the simulation to study the propagation of the transient through different components of the TVSS which will include a study of the total energy absorbed by each of the components over time. This type of study would be helpful in determining the most vulnerable components of the system for failure in the event of a high energy transient.

First, we will examine the behaviour of the MOV during surge propagation. The simulated voltage across the device along with the current through the MOV is presented in Fig. 7.13A and B, respectively. We see that the simulations compare well with the validations with respect to both shape and magnitude in spite of the ringing in the simulations not being damped as much as the validations. In Fig. 7.13B and C we have shown the power variation and energy absorption of the MOV during the period of surge propagation. We see that the energy absorption saturates prior to about 30 μs. Very good agreement is observed between the simulation and validation for the total energy absorbed by the MOV, in spite of the mismatch in the amplitudes of the ringing peaks in the instantaneous power curves of Fig. 7.13B. This is encouraging, especially since we have added two new components to the circuit under test.

Now we will study how the same surge propagates through the capacitor C_f of the line filter. The simulated voltage across the capacitor and the current through this component is presented in Fig. 7.14A and B, respectively, along with the experimental curves for a surge of 1 kV. Again we see that the simulations compare reasonably well with the validations with respect to both shape and magnitude in spite of the ringing in the simulations not being damped as much as the validations. In Fig. 7.14C and D we have shown the power variation and energy absorption in the capacitor C_f during the period of surge propagation. Approximately 5% error is observed between the simulation and validation for the energy absorbed by the capacitor. We see that the resonance frequency ($f_r = 1/2\pi\sqrt{L_fC_f} \approx 141$ kHz) in the simulation of capacitor current matches with the resonance frequency of the line filter components L_f and C_f, whereas the experimental curves exhibit a frequency ~5% higher. We believe that parasitic capacitances and inductances present in the experimental set-up due to connecting cables and proximity of components are responsible for this small deviation in frequency.

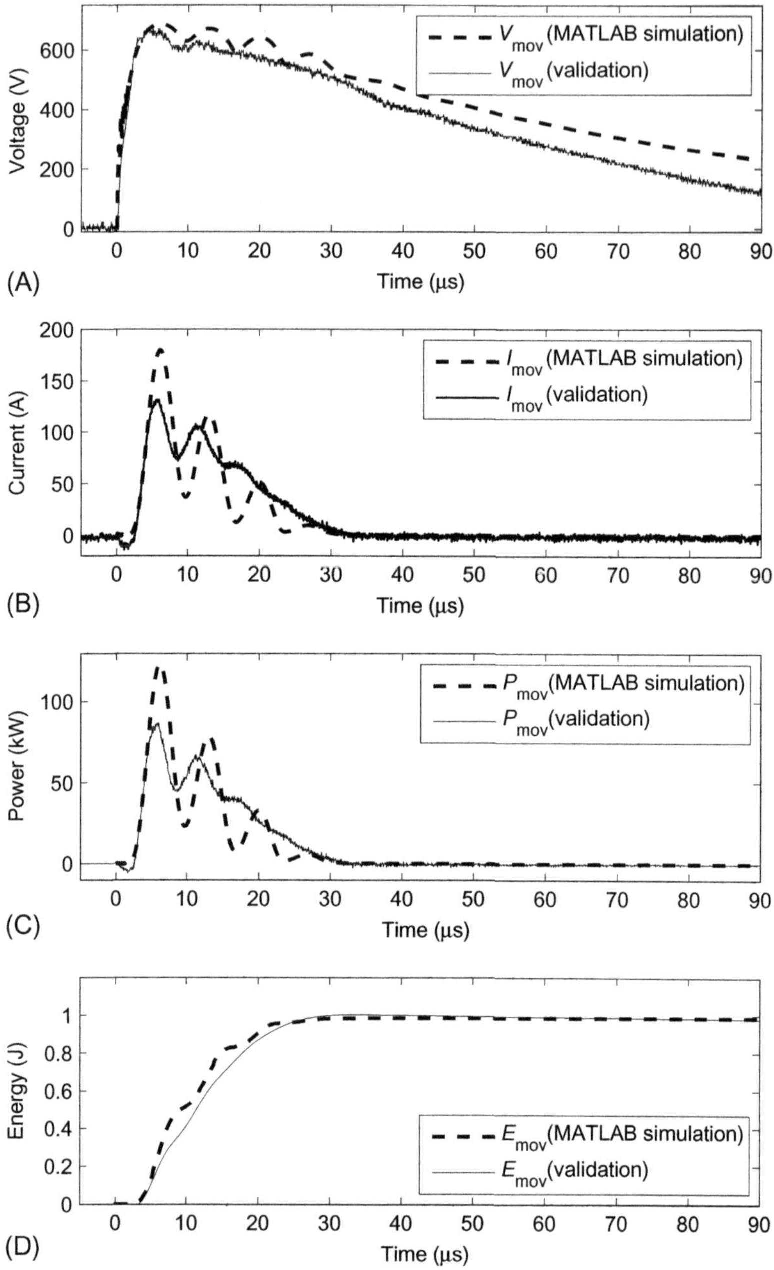

FIG. 7.13 Numerical simulation and validations for the MOV of the TVSS: (A) MOV voltage (B) MOV current (C) power variation in the MOV (D) energy absorption in the MOV.

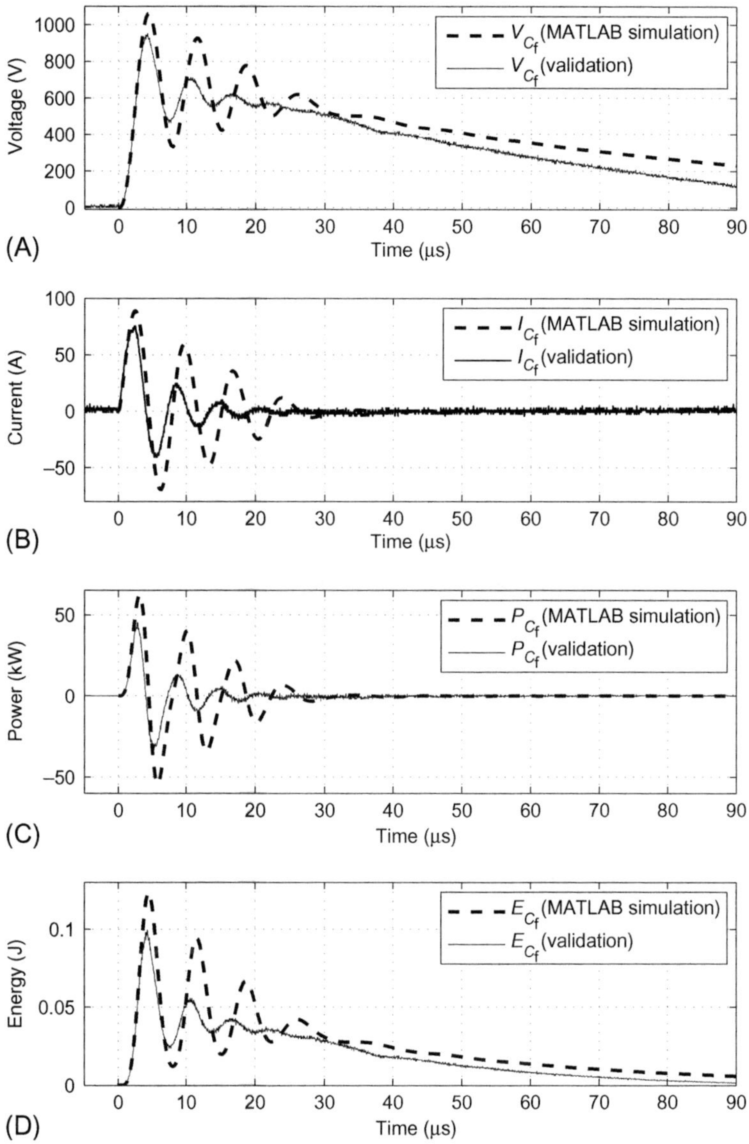

FIG. 7.14 Numerical simulation and validations for the capacitor C_f of the TVSS (A) capacitor current (B) capacitor voltage (C) power variation in the capacitor (D) energy absorption in the capacitor.

The maximum energy absorption in the capacitor takes place ~4 μs after the transient and we see that the maximum energy shown by the simulation at this point is about 20% higher than the experimental result. This energy, which is about 100 mJ, is much smaller than the energy absorbed by the MOV, which is in the range of 1 J.

Finally to complete the discussion of this circuit, let us look at how the surge propagates through the inductor L_f of the line filter in Fig. 7.15. Again we have similar wave shapes although the peaks of ringing in the simulation tend to be higher than in the validating curves. The maximum energy absorption in the

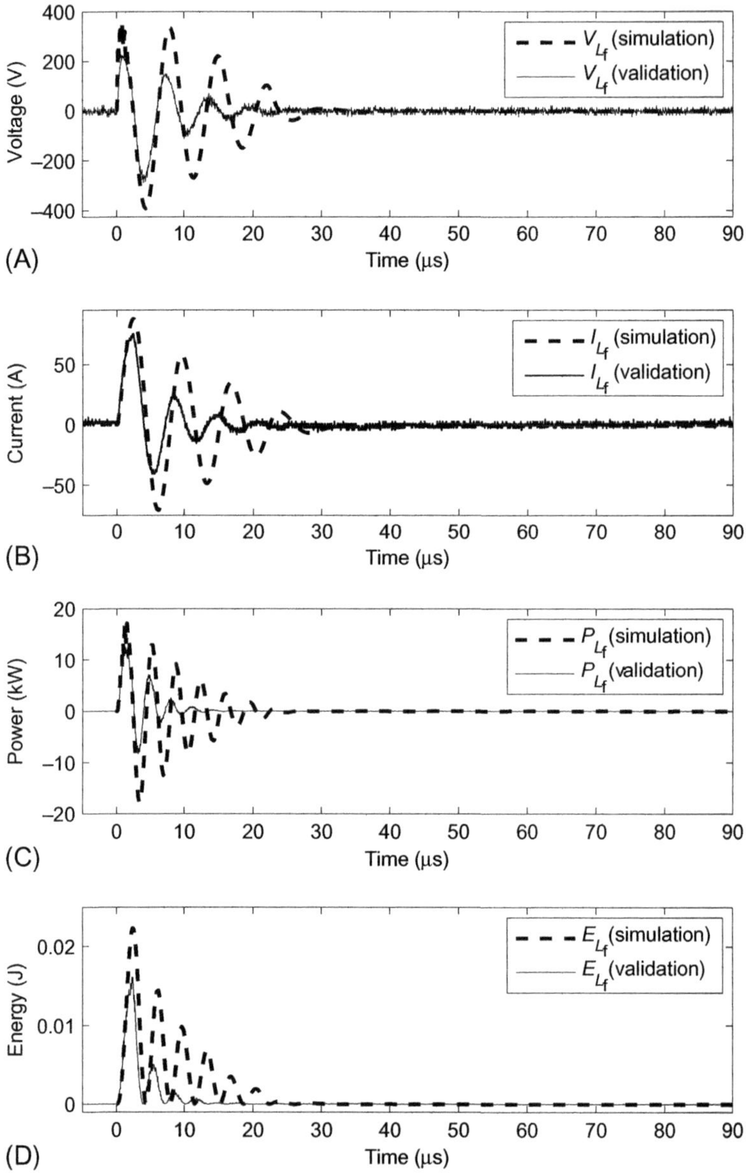

FIG. 7.15 Numerical simulation and validations for the inductor L_f of the TVSS (A) inductor current (B) inductor voltage (C) power variation in the inductor (D) energy absorption in the inductor.

inductor can be noted in Fig. 7.15D and takes place ~2.5 μs after arrival of the surge. Although the maximum energy levels of the simulation and validation differ by about 25% the amount of energy absorbed in this case is even smaller and is in the range of 15–20 mJ.

7.4 Investigation of Surge Propagation Through a Complete 2-Wire Category A/B TVSS

Having successfully investigated a two-stage TVSS consisting of both linear and nonlinear components in the last section, the investigation is taken a step further by adding a second nonlinear element: a TVS diode. This provides a second level of protection, thereby making a superior protector having two nonlinear and two linear components (the two resistors, small in value, are added only for current measuring purposes) as shown in Fig. 7.16.

In order to appreciate the investigation of surge propagation through the circuit shown in Fig. 7.16, it is important to understand how this circuit operates whenever a HV surge arrives at its input terminals. While the operation is similar to that of the 3-wire version of this circuit discussed briefly in Section 1.4.3, it must be mentioned that the nonlinear surge protective devices used must be properly coordinated for effective functioning of the overall circuit in the case of a HV surge; if properly coordinated the MOV will always conduct before the TVS diode.

In this 2-protective-level TVSS, the first level of protection is provided by the MOV; because of its relatively high transient energy absorption capability most of the surge energy is absorbed by this component in the case of a HV transient. As we saw in the previous section, the sudden rush of current through the MOV gives rise to a ringing voltage and current in the line filter elements L_f and C_f. As a result, a HV surge can develop damagingly high voltage peaks across the filter capacitor that may be let through to the "protected" equipment. This is where the second-level protector, the TVS diode, comes into play, by conducting whenever the voltage across C_f attempts to go beyond the *clamping voltage* of the TVS diode. This situation will prevail for all stress currents up to

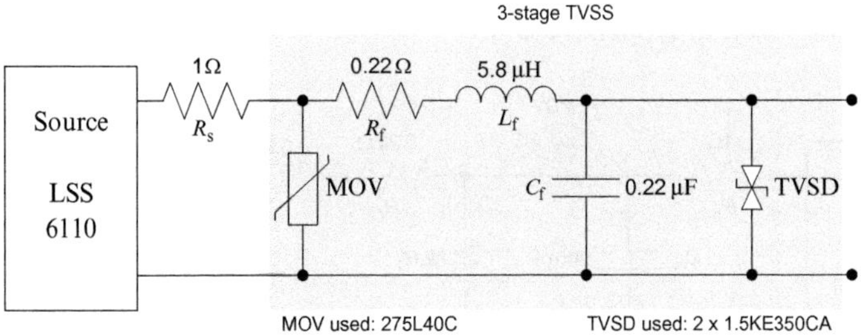

FIG. 7.16 LSS driving a complete 3-stage category A/B protection unit. The protection unit is shown in the shaded area.

the failure point of the TVSD. If this level is reached, the TVSD will fail to a short circuit, clearing the protection fuse that would be present in a commercial or industrial set-up [2]. If the transient energy is very large and of short duration like in the case of an extra high voltage lightning transient, there can be an open circuit due to the silicon chip itself exploding [3].

For the reasons indicated before, for this 2-level protector to function correctly, the breakdown voltage of the TVSD must be higher than the firing voltage of the MOV. In the design we have considered here the 275L40C MOV has a clamping voltage just over 600 V and we used two 1.5KE350CA-type TVS diodes in series to get a breakdown voltage around the 700 V mark.

Let us now develop the equivalent circuit for the set-up given in Fig. 7.16 in order to develop the set of differential equations required to numerically simulate the propagation of an HV transient through it. The equivalent circuit resulting from replacing the MOV and the TVS diode by their nonlinear resistance models is shown in Fig. 7.17. The first two loops belonging to the equivalent circuit of the LSS are not explicitly shown here since they result in the same two state equations as developed in the earlier cases.

In order to formulate the set of state equations required to simulate the previous circuit, six loop currents i_1, i_2, i_3, i_4, i_5, and i_6 are considered. The first two loop currents i_1 and i_2 are flowing within the LSS equivalent circuit and are not explicitly shown in Fig. 7.17. In addition, we will consider the nodal voltages v_3 and v_5 across the nonlinear resistances NLR_1 and NLR_2. The power-law-based mathematical model is used for the MOV and TVSD nonlinear resistances. For the current and voltage variables of Fig. 7.17, these models can be represented by Eq. (7.16) for the MOV and by Eq. (7.17) for the TVS diode.

$$i_4 - i_5 = k_m {v_3}^{\alpha_m} \tag{7.16}$$

$$i_6 = k_t {v_5}^{\alpha_t} \tag{7.17}$$

The set of state equations formulated for the simulation of the circuit of Fig. 7.17 is given as follows. The loop currents i_1, i_2, i_3, and i_5; the nodal voltages v_3 and v_5; and the derivatives of the loop currents $i_1$and i_5 were selected to be the state variables. The complete derivation of the state equations (Eqs 7.18–7.25) is included in Appendix D.6.

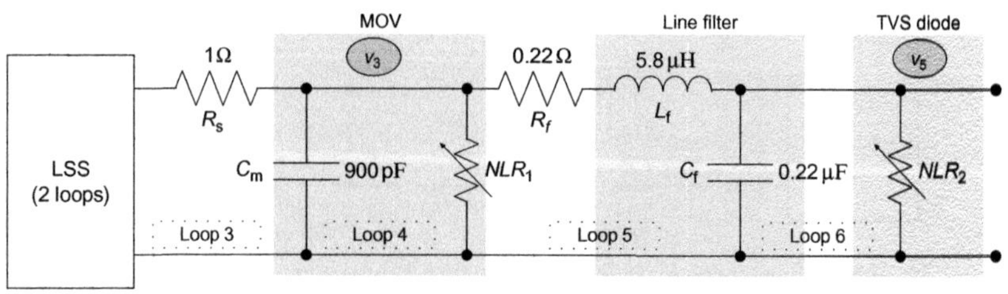

FIG. 7.17 MOV and TVS diode models used to redraw the circuit of Fig. 7.16.

$$\frac{di_1}{dt}=j_1 \tag{7.18}$$

$$\frac{dj_1}{dt}=-aj_1-(b-ac)i_1-adi_2+aei_3 \tag{7.19}$$

$$\frac{di_2}{dt}=ci_1-di_2+ei_3 \tag{7.20}$$

$$\frac{di_3}{dt}=cfi_1-dfi_2+(ef-g)i_3+gi_5+gk_\mathrm{m}v_3^{\alpha_\mathrm{m}} \tag{7.21}$$

$$\frac{dv_3}{dt}=h\left(i_3-i_5-k_2v_3^{\alpha_2}\right) \tag{7.22}$$

$$\frac{di_5}{dt}=j_5 \tag{7.23}$$

$$\frac{dj_5}{dt}=ki_3-(k+l)i_5-kk_\mathrm{m}v_3^{\alpha_\mathrm{m}}+lk_\mathrm{t}v_5^{\alpha_\mathrm{t}}-nj_5 \tag{7.24}$$

$$\frac{dv_5}{dt}=mi_5-mk_\mathrm{t}v_5^{\alpha_\mathrm{t}} \tag{7.25}$$

where $a=R_1/L_1$, $b=1/L_1C_1$, $c=R_1/L_2$, $d=(R_1+R_2+R_3)/L_2$, $e=R_3/L_2$, $f=R_3/(R_3+R_S)$, $g=1/C_\mathrm{m}(R_3+R_S)$, $h=1/C_\mathrm{m}$, $k=1/L_\mathrm{f}C_\mathrm{m}$, $l=1/L_\mathrm{f}C_\mathrm{f}$, $m=1/C_\mathrm{f}$ and $n=R_\mathrm{f}/L_\mathrm{f}$.

The initial conditions for the simulation are $i_1(0)=0$, $j_1(0)=V_{C1}(0+)/L$, $i_2(0)=i_3(0)=v_3(0)=i_5(0)=j_5(0)=v_5(0)=0$. As before, the value given for $j_1(0)$ results due to the initial voltage $V_{C1}(0+)$ of the capacitor C_1. A complete listing of the code used for the simulation LSS_TVSS.m is given in Appendix D.7. The values of $K_\mathrm{m}=2.4496\times10^{-50.15}$ and $\alpha_\mathrm{m}=18.3338$ for the "on" model of the *nonlinear resistance of the* 275L40C MOV are taken from the improved model discussed in Section 7.2. The values of $K_\mathrm{t}=4.0638\times10^{-161}$ and $\alpha_\mathrm{t}=56.3468$ for the "on" model of the series combination of two 1.5KE350CA TVS diodes were obtained as discussed in Sections 5.5.2.2 and 5.5.2.3. The "off" models of the MOV and the TVS diode have not been considered here due to the nature of the HV input transients.

The simulation is used to study the propagation of the transient through various levels of the TVSS and compare the simulations with results obtained experimentally.

Again most of the energy delivered by the HV transient to this circuit will be absorbed by the MOV, which provides the first level of protection. The simulated voltage across the MOV and the current through it are presented in Fig. 7.18A and b, respectively. We see that the simulations compare well with the validations with respect to both shape and magnitude. The ringing that was present, in the absence of the TVS diode Fig. 7.13 has almost disappeared. In Fig. 7.18B and C we have shown the power variation and energy absorption of

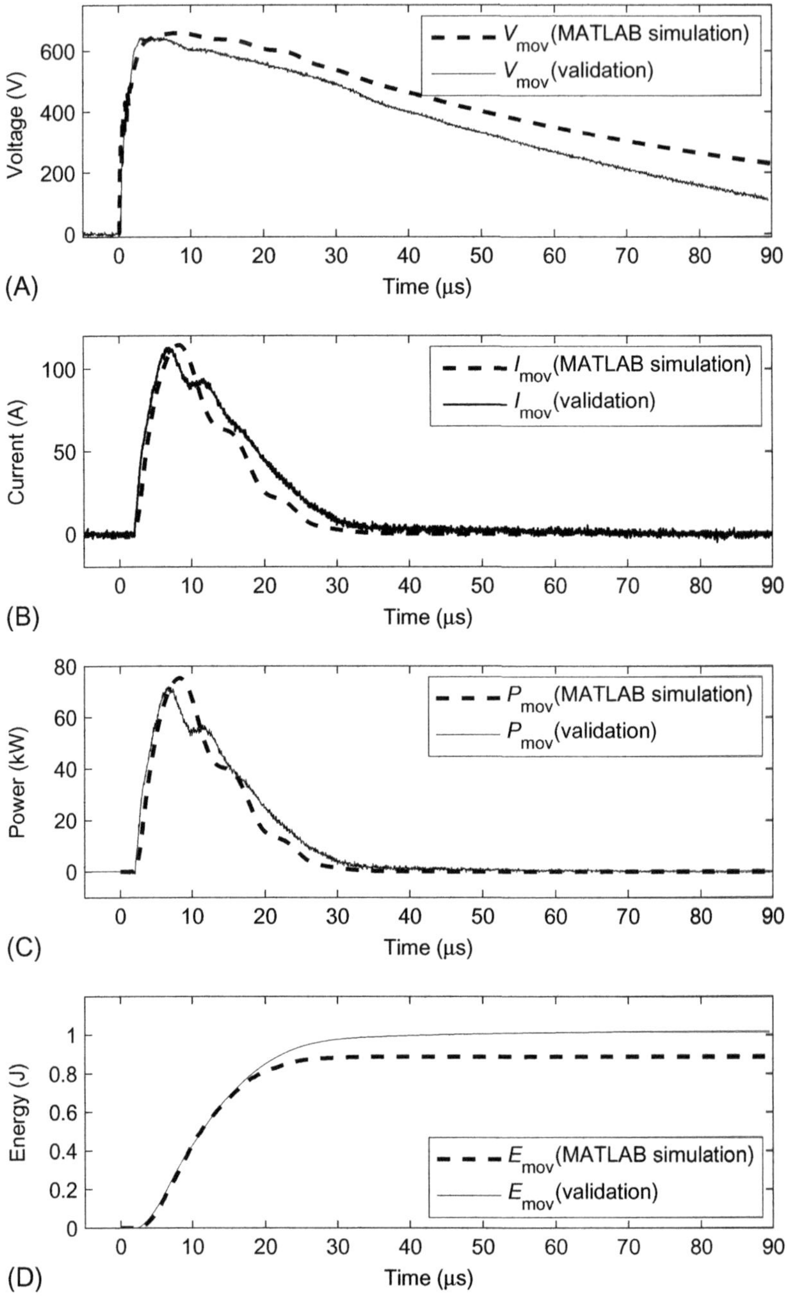

FIG. 7.18 Numerical simulation and validations for the MOV of the TVSS (A) MOV voltage (B) MOV current (C) power variation for the MOV (D) energy absorption in the MOV.

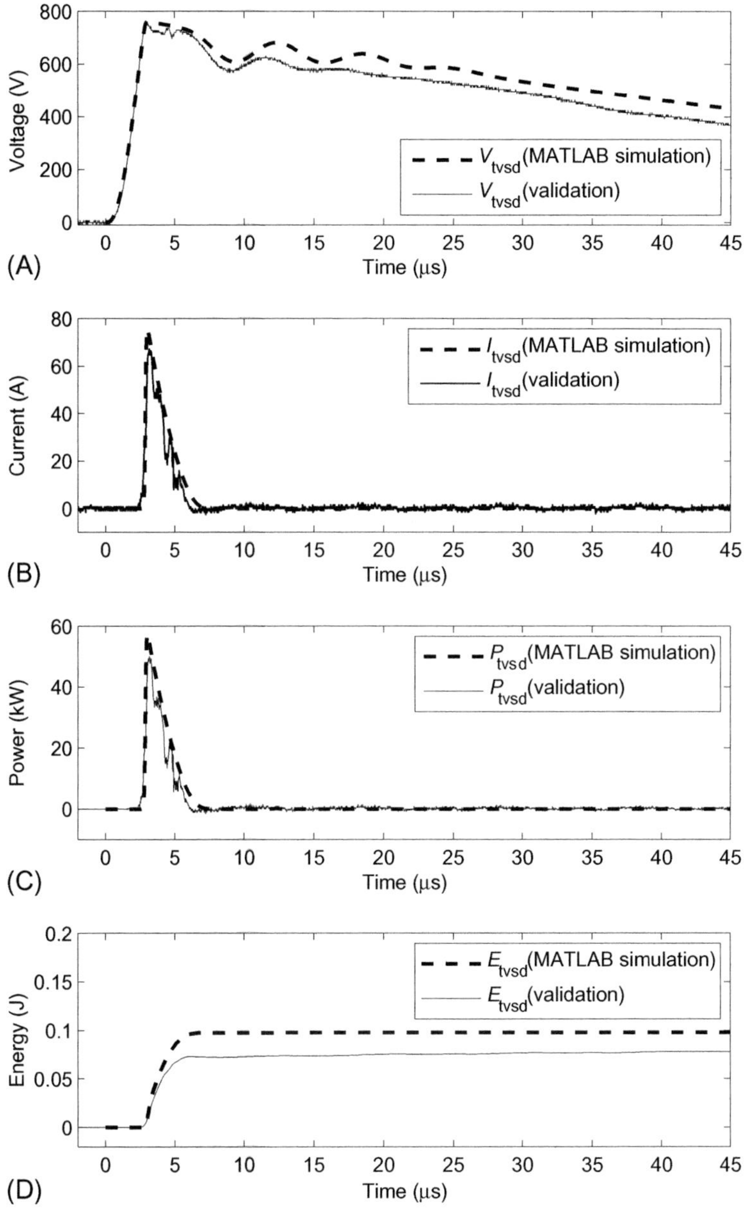

FIG. 7.19 Numerical simulation and validations for the TVS diode of the TVSS (A) TVS diode voltage (B) TVS diode current (C) power variation in the TVS diode (D) energy absorption in the TVS diode.

the MOV during the period of surge propagation. Again there is good agreement between simulation and validation for the total energy absorbed by the MOV, the percentage error being ~12%.

Let us now look at how the same surge propagates through the TVS diode. An observation of the voltage and current curves for this device will clearly explain how this device operates and provides a second level of protection for downstream electronics that would be connected to a TVSS. The simulations and experimental validations for the voltage across the diode and the current through this component are presented in Fig. 7.19A and B, respectively, for a surge of 1 kV. Again we see that the simulations compare well with the validations with respect to both shape and magnitude in spite of a minor mismatch in the current curves. It must be noted that the TVS diode voltage gets clipped as soon as the filter capacitor tries to charge beyond the breakdown voltage of the TVS diodes. Further evidence to this fact can be seen if we compare the TVS voltage curve with the line filter capacitor voltage curve shown in Fig. 7.14A, where it rises to almost 900–1000 V. Both the simulation and the validation show the same time instant (around the 3 μs mark) for the point of switching of the TVS diode from its "off" state to its "on" state.

In Fig. 7.19C and D we have shown the power variation and energy absorption in the TVS diode during the period of surge propagation. Approximately a 20% error is observed between the simulation and validation for the energy absorbed by the TVS diode. This energy, which is <100 mJ, is much smaller than the energy absorbed by the MOV, which is in the range of 1 J.

References

[1] Littlefuse, Varistor SPICE Models, Retrieved from: http://www.littelfuse.com/technical-resources/spice-models/varistor-spice-models.aspx, 2003. (Accessed May 2013).

[2] K.H. Billings, Switchmode Power Supply Handbook, McGraw-Hill, New York, 1999.

[3] Vishay General Semiconductor, Failure modes and fusing of TVS devices, Application Note 88440. Retrieved from: http://www.vishay.com/docs/88440/failurem.pdf, 2007. (Accessed October 2013).

Chapter 8

Design of Surge Protectors With a Practical Approach

8.1 Introduction

So far we have discussed the essential standards, necessary components, and their characterization as applied to surge protection systems. Practical circuit of a surge protector does not carry any transistors, or integrated circuits but it usually carries few nonlinear devices (NLD), such as metal-oxide varistors (MOVs), BBDs together with simple linear components, such as inductors and capacitors. In this chapter and Chapter 9 we present concepts useful in selecting NLDs and essential design calculations required to build a packaged and testable surge protector device (SPD), dividing the contents into design of (i) a traditional version and (ii) a supercapacitor-based version, usable for 230 V, 50 Hz input AC supply side of an electronic product or a system.

The discussion commences with guidelines on how to use the relevant component datasheets and selecting suitable NLD components, and then combining them into a complete circuit, which usually acts as a filter for removing the harmful effects of the transient high-voltage, high-energy surge from the energy supply side as well as the essential components to block any high-frequency components (usually known as RFI/EMI) generated within the protected electronic system flowing towards the other electrical systems powered by the same AC distribution circuit. The examples we use in the chapter are for Class A or Type 3 surge protector, which is used to protect only a single electrical product or a system, where the protector unit is placed between the electrical socket outlet and the protected electrical product.

8.2 Simple and Traditional Surge Protector Useful for 230 V, 50 Hz AC Socket Outlet With Some Background on How Typical Surges Propagate Into an Electrical Product

Transient overvoltages are distinguished into different categories based on their origination. Internal transients originate within the actual system to be protected. And they can be divided into:

Design of Transient Protection Systems. https://doi.org/10.1016/B978-0-12-811664-7.00008-2

- inductive load switching;
- arcing;
- direct contact with a higher voltage;
- mutual coupling (capacitive or inductive); and
- electrostatic discharge.

In many situations of internal causes, the worst-case conditions can be estimated or traceable by a test circuit, and the designers can optimize the transient overvoltage protective circuit. External overvoltages are usually caused by

- lightning;
- strong magnetic fields (electromagnetic coupling based); and
- line interference.

In these external situations, waveform, peak amplitude, and the frequency of occurrence of the transient cannot be predicted or estimated and it makes it difficult to develop a reliable SPD. Whilst frequent attempts were made by electrical distribution companies and local authorities to define the overvoltage vulnerability of typical supply systems, such as industrial, municipal, and rural, scale of local differences makes them a subject of serious uncertainty.

Given this situation, for reliable protection a certain degree of overdesign should be considered. As an example, in many equatorial belt countries where the monsoon season lightning could be severe, many commercial SPDs installed to protect critical systems do not do well. In Sri Lanka, several of the authors of this work have seen frequent 'disasters' where both the SPD and the system under protection were both severely damaged. Given this uncertainty of predicting the transient peak voltage values, in 230 V, 50 Hz power line situations following approximate guidelines are used for reasonable levels of protection:

- transient peak amplitudes up to 6 kV and
- pulse duration varying from 0.1 μs to 1 ms.

8.3 External Activities Which Create Superimposed High-Voltage Transients: A Hypothetical Example With a GDT, an MOV, and a BBD

Fig. 8.1 depicts a case of a lightning type transient voltage getting induced on the live wire with respect to the earth on an electrical supply connection into the premises where multiple surge absorption components could be installed at different locations. In this figure the transient superimposed will have a shape similar to the 1.2 μs/50 μs combined voltage surge waveform discussed in Chapter 2. The circuit shown in Fig. 8.1 has a gas discharge tube, two inductors, an MOV, and a BBD placed along the path [1].

As discussed in Chapter 3 there is no ideal component that can fulfil all the technical requirements of surge protection equally effectively. Therefore, we

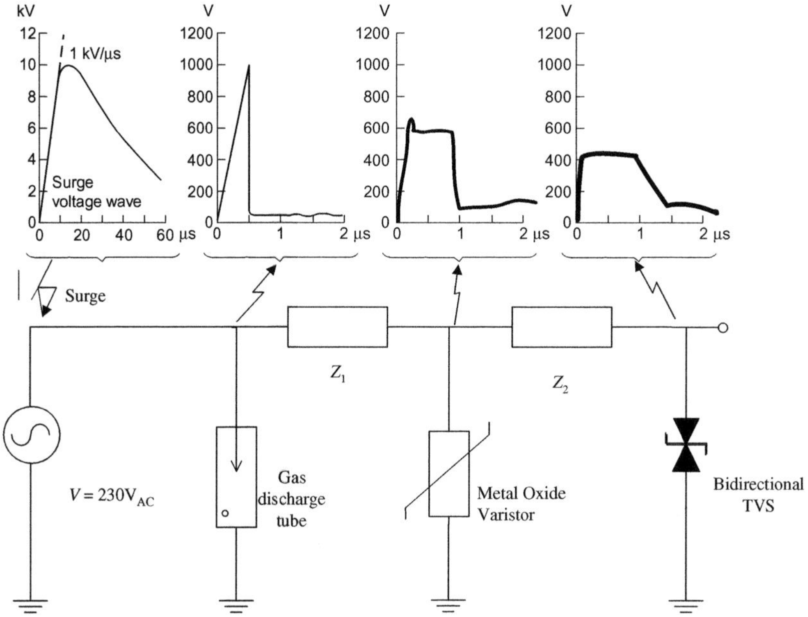

FIG. 8.1 A power flow path consisting of several different surge absorbing or attenuating components including GDT, MOV, BBD, and two inductors. *(Adapted from Weidmüller, The Basics of Lightning and Surge Protection, 2013.)*

use a variety of components with different physical methods of operation that complement each other. These posses distinct protective effects, namely:

- superfast reaction time;
- high instantaneous current carrying capacity;
- lower residual voltage; and
- long service life.

Fig. 8.1 is used as a case to show how these effects could be combined into one single overall circuit, based on GDT, MOV, and BBD together with one or two inductors.

If the instantaneous peak voltage of the combined surge is above the firing voltage of the GDT, the device fires and enter into conduction, by which the terminal voltage across the GDT becomes similar to the case of a rising triangular shape, which suddenly collapses at the instant of firing (ignition) of the spark gap of GDT shown in Fig. 8.1. This effect is usually called a crowbar effect, during which the AC line's energy delivery may be also compromised due to its short circuiting path generated across the two terminals of the GDT. Because GDT is a relatively slow device, before the GDT firing occurs, the voltage across the MOV keeps growing as shown in the relevant time graph, with a flattened start-up period. When the GDT devise remain fired, the lower voltage clamped at the terminals now causes the firing of the MOV into conduction, as shown by the second part of the relevant time graph. In this process the effect of the series inductor creates a voltage drop due to the *Ldi/dt* value, which creates a separate current loop created by the Z1 and the MOV. Sharp drop within the time graph at the MOV terminal is due to the firing of the MOV. Similar case occurs at the BBD where the clamping voltage (breakdown voltage) of the BBD is lower than the MOV's clamping voltage and reaction time lies in the picoseconds range [2].

8.4 Subsets of Circuits Useful for Class a Protectors

Fig. 8.2 depicts several subsets of circuits, which could be combined into a complete SPD which performs two basic functions namely:

- filtering of the transient high-voltage high-energy surge, which could enter into the protected unit; and
- filtering of the radio-frequency components, which could be generated within the protected electrical product.

Fig. 8.2A is one of the simplest possible surge protection circuits, where an MOV and a simple fuse are combined. Under normal AC mains conditions such as 230 V, 50 Hz, where the peak voltages are $\pm 230\sqrt{2}$V which is approximately $\pm$ 325 V, MOV is selected with a firing voltage above this 325 V by about 20%. If a transient gets superimposed on the mains waveform above this value, the MOV fires and creates a low impedance path for the surge energy-related transient signal. This dissipates the surge energy within the low impedance path of the MOV, by which the connected load can be protected. One important practical property of MOVs is that with every event of firing certain amount of material degradation occurs and due to the nature of the material, which is beyond the subject of this chapter. With this situation in most MOVs firing voltage could be gradually reduced and when this keeps happening due to repeated surges the MOV will start firing due to RMS voltage fluctuations continuously, which in turn will burn the fuse. Another case is that when repeated surges occur, heating within the MOV could exceed its thermal capacity and the MOV device could get destroyed.

Fig. 8.2B is a simple inductor and capacitor added to the case in circuit of Fig 8.2A, which could also help creating a short circuit path for the surge, and to dissipate surge energy. When the surge voltage occurs, a surge current through the inductor will cause a voltage drop determined by the *Ldi/dt* effect, and that causes a lower effective instantaneous surge voltage at the protected load. At the same time, due to the capacitance between the two rails, the voltage rise will be retarded. Another useful property of this LC filter is its high-frequency filtering property to stop transmitting interference between the two sides of the filter.

Fig. 8.2C is a more advanced version of an SPD where the above two cases and a BBD are combined with the above two cases. One additional inductor is

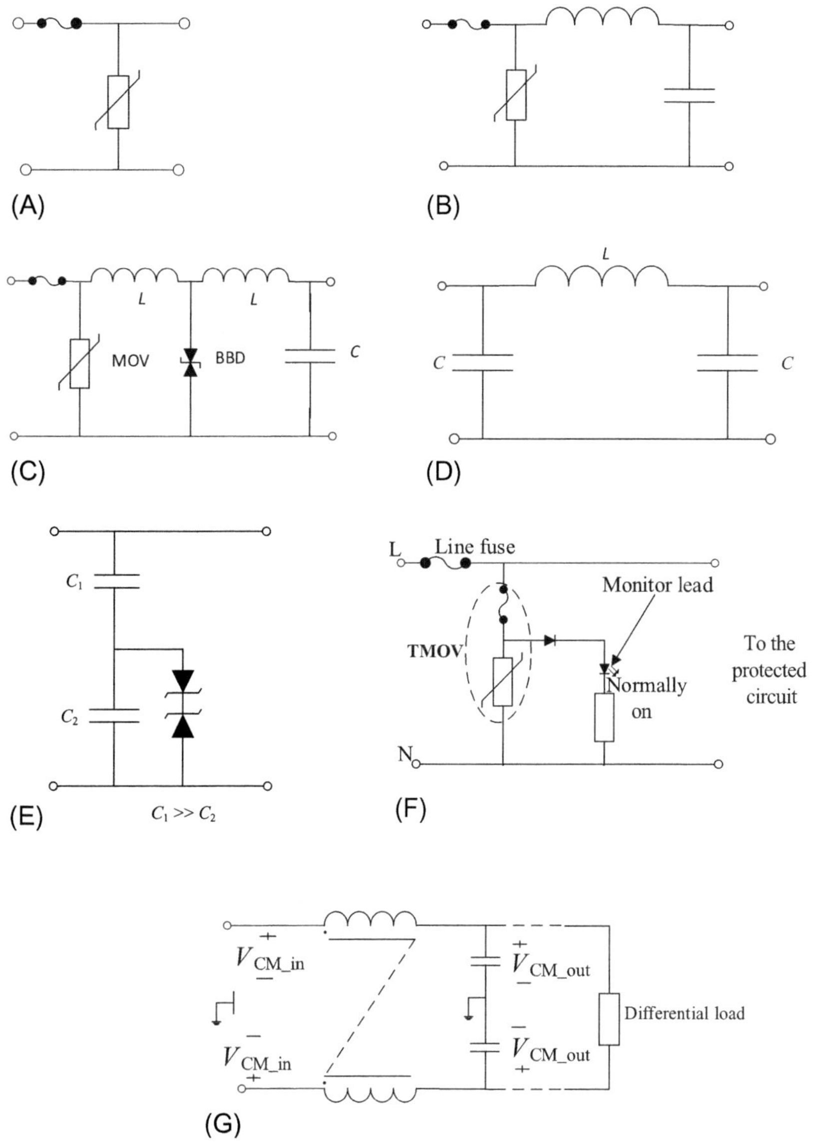

FIG. 8.2 Several subsets of a Class A surge protector. (A) Simple MOV circuit, (B) simple inductor and capacitor filter, (C) advance surge protector, (D) typical π-type filter, (E) capacitor and diode combination for surge filtering, (F) TMOV circuit, and (G) common mode surge protection.

inserted to form a T-type filter, where the BBD is in the middle of the T section. The property of the BBD is to absorb the excessive transient-related energy at a faster speed than an MOV, by which it makes a much better transient energy absorption solution. BBD is usually selected with a firing voltage value that about 1.1 times the peak of the AC peak.

Fig. 8.2D is a typical π-type filter, which is frequently used in high-performance SPDs, where inductor voltage drop and the capacitor's low impedance at high frequencies are effectively used.

Fig. 8.2E is another practical surge absorbing circuit based on two capacitors and two diodes combined. In this case C_1 is a much larger capacitor than the C_2, which forms a capacitive divider where a higher instantaneous voltage across C_2 gets short circuited by the action of the two diodes connected in inverse fashion.

Fig. 8.2F is based on a much never type of MOV, which is called thermally protected MOV (TMOV), where a built-in internal fuse is in series with the varistor. These are usually three pin devices, where the additional pin is used to drive an LED or other type of an indicator, which provides useful information to the user of the varistor's status. Indicator can be connected to either live or neutral to facilitate the relevant indication. Fig. 8.2F is for the specific case of indicator lighting up when the TMOV is having a faulty fuse, due to the failure of the varistor part due to excessive energy dissipated within the device. The family of TMOV devices from manufacturers, such as Littelfuse, is due to the specifications proposed in the third cdition of UL 1449 Standard. For more details, Ref. [2] is suggested.

Fig. 8.2G shows differential mode coupled inductors, which are typically used in more expensive SPDs with a pair of capacitors or where mid-point is connected to the ground.

Whilst a commercial SPD will carry both subset of components, namely the surge absorbing part and the high-frequency filtering part, which are carefully combined to meet the standards of surge protection and EMC filtering; in this discussion, we will present details only for estimating the transient surge absorption portion of the overall circuit and its laboratory testing of the final design for confirmation of its capability.

8.5 Principle of Protection Based on the Characteristic Impedance: Simplified Case

In Chapter 1 we have discussed how the general concept of the impedance dividers work in a transient protection system referring to Fig. 1.7. Now we can extend these concepts practically into the design and development of an SPD based on MOVs or BBDs, which could be placed in parallel to the load to be protected.

Fig. 8.3 depicts the case of the a surge superimposed on a power line or telecom line, where surge is represented by V_{surge}, which is voltage source lasting only microsecond order times, and the series impedance shown by Z_{source}, which represents any of the following or their combinations:

- ohmic resistance of the connection cables, including the effect of distribution systems;

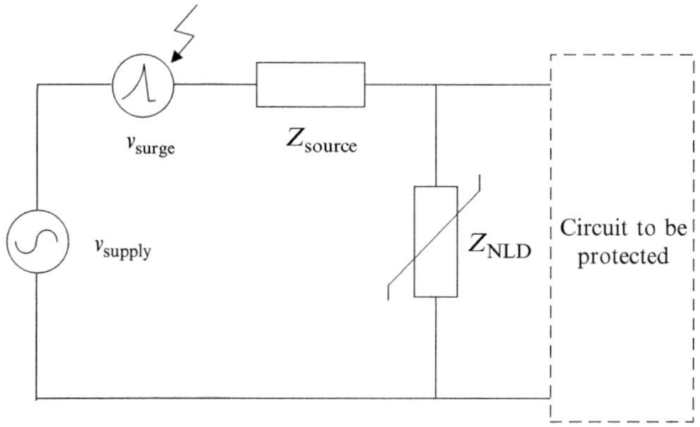

FIG. 8.3 Simplified representation of the voltage dependent resistance of the NLD and the voltage independent impedance of the source.

- inductive reactance of a coil in series; and
- complex characteristic impedance of a transmission line.

As shown in Fig. 8.3 when the transient occurs,

$$V_{NLD} = \frac{Z_{NLD}}{Z_{NLD} + Z_{source}}. \tag{8.1}$$

If the surge voltage does not exceed the firing voltage of the NLD, Z_{NLD} becomes very large compared with the series impedance of the surge path. In Eq. (8.1), we assume that the load to be protected is also represented by the overall value of Z_{NLD}.

Fig. 8.4 depicts the actual case of an NLD used for protection, such as an MOV, together with the actual AC power source also coming in series with the surge source. In this case two voltage sources connected in series (V_{supply} and the V_{surge}), Z_{source} (which could be in the range of fractional ohms to few ohms) and Z_{NLD} works in a series loop.

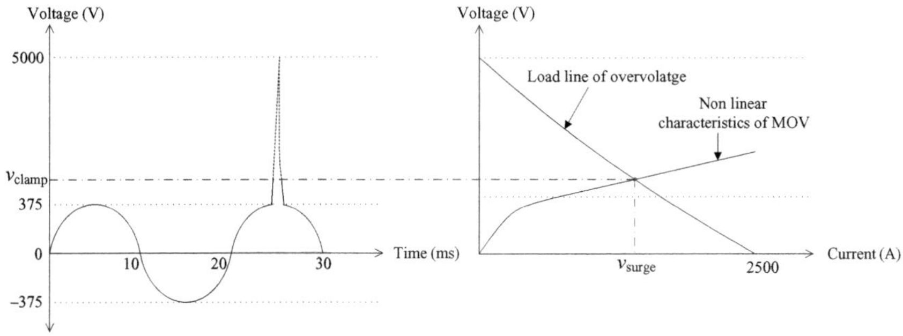

FIG. 8.4 Principle of overvoltage protection by an NLD such as an MOV or a BBD.

Fig. 8.4 depicts the operation of the circuit in Fig. 8.3, where the nonlinear behaviour of the MOV (or BBD) is represented with the load line assuming that impedances are simplified to be resistors, and the transient peak of the surge is creating a maximum peak voltage value of 5 kV within a 230 V, 50 Hz supply where sine wave peak voltage is approximately ± 375 V (allowing for an occasional overvoltage) and the Z_{source} is 2Ω. Right-hand side of the figure is similar to the common transistor amplifier load lines showing the bias conditions. This graph represents the maximum possible short circuit current of 5000/2 = 2500 A. Because the two resistances are in series, and one being an NLD, loop current due to the surge voltage creating the sum of two voltages across resistors being equal to the sum of the two input voltages, the combination circuit is expected to operate at the intersection point of the two graphs in the right-hand side of Fig. 8.3. When the MOV fires, and move into the lower-resistance (in the order of few ohms, high current in the order of few 100 to few 1000 A) region, resistor divider action ultimately creates the case of MOV achieving its clamping voltage across the MOV, protecting the critical load connected.

8.6 A More Realistic Case

The preceding discussion is only a very much simplified case where we consider the surge waveform is considered having a constant amplitude over time. However, the surge voltage being non-DC or nonsinusoidal case (of single frequency), the typical surge current waveform contains a large portion of frequencies within the range of kilohertz to megahertz. For different frequencies, the connection cables and leads, etc., will have different characteristic impedances where the straight line in the right-hand side of Fig. 8.4 will have different slopes at different frequencies due to combination effects of multiple elements. Fig. 8.5 shows the approximate behaviour of a characteristic impedance of a power supply line at different high frequencies in the order of 100 kHz–100 MHz. To avoid complexity of calculations, characteristic impedance could be considered as a constant value of 50Ω, which is shown in figure at 1 MHz.

Whilst we have used a constant series source impedance of 2Ω in Fig. 8.3, in real-world situations, the connecting power line may present different characteristic impedances for various high-frequency components of the transient over voltage, the analysis will have to use these different values, if one needs to get a better idea of the clamping voltage developed across the MOV or the BBD. More discussion on this is available in [3]

8.7 Selection of Commercial MOVs for an SPD

Referring to the case in Fig. 8.2A we can use commercially available devices from different manufacturers in configuring an SPD. In this chapter we have used devices from companies, such as TDK-EPCOS and Littelfuse, in the specific example, for convenience. In complying with new standards, such as

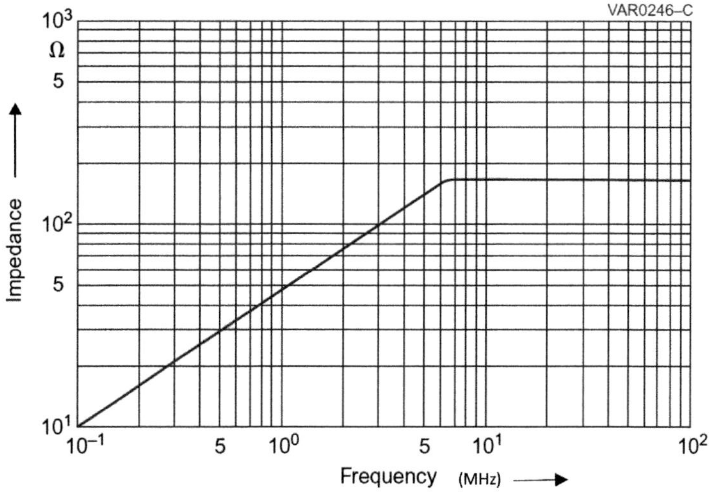

FIG. 8.5 Impedance of a supply line for high-frequency components of a transient overvoltage. *(Adapted from EPCOS AG, SIOV Metal Oxide Varistors, 2018.)*

UL 1449-Edition 4, it may be compulsory in some countries to use newer types of NLDs, such as thermally protected MOVs. This is to minimize the risk of fire hazards within an SPD product for increased safety requirements.

Let us start with a sample set of specifications such as the following: AC power circuit conditions and requirements: 230 V rms AC line-voltage current/voltage waveforms for surge: IEEE Combination (current) waveform of 8×20 μs; voltage waveform of 1.2×50 μs; peak current during the surge: 3000 A; endurance requirement—survive up to 40 surges; and other components (transformer, capacitors, etc.): rated to withstand 1000 V maximum. Let us develop this protector based on components from a reputed MOV manufacturer, such as Littelfuse, based on their device datasheets and guidelines given in [4]. Table 8.1 is an extract from the relevant datasheet from Littelfuse, Inc. for simple MOVs where thermal protection is not built-in to the device [5].

In selecting the device, we have to allow the rms voltage (for the 230 VAC rms) to occasionally to rise by 20%, which makes us select a device with an AC rms rating of 230 VAC $\times$ 1.2 = 276 VAC. This allows for 20% head room to take into account occasional voltage swells. Lets select the 20-mm disc with the clamping voltage of 275, V275PA40C in Table 8.1. This is having less clamping voltage than the V275PA40A. Manufacturer provides the pulse rating curve in their datasheet as depicted in Fig. 8.6 [2]. The graphs in Fig. 8.6 show the repeated-pulse endurance limits of the 20-mm MOV device family for different peak current ratings versus pulse duration. Longer the pulse duration, maximum peak current withstand capability reduces. Top graph shows a case where device will survive only one hit, whilst the lowest graph shows the capability to

TABLE 8.1 Twenty-Millimetres Simple MOV Devices From Littelfuse [2]

Part Number and Device Branding	Maximum Ratings (85°C)				Specifications (25°C)					
	Continuous		*Transient*		*Varistor Voltage (at 1 mA DC Test Current)*			*Max Clamping Volt V_c (at Test Current (8/20 μs))*		*Typical Capacitance*
	V_M *(AC)* *(V)*	V_M *(DC)* *(V)*	W_{TM} *(J)*	I_{TM} *(A)*	*Min* *(V)*	V_N *(DC)* *(V)*	*Max* *(V)*	V_C *(V)*	I_P *(A)*	f = 1 *MHz* *(pF)*
V130PA20A	130	175	70	6500	184	200	243	360	100	1900
V130PA20C	130	175	70	6500	184	200	220	325	100	1900
V150PA20A	150	200	80	6500	212	240	284	420	100	1600
V150PA20C	150	200	80	6500	212	240	243	360	100	1600
V250PA20A	250	330	130	6500	354	390	453	675	100	1000
V250PA20C	250	330	130	6500	354	390	413	620	100	1000
V275PA40A	275	369	140	6500	38 9	430	494	740	100	900
V275PA40C	275	369	140	6500	389	430	453	680	100	900
V320PA40A	320	420	160	6500	462	510	565	850	100	750
V320PA40C	320	420	160	6500	462	510	540	800	100	750
V350PA40A	350	460	165	6500	500	559	618	910	100	700
V350PA40C	350	460	165	6500	500	535	570	840	100	700

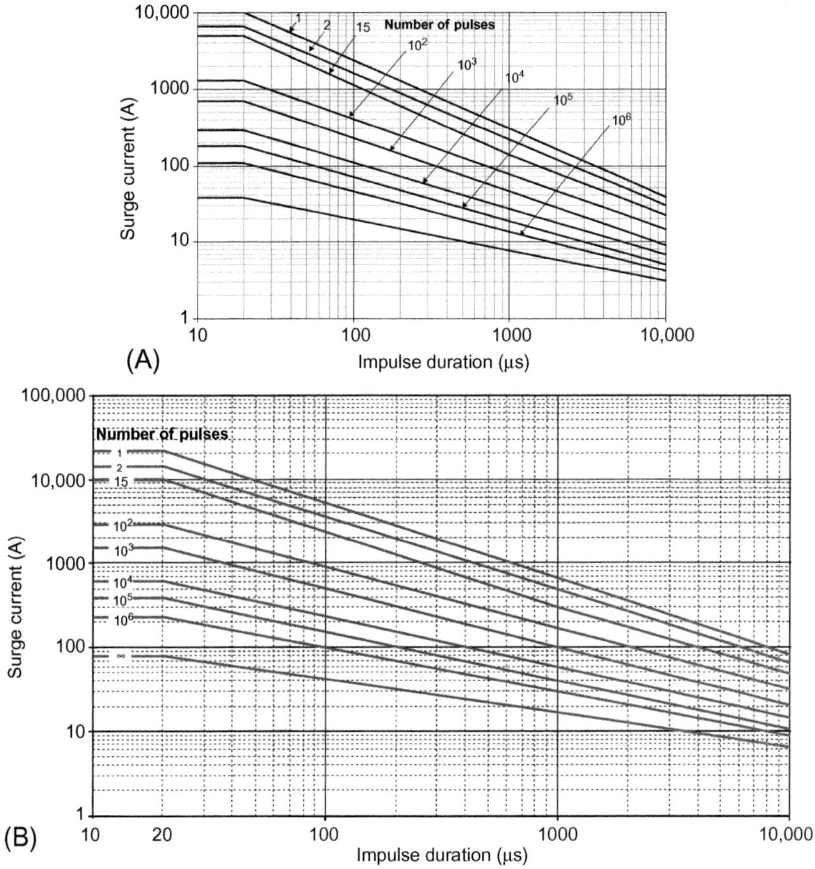

FIG. 8.6 Pulse rating curves for Littelfuse [5] (A) for 20-mm disc sizes in MOV family and (B) for 25-mm TMOV family [2].

withstand an indefinite number of hits. However, in this case, the maximum peak current at 20 μs duration is much below 32 A approximately. For 20 μs duration pulse, even at 2000 A peak current, the device cannot withstand anything above 10 hits. So this device we consider is not usable because the specification is for 3000 A peak current.

Now let us therefore consider a different device family, such as the 25 mm TMOV family from the same manufacturer, where the device impulse duration versus surge current graph is shown in Fig. 8.6B. From this graph, for an impulse duration of 20 μs with a peak current of 3000 A, we can see the endurance is over 100 hits. Now we have to select the particular device from the family, which meets the rms value of 275 VAC. Table 8.2 depicts the data for this TMOV family, which even satisfy the UL 1449 Ed 3 requirement for thermal protection. Now let us select a device such as TMOV25SP275M from Table 8.2

TABLE 8.2 TMOV and iTMOV Devices From Littelfuse [2]

Absolute Maximum Ratings

- For ratings of individual members of a series, see Device Ratings and Specifications chart.

	TMOV 25S Varistor Series	Units
Continous:		
AC Voltage Range ($V_{M(AC)RMS}$)	115 to 750	V
Transient:		
Peak Pulse Current (I_{TM})		
For 8x20μs Current Wave, single pulse	20,000	A
Single-Pulse Energy Capability		
For 2ms Current Wave	170 to 670	J
Operating Ambient Temperature Range (T_A)	-55 to +85	°C
Storage Temperature Range (T_{STG})	-55 to +125	°C
Temperature Coefficient (αV) of Clamping Voltage (V_C) at Specified Test Current	<0.01	%/°C
Hi-Pot Encapsulation (COATING Isolation Voltage Capability)	2,500	V
Thermal Protection Isolation Voltage Capability (when operated) *See notes under Device Ratings & Specifications section for more information	600*	V
COATING Insulation Resistance	1,000	MΩ

CAUTION: Stresses above those listed in "Absolute Maximum Ratings" may cause permanent damage to the device. This is a stress only rating and operation of the device at these or any other conditions above those indicated in the operational sections of this specification is not implied.

Device Ratings & Specifications

2 Leaded Device - Without Indicator Lead		3 Leaded Device - With Indicator Lead Option		Model Size Disc Diameter (mm)	Maximum Rating (85°C)				Specifications (25 °C)			
					Continuous		Transient		Varistor Voltage at 1mA Test Current		Clamping Voltage at 100A Current 8/20μs	Typical Capacitance (f=1MHz)
					AC Volts	DC Volts	Energy 2ms	Peak Current 8/20μs				
Part Number	Branding	Part Number	Branding		$V_{M(AC)RMS}$	$V_{M(DC)}$	W_{TM}	I_{TM} 1 × Pulse	$V_{N(DC)}$ Min	$V_{N(DC)}$ Max	V_C	C
					(V)	(V)	(J)	(A)	(V)		(V)	(pF)
TMOV25SP115E	P25T115E	TMOV25SP115M	P25T115M	25	115	150	170	20000	162	198	295	3200
TMOV25SP130E	P25T130E	TMOV25SP130M	P25T130M	25	130	170	190	20000	184.5	225.5	335	2800
TMOV25SP140E	P25T140E	TMOV25SP140M	P25T140M	25	140	180	210	20000	198	242	355	2500
TMOV25SP150E	P25T150E	TMOV25SP150M	P25T150M	25	150	200	220	20000	216	264	390	2300
TMOV25SP175E	P25T175E	TMOV25SP175M	P25T175M	25	175	225	250	20000	243	297	450	1900
TMOV25SP200E	P25T200E	TMOV25SP200M	P25T200M	25	200	265	270	20000	283	345	530	1700
TMOV25SP230E	P25T230E	TMOV25SP230M	P25T230M	25	230	300	300	20000	324	396	585	1500
TMOV25SP250E	P25T250E	TMOV25SP250M	P25T250M	25	250	320	330	20000	351	429	640	1400
TMOV25SP275E	P25T275E	TMOV25SP275M	P25T275M	25	275	350	350	20000	387	473	700	1250
TMOV25SP300E	P25T300E	TMOV25SP300M	P25T300M	25	300	385	370	20000	423	517	765	1150
TMOV25SP320E	P25T320E	TMOV25SP320M	P25T320M	25	320	420	390	20000	459	561	825	1080
TMOV25SP385E	P25T385E	TMOV25SP385M	P25T385M	25	385	505	430	20000	558	682	1010	900
TMOV25SP420E	P25T420E	TMOV25SP420M	P25T420M	25	420	560	460	20000	612	748	1100	820
TMOV25SP440E	P25T440E	TMOV25SP440M	P25T440M	25	440	585	470	20000	643.5	786.5	1160	790
TMOV25SP460E	P25T460E	TMOV25SP460M	P25T460M	25	460	615	490	20000	675	825	1220	750
TMOV25SP510E	P25T510E	TMOV25SP510M	P25T510M	25	510	670	520	20000	738	902	1335	680
TMOV25SP550E	P25T550E	TMOV25SP550M	P25T550M	25	550	745	550	20000	819	1001	1475	630
TMOV25SP625E	P25T625E	TMOV25SP625M	P25T625M	25	625	825	600	20000	900	1100	1625	550
TMOV25SP750E	P25T750E	TMOV25SP750M	P25T750M	25	750	970	670	20000	1080	1320	1950	460

Notes:
Average power dissipation of transients should not exceed 1.5 watts.

where under all possible pulse duration conditions maximum peak current is at 20, 000 A. We have to ensure that the device clamping voltage should be lower than 1000 V, for a peak current of 3000 A. From the graphs of Fig. 8.6A at RMS capability of 275 V we see that maximum current capability at 3000 A is less than about 800 V. This clamping voltage is able to protect all devices, which are designed from a peak transient of 1000 V.

8.8 Combined MOV Varistor and TVS Scenarios

For a better reliability and safer-operation guarantee, most advanced SPDs combine MOVs and BBDs (which are generally known as transient voltage surge devices) in the same circuit as shown in Fig. 8.7. Fig. 8.7A is a simpler version with an MOV and a BBD separated by an inductor. In Fig. 8.7B it is a more advanced version with two inductors and capacitors are inserted as an additional filter.

8.9 Surge Simulator-Based Testing of Practical Circuits

When a surge protector is configured with the simple guidelines earlier, it is very important to test it using a lightning surge simulator (LSS) such as the ones shown in Fig. 8.8. Fig. 8.8A is a photograph of an LSS from Noise Laboratories, Japan, which is capable of reaching IEEE Standard voltage waveforms up to 6.6 kV peak value. Fig. 8.8B is more advanced version, Model LSS F03 series from the same equipment manufacturer, which could reach peak voltage waveforms of up to 15 kV.

These types of test instruments are capable of adjusting the peak value in steps of 100 V and with repeated pulsing with either 20 or 10 s apart. In the next few paragraphs, we will show various versions of SPDs developed from a

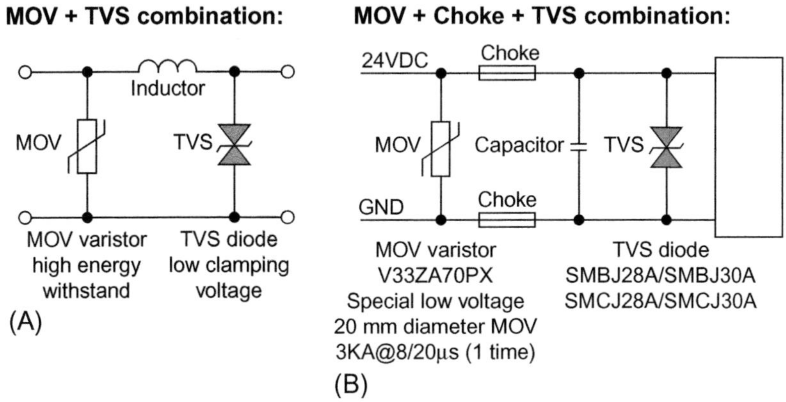

FIG. 8.7 MOV-TVS combined SPD configurations: (A) simpler version based on three components; and (B) with an L-C filter replacing the inductor for better RFI/EMI filtering and lower clamping voltage at the critical load end.

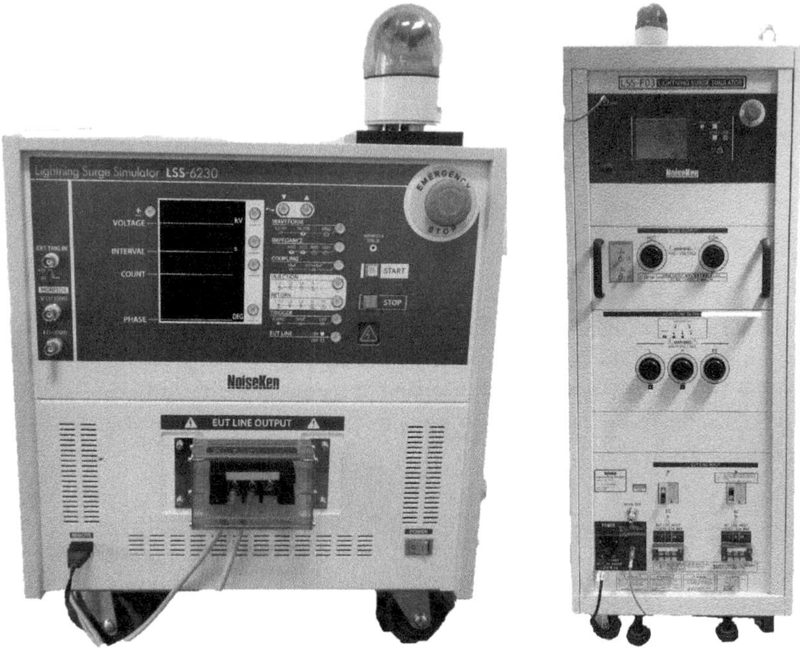

(A) (B)

FIG. 8.8 Different types of surge simulators. (A) Noiseken LSS 6230 and (B) Noiseken LSS F03.

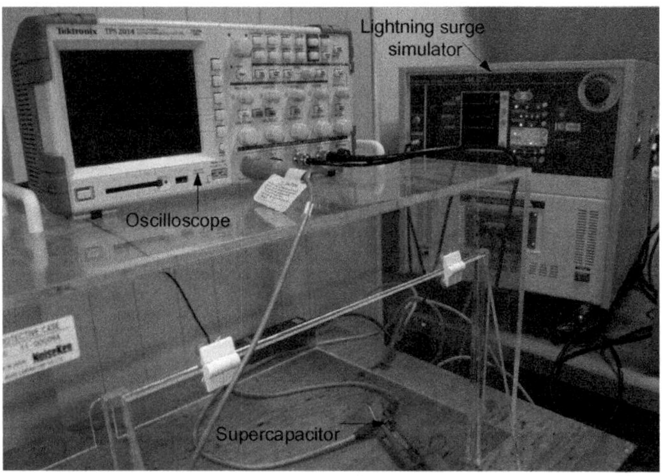

FIG. 8.9 Surge test setup.

simple MOV to much more advanced and complex versions. Fig. 8.9 depicts the standard connections in a typical test setup, with an oscilloscope connected to measure input surge voltage waveform, surge current from the LSS into the SPD input and the output waveform at the load end of the SPD. An LSS typically has two monitoring channels to feed the voltage and the current to the oscilloscope inputs, via 1000:1 (for 6.6 kV model) attenuation ratio. A 15 kV model usually comes with a voltage monitoring channel with a 2000:1 attenuation ratio. Appendix A provides an overview of a typical measurement set up, with the readings observed at different test points, together with a simplified equivalent circuit of LSS indicating its measuring points.

8.10 Possible Circuit Configurations and Test Waveforms

Table 8.3 is a compilation of useful test results for different SPD configurations, in order to demonstrate the case of selecting useful components for an SPD and to see their surge absorbing capability based on a surge created by an LSS. All these cases were subjected to a 2 kV peak-based IEEE Standard 1.2/50 μs combined surge waveform generated by a Noiseken LSS 6230. In figures in Table 8.3, scope's Channel 1 (Ch 1) shows the voltage waveform at the input side MOV of the SPD. Oscilloscope's Channel 2 (Ch 2) indicates in relevant cases the voltage appearing at the critical load protected by the SPD, and Channel 4 is depicting the current drawn from the LSS monitored via the current monitoring channel of the LSS.

8.10.1 Simple SPDs Based on a Single MOV With Different Clamping Voltages

Case 1 in Table 8.3 is possibly the simplest possible case of an MOV (of Type TMOV14R275E from Littelfuse) with a clamping voltage of 710 V subjected to 2 kV peak surge. As seen from the relevant oscillographs, output voltage gets clamped at a peak value of 736 V with a surge current created in the loop of 568 A.

Case 2 in Table 8.3 is where a slightly higher clamping voltage-based MOV (of Type TVT 14471 from Thinking Electronic Industrial Co., Ltd) is used. In this case with the higher clamping voltage MOV of value 775 V we see the output clamping voltage rising to 784 V with a loop surge current of 536 A.

8.10.2 MOV and Capacitor-Based Version

Case 3 in Table 8.3 is with the same MOV as in Case 2, but supplemented with a 0.047μF capacitor (400 V). In this case the clamping voltage at the output does not change, but loop surge current changes slightly.

TABLE 8.3 Variation of Waveforms With Different Arrangements of Varistors, Inductors, and Capacitors

Number	Configuration	Oscillogram	Comments
1	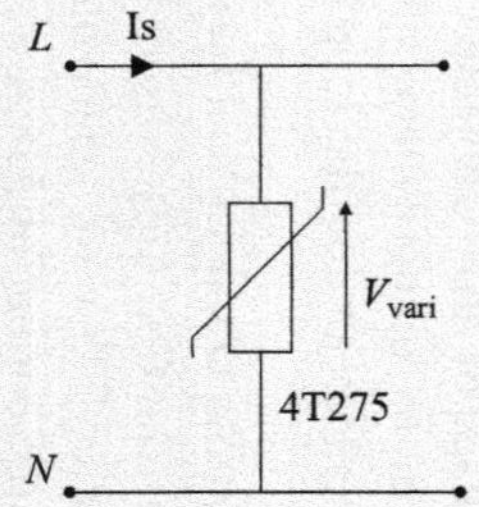	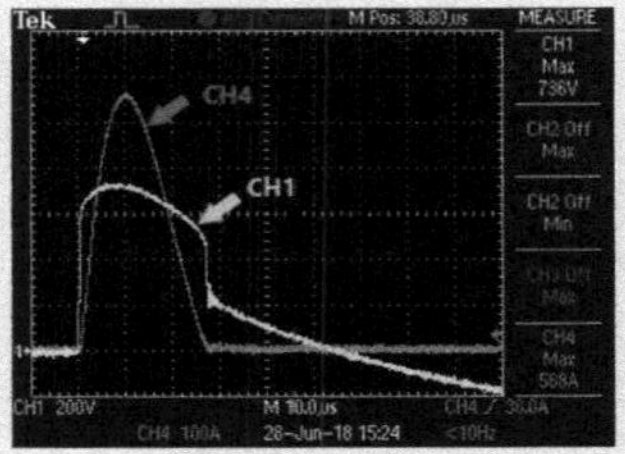	Note the output surge voltage collapsing at the time surge current dropping to zero
2	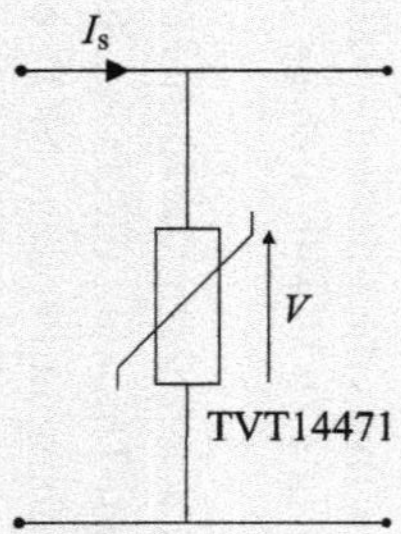	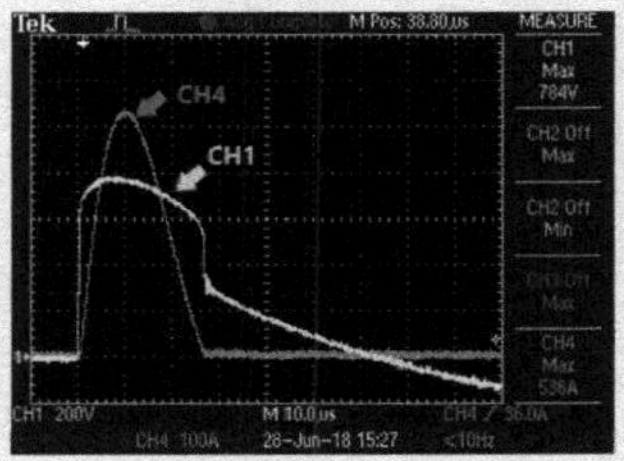	Note the output surge voltage collapsing at the time surge current dropping to zero as in case 1
3	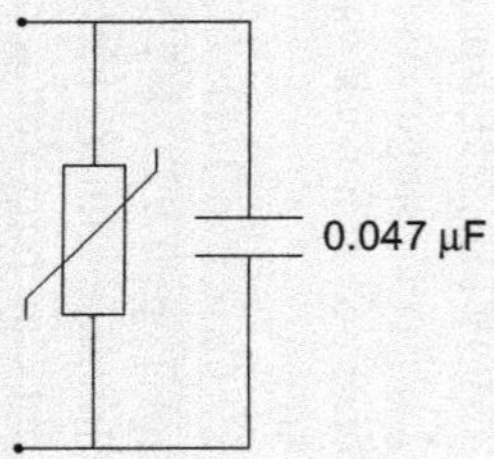	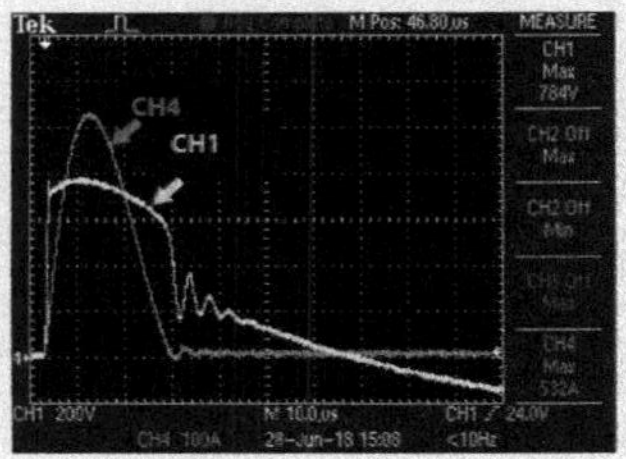	Note the output surge voltage with some resonance-based ringing effect, due to the line inductances resonating with the capacitor, at the time the loop current collapses to its minimum

4	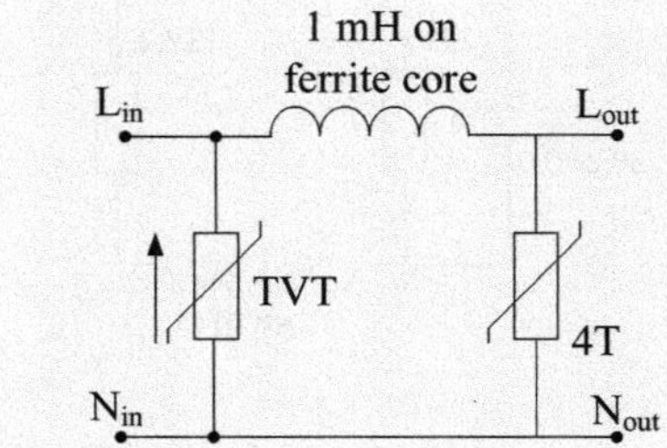	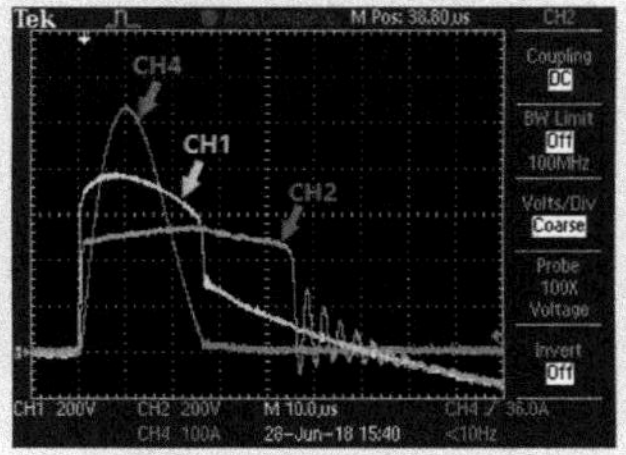	Note the output surge voltage with some resonance-based ringing effect, due to the series inductances resonating with the capacitance of the fired MOV after the loop current collapses to its minimum
5	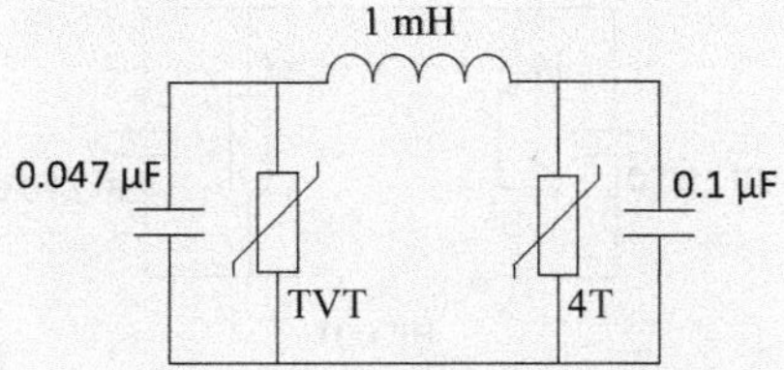	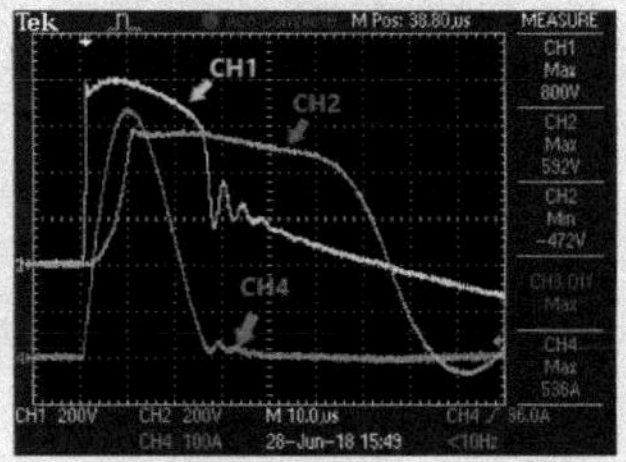	In this case resonance effect is shown at the input MOV voltage, compared to Case 4 where ringing occurs at output end
6	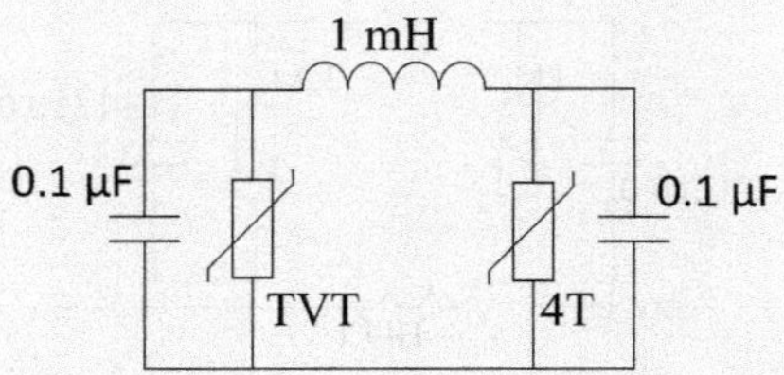	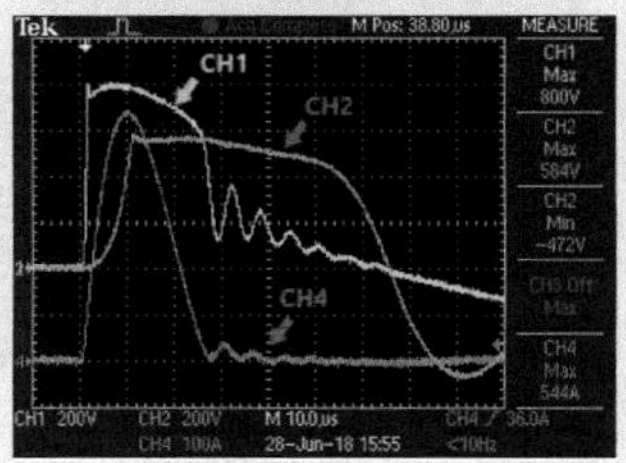	In this case resonance effect is still shown at the input MOV voltage, but at a different frequency

Continued

TABLE 8.3 Variation of Waveforms With Different Arrangements of Varistors, Inductors, and Capacitors—cont'd

Number	Configuration	Oscillogram	Comments
7	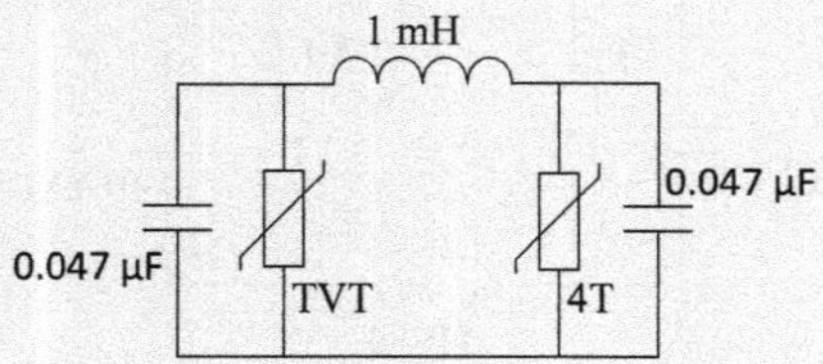	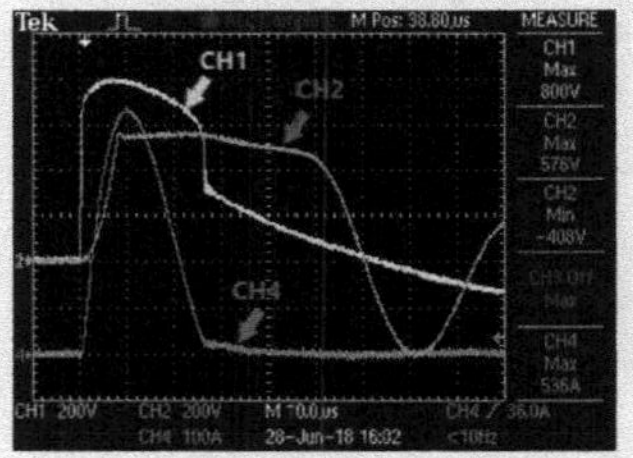	No resonating effect at both ends
8	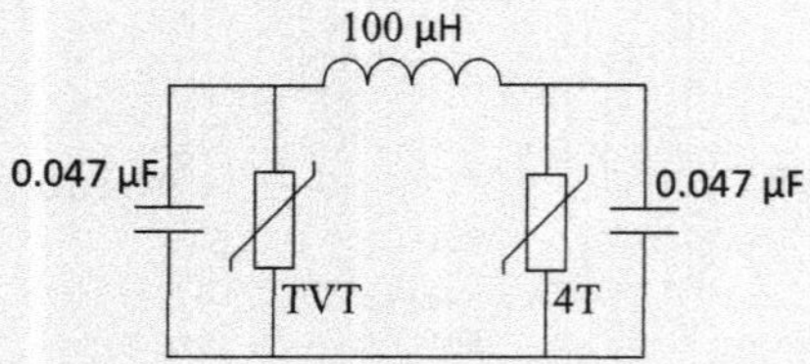	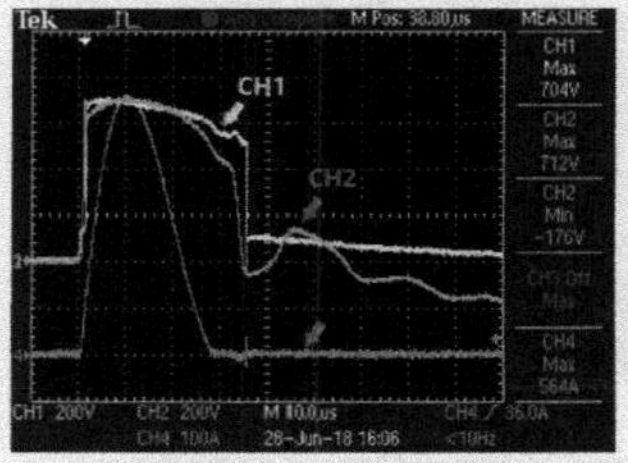	Negative part of the output MOV voltage is reduced significantly
9	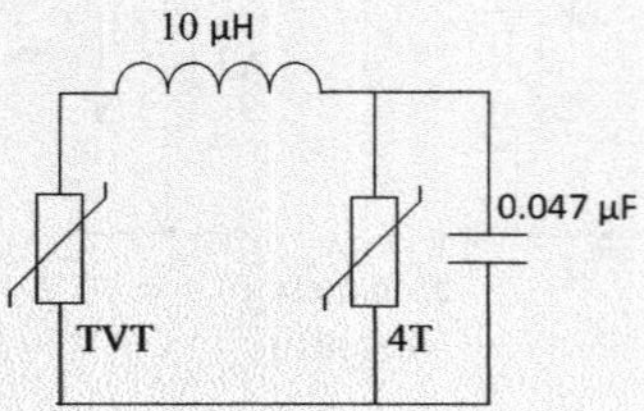	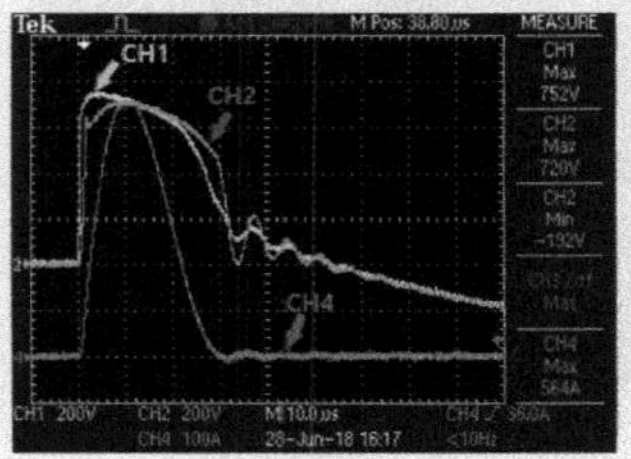	Output and input waveforms following similar shapes

10	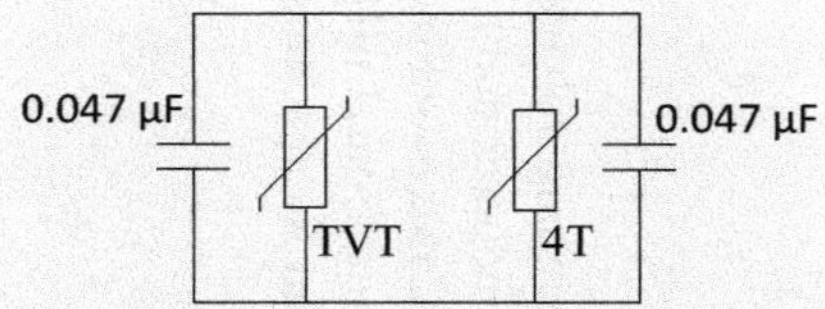	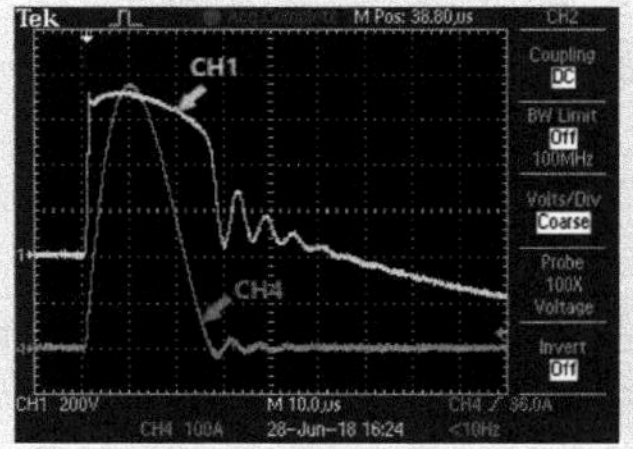	Input and output MOVs are showing the same waveform
11	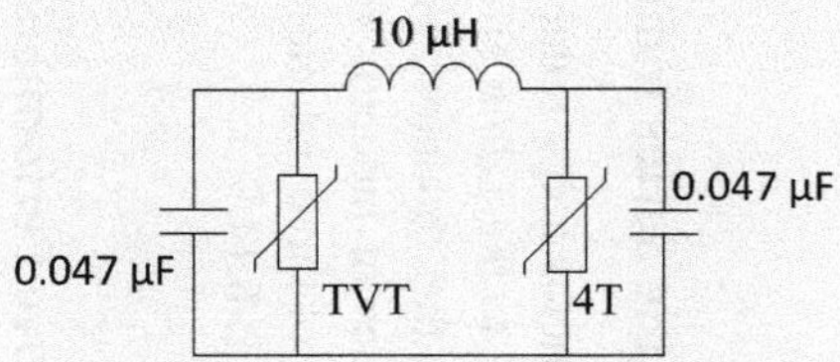	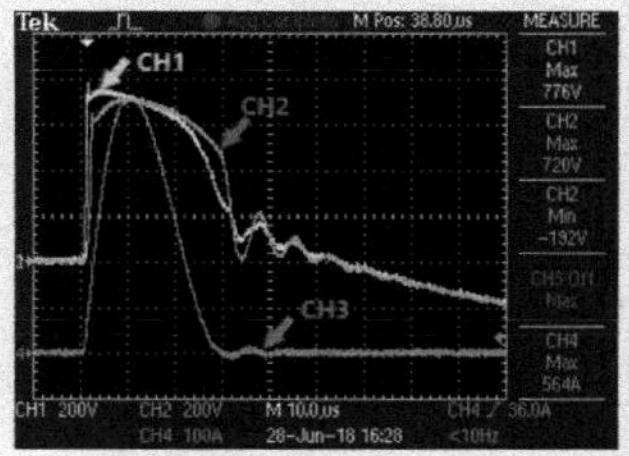	Nearly same as in case 9

8.10.3 Two MOVs and a Series Inductance of 1 mH Wound on a Ferrite Core

Case 4 in Table 8.3 depicts a case of two MOVs we had in Cases 1 and 2 combined with an inductor in series. Voltage waveform at the output end is lower than the voltage at the input end, with a similar case of current as in previous cases.

8.11 Adding X-Type Capacitors to an SPD

Case 5 in Table 8.3 is the case of two capacitors enhancing the circuit configuration in case 4. Here the current peak is around 536 A, but the output voltage across the second MOV goes from positive 592 to 472 V. Input side MOV clamping voltage peaks at 800 V.

Case 6 shows the circuit when you change the input MOV end's capacitor is changed to a higher value. Only difference we see is the resonance occurring at a slightly different frequency. Peak voltages are changed only by a small margin.

Case 7 shows the effect of two capacitors reduced to 0.047 with other three parts remaining the same. In this case you see hardly any ringing, but with the output MOV waveform negative peak reduced slightly.

Case 8 is same as in case 7 but the inductance reduced to 100 μH. Now we see the peak clamping voltages at each of the MOV are nearly the same, because the value of L reduced by a factor of 10.

Cases 9–11 are with inductance reduced to 10 μH or to 0 (case 10) with one or two capacitors still connected. In all these resonant part of the output waveform is slightly changed.

8.12 Changing the Value of the Series Inductor

In a surge protector the series inductor helps reducing the clamping voltage depending on its value. This is due to the Ldi/dt effect of the inductor due to the surge current driven through the inductor. Figures in Cases 4–11 in Table 8.3 depict three possible cases of inductor values varying from 1 mH to 10 μH. Fig. 6.15 depicts the case of three inductors with added X capacitors into the basic versions shown in Fig. 6.14.

8.13 Adding a BBD to Improve Performance

A BBD of a suitable rating can also improve the surge performance of an SPD, but it adds a significant cost.

8.14 Surge Protector to Protect Against Both the Differential Mode and the Common Mode

MOVs are the most widely used device for protecting sensitive electronics from damaging overvoltage transients. To date, most consider MOVs to be sacrificial devices, expected to disconnect (open) in the event of a failure. Unfortunately, this is not the case. MOVs used in SPDs can fail explosively when subjected to sustained steady-state power frequency over voltages. Traditional MOVs are highly susceptible to damage from sustained/temporary overvoltage conditions. During these overvoltage conditions, MOVs enter a partial conductive state where they absorb the associated energy—eventually generating enough heat to force the component to rupture and initiate a short circuit condition. For the preceding reasons, many design engineers have pursued alternatives to include an MOV-based suppressor in their design—adding series components, extra costs, and engineering time. However, the thermally protected MOV (TPMOV) addresses the inadequacies of the traditional MOV, making this technology a viable consideration.

References

[1] Weidmüller, The Basics of Lightning and Surge Protection, 2013.

[2] Littelfuse, Inc., TMOV and iTMOV Varistor Series, 2016.

[3] EPCOS AG, SIOV Metal Oxide Varistors, 2018.

[4] Littelfuse, Inc., MOVs and Fuses: Selecting the Appropriate Circuit Protection Component, 2017.

[5] Littelfuse, Inc., Metal-Oxide Varistors (MOVs), Radial Lead Varistors > TMOV and iTMOV Varistor Series, 2017.

Chapter 9

Development of the Supercapacitor-Assisted Surge Absorber (SCASA) Technique

9.1 Introduction

In Chapter 4 surge absorption capability of SCs was discussed. SCs are able to absorb short duration transient surges with multiple repetitions. A 5-F supercapacitor fully charged to 2.7 V can store 18.2 J where as a 2-kV, 1-kA combination wave has 9 J of energy as shown in Table 2.2. When comparing the preceding two values, the 5-F SC can easily store twice the energy content of the high-voltage source. Despite the SC's low ESR, the large capacitance value creates a significantly long-time constant. When the circuit path resistances are added, the circuit time constant becomes very much larger, usually in the order of fractional seconds to several seconds. Because an induced voltage due to a transient event does not usually last more than 100 μs, a SC will not charge significantly. However, the continuous DC voltage rating of SC is only few volts.

The low DC voltage rating of SCs is the main factor, which limits SC being used in mains power applications as a direct replacement to typical surge absorbers. In most traditional mains operated appliances, the surge protection device comes directly in parallel to the AC mains supply at the source side input. Because the impedance of a SC at 50 Hz is extremely low, an SC cannot be a direct replacement for a traditional surge absorber device, such as MOV or a BBD. Given the preceding characteristics of a SC, it is necessary to develop a circuit in such a way that the SC is placed in a low-voltage circuit during normal AC operation. Therefore, if a SC is to be used as the surge absorbing device, a circuit arrangement should be implemented to

- trigger the SC circuit path only at the event of a high-voltage surge; and
- transfer the surge energy into a subcircuit where a SC absorbs the surge energy or diverting the surge energy into a different part of the surge protector circuit.

Design of Transient Protection Systems. https://doi.org/10.1016/B978-0-12-811664-7.00009-4

In the SCASA technique [1–4], a coupled inductor is used. During the event of a high-voltage surge, a varistor is set to fire and trigger the coupled inductor circuit. By the overall action of the SC-based circuit configuration, surge energy is dissipated in resistive parts of the circuit, whilst the resistor and the capacitor assist the dissipation with a useful delay.

Inductors are already found in surge protection circuits mainly in series to the load as seen in Fig. 9.1. An inductor placed in series with the surge absorber acts as a filter by opposing the quick rise of current through the circuit. During the event of a power-line surge when surge current pulse, i, passes through the inductor, a voltage proportional to di/dt will appear across the inductor opposing the source voltage. Therefore, a typical surge protector circuit cannot reduce the load voltage below the varistor clamping voltage, v_{var}. Further the energy stored in the inductor will cause oscillations, affecting the load power quality.

In the SCASA technique, the coupled inductor-secondary winding, polarity and placement are arranged in such a way that the instantaneous load voltage is lower than the varistor clamping voltage.

By the inductor placement and polarity arrangement of the coupled inductor, as shown in Fig. 9.2, it is possible to

- trigger the coupled inductor circuit only at the event of a high-voltage transient surge;

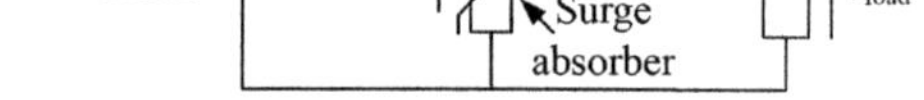

FIG. 9.1 Series inductor with high-voltage clamping device.

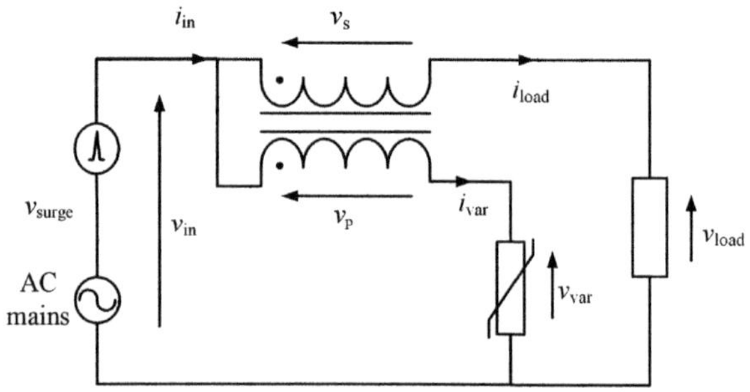

FIG. 9.2 A coupled inductor, placed in series with the load to reduce the load voltage during the event of a surge.

- reduce the load voltage during the event of a high-voltage transient surge; and
- store a part of the surge energy in the coupled inductor.

To analyze the coupled inductor circuit, assume the perfect coupling to describe the transformer turns ratio, which allows the desired operation. From varistor current, i_{var}, and load current branches in Fig. 9.2:

$$v_{\text{in}} = v_{\text{var}} + v_{\text{p}} = v_{\text{s}} + v_{\text{load}}. \tag{9.1}$$

If the transformer has n_{p} primary turns and n_{s} secondary turns, turns ratio, $n = n_{\text{s}}/n_{\text{p}}$, then $v_{\text{s}} = nv_{\text{p}}$. Hence, the load voltage can be expressed as:

$$v_{\text{load}} = v_{\text{var}} + v_{\text{p}} - nv_{\text{p}} = v_{\text{var}} - (n-1)v_{\text{p}}. \tag{9.2}$$

Because the design should lower the voltage seen by the load to a voltage less than the varistor clamping voltage.

$$v_{\text{load}} < v_{\text{var}} \Rightarrow (n-1)v_{\text{p}} > 0.$$

Because $v_{\text{p}} > 0$ for a positive transient,

$$n - 1 > 0 \Rightarrow n > 1.$$

This means that the coupled inductor should be a step-up winding to lower the load voltage below the varistor clamping voltage. If the turns ratio becomes too large the induced voltage will make the transient load voltage negative. Further adding a large inductor will show a high impedance to the load current. Detailed design guide on the coupled inductor will be discussed in Section 9.3.

By inserting the coupled inductor into the circuit with the surge absorber, during the event of a surge, a part of the surge energy will get stored in the two inductors. To avoid this stored energy propagating to the load side, a closed loop should be formed connecting the two inductors to dissipate the stored energy. This scenario is depicted in Fig. 9.3, where the two coils are connected together by a subcircuit.

Applying KVL to Loop 1 in Fig. 9.3:

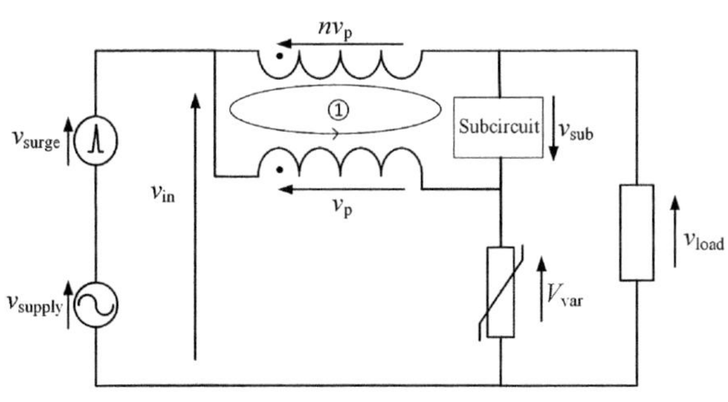

FIG. 9.3 A subcircuit is used to form a closed loop with the primary and secondary coils.

$$v_{\text{sub}} = (n-1)v_{\text{p}} = (n-1)(v_{\text{in}} - v_{\text{var}}).$$

For the worst-case analysis, lets consider the case of peak-surge voltage, $v_{\text{surge}}^{\text{max}}$ occurring aligned with the peak-line voltage, $v_{\text{supply}}^{\text{max}}$. Then the preceding equation can be rewritten as

$$v_{\text{sub}} = (n-1)\left(v_{\text{supply}}^{\text{max}} + v_{\text{surge}}^{\text{max}} - v_{\text{var}}\right).$$

Because $n > 1$ and $v_{\text{surge}}^{\text{max}} >>> v_{\text{supply}}^{\text{max}} \approx v_{\text{var}}$, the subcircuit should be capable of handling a very high voltage, similar to that of a lightning surge. As it was discussed earlier in Chapter 4, SCs are capable of withstanding lightning strikes and their associated energies. Therefore, a SC can be considered to be the device in the subcircuit, which will dissipate/store the transient lightning surge energy stored in the two inductors. Though the subcircuit is exposed to a high voltage during the event of a surge, during normal AC operation the subcircuit will not carry current other than small leakage currents caused by component parasitics. The following section will quantify the size of the subcircuit SC explaining its operation in the coupled inductor loop.

9.2 Subcircuit Operation

For analyzing the subcircuit, consider the coupled inductor as a pair of perfectly coupled inductors: primary inductance, L_{p}, and secondary inductance, L_{s}. When a high-voltage transient surge enters the circuit, a part of the surge energy is stored in the two inductors whilst dissipating the rest in the surge absorber. The stored energy in the inductors, L_{p} and L_{s}, should be dissipated within the inductor loop without allowing any propagation to the critical load. This is done through the subcircuit SC. When the stored energy in the inductors circulates through the subcircuit, it creates oscillations. In order to avoid undesirable oscillations reaching the load, the resonance frequency of the inductor-SC loop should be equal to the line frequency. Eq. (9.3) describes the resonance frequency, f_{r}, of an LC circuit.

$$f_{\text{r}} = \frac{1}{2\pi\sqrt{LC}}. \tag{9.3}$$

For the coupled inductor-subcircuit loop, using L_{p}, L_{s}, and the AC line frequency, the subcircuit capacitance, C_{sub}, can be obtained. The detailed analysis is in the next section. With the optimum capacitor and by managing other component parasitics the proposed circuit could also act as an electromagnetic interference (EMI) filter. This discussion is beyond the scope of the chapter. However, a basic filter analysis is presented in Section 9.4.

9.3 SCASA Circuit Analysis

This section analyzes the operation of the SCASA circuit under a surge, to evaluate the primary inductance, L_p, coupled inductor turns ratio, n, and subcircuit SC value. As shown in Fig. 9.4 the input voltage v_{in} with a superimposed high-voltage spike will fire the varistor and it will get clamped at V_{var}. The primary and secondary inductor voltages are v_p and v_s, respectively. For the worst-case analysis surge is considered to be superimposed with the AC mains voltage at a 90 degrees phase shift. That is, the maximum input surge voltage, v_{surge}^{max}, occurs at the maximum AC mains voltage, v_{supply}^{max}. Further the load is considered to be an open circuit, so that the varistor will have to handle the total surge current and the effect of transformer mutual inductance is nullified.

From the primary winding of the transformer,

$$V_{supply}^{max} + V_{surge}^{max} - V_{var} = v_p = Lp\frac{di_{var}}{dt}. \tag{9.4}$$

Because the varistor carries the total surge current at no load condition, $i_{var} = i_{surge}$. An estimate for the rate of charge surge current can be obtained by estimating the standard lightning surge current by a linear rise from 0 to i_{surge}^{max} within 8 μs as shown in Fig. 9.3. Hence Eq. (9.4) can be rewritten as:

$$L_p = \left(V_{supply}^{max} + V_{surge}^{max} - V_{var}\right)\frac{8\mu s}{I_{surge}^{max}}. \tag{9.5}$$

According to the transformer winding arrangement shown in Fig. 9.4, the voltage across the secondary winding will be opposing the input surge voltage, hence the load voltage, v_{load}, will be $v_{in} - v_s$.

Now the transformer turns ratio, n, should be selected such that the voltage across the secondary winding will maintain the load voltage, v_{load}, at V_{supply}^{max}. From Eq. (9.4),

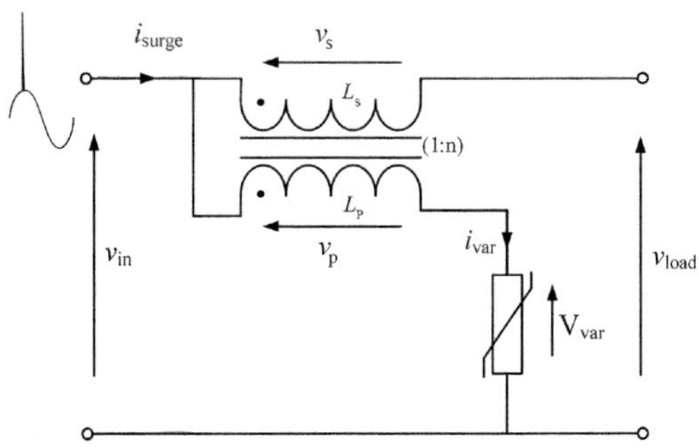

FIG. 9.4 Coupled inductor under an input voltage surge.

$$\begin{aligned} v_{\text{load}} &= V_{\text{supply}}^{\max} + V_{\text{surge}}^{\max} - v_{\text{s}} = V_{\text{supply}}^{\max}, \\ \Rightarrow v_{\text{s}} &= V_{\text{surge}}^{\max}. \end{aligned} \tag{9.6}$$

Using v_{p} and v_{s} values from Eqs (9.4), (9.6), transformer turns ratio, n, can be obtained as given in Eq. (9.7):

$$n = \frac{v_{\text{s}}}{v_{\text{p}}} = \frac{V_{\text{surge}}^{\max}}{V_{\text{supply}}^{\max} + V_{\text{surge}}^{\max} - V_{\text{var}}}. \tag{9.7}$$

Assuming an ideal transformer, the ratio between L_{p} and L_{s} is described in Eq. (9.8):

$$L_{\text{s}} = n^2 L_{\text{p}}. \tag{9.8}$$

Now the stored surge energy in the coupled inductors should be dissipated without affecting the 50 Hz AC mains supply voltage waveform. Therefore, when a capacitor is introduced into the system as shown Fig. 9.5A, resulting inductor-capacitor loop should be made to resonate at the mains line frequency. The mutual inductance of the coupled inductor is $M = k\sqrt{L_{\text{p}}L_{\text{s}}}$, where k is the coupling coefficient. Fig. 9.5B illustrates the LC loop with uncoupled inductors.

The value for C can be obtained by equating the resonance frequency of the LC loop to the line frequency, f_{L}, as given in equation:

$$f_{\text{L}} = \frac{1}{2\pi\sqrt{(L_{\text{p}} + L_{\text{s}} - 2M)C}}, \tag{9.9}$$

$$C = \frac{1}{4\ f_{\text{L}}^2\ \pi^2 (L_{\text{p}} + L_{\text{s}} - 2M)}. \tag{9.10}$$

9.4 Normal AC Operation

In normal AC operation the varistor will be in 'off' state. The varistor equivalent circuit in 'off' state is shown in Fig. 9.6. The varistor junction capacitance, C_{var}, is the dominant component in the equivalent circuit. Further, varistor lead inductance, L_{var}, is negligible compared with L_{p} and L_{s}.

Hence, the equivalent LC network formed by L_{p}, L_{s}, C, and C_{var} during normal AC operation is given in Fig. 9.7A. The L_{p}, L_{s}, and C are forming a delta

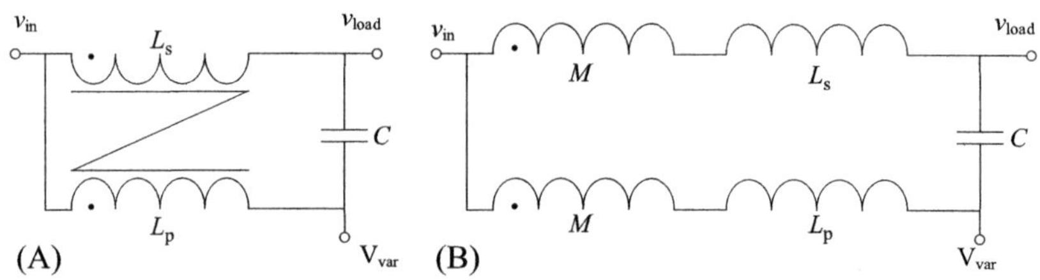

FIG. 9.5 The coupled inductor-SC loop: (A) coupled circuit and (B) uncoupled circuit.

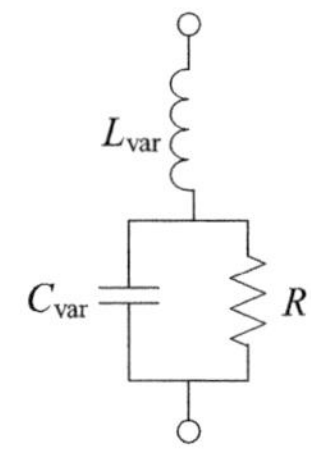

FIG. 9.6 'Off' state equivalent circuit of a varistor.

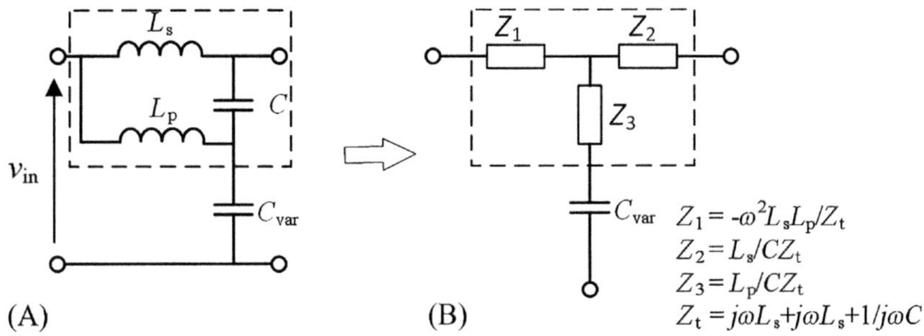

FIG. 9.7 Normal AC operation of SCASA circuit: (A) simplified circuit for filtering and (B) delta to star transformation.

combination, which can be transferred into a wye combination as per Fig. 9.7B. This indicates that the circuit simplifies to a T-type filter, which may be further tuned to improve the filtering properties of the SCASA topology to deal with RFI/EMI. However, this is beyond the scope of this chapter.

Fig. 9.8 is a snapshot of the input and output waveforms at 50 Hz mains. This shows that the filter does not cause a drop in the nominal 50 Hz mains voltage to the load.

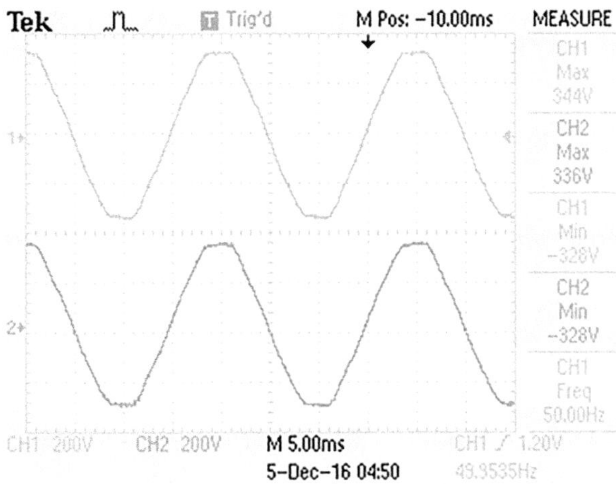

FIG. 9.8 50 Hz mains line operation of SCASA circuit, CH1-input voltage, and CH2-output voltage.

9.5 Sample Design Calculation

Based on industrial surge protector design guidelines given by Littelfuse, Inc. [5] choose rms varistor voltage to be at least 110% of the rms AC mains voltage, 230 V. Varistor V20E275P by Littelfuse, Inc. was chosen with nominal let—through voltage of 275 V which is 120% of 230 V line voltage. The chosen varistor has 710 V peak clamping voltage at the highest possible single pulse surge current of 10,000 A. According to the UL Standard [6], the worst-case surge is considered to have $V_{\text{surge}}^{\max} = 6000\text{V}$ and $I_{\text{surge}}^{\max} = 3000\text{A}$. The same standard states that varistor clamping voltage should be at least 700 V for a 3 kA surge. Therefore, 710 V will be chosen as V_{var} for the worst-case design calculations. From 230 V AC mains, $V_{\text{supply}}^{\max} = 230\sqrt{2} = 325$ V. From Eq. (9.5) the coupled inductor primary coil inductance can be calculated as:

$$L_{\text{p}} = (325\text{V} + 6000\text{V} - 710\text{V})\frac{8\times10^{-6}\text{s}}{3000\text{A}} = 14.97\,\mu\text{H}. \tag{9.11}$$

Substituting the same parameters in Eq. (9.7), yields the coupled inductor turns ratio, n:

$$n = \frac{6000\text{V}}{325\text{V} + 6000\text{V} - 710\text{V}} = 1.07. \tag{9.12}$$

Hence from Eq. (9.8), the secondary inductor value is calculated as:

$$L_{\text{s}} = (1.07)^2 \times 14.97\,\mu\text{H} = 17.14\,\mu\text{H}. \tag{9.13}$$

For surge applications the transformer-coupling coefficient is around 0.7. Hence the mutual inductance will be:

$$M = 0.7\sqrt{14.97\times17.14}\,\mu\text{H} = 11.2\,\mu\text{H}. \tag{9.14}$$

Now select the suitable capacitor value for the subcircuit by equating the inductor-capacitor resonant frequency to 50 Hz line frequency. Using Eq. (9.10) the subcircuit capacitor value, C, can be found to be:

$$C = \frac{1}{10^4\times\pi^2(14.97\,\mu\text{H} + 17.14\,\mu\text{H} - 2\times11.2\,\mu\text{H})} = 1.04\text{F}. \tag{9.15}$$

The closest practically available capacitor is a 1 F single cell. From the varistor datasheet it shows 900 pF equivalent capacitance, C_{var}, during normal AC mode of operation. Hence the resonant frequency of the obtained inductor capacitor filter arrangement according to Eq. (9.3):

$$f_{\text{r}} = \frac{1}{2\pi\sqrt{(14.97\,\mu\text{H} + 17.14\,\mu\text{H} - 2\times11.2\,\mu\text{H})(1 + 900\times10^{-12}\text{F})}} = 51.1\text{Hz}. \tag{9.16}$$

9.6 Designing of Coupled Inductor

The coupled inductor used in SCASA circuit acts as a transformer during a high-voltage surge, that is, when the varistor is conducting. It will be in series with the load. To design the transformer, consider its equivalent circuit in order to account for the transformer's leakage present in the primary and the secondary windings. Fig. 9.9A shows a $1{:}n$ transformer with L_1 and L_2 magnetizing inductances, and l_1 and l_2 leakage inductances. By referring the secondary to the primary, an equivalent circuit can be drawn as shown in Fig. 9.9B.

As shown in Fig. 9.9B the total inductance seen by the surge in the primary path is $L_2/n^2 + l_2/n^2 + L_1 + l_1$. Magnetizing inductances, L_1 and L_2, correspond to the flux coupled through the magnetic materials. Leakage inductances l_1 and l_2 are for the flux leaked through air. According to [7], an empirical estimate for the permeance of leakage through air, Λ_σ, for a toroid, in μH can be found as:

$$\Lambda_\sigma = A\, d_{\text{int}}, \tag{9.17}$$

where constant $A = 2.3$ μH/m is an empirical constant and d_{int} is the inner diameter of the toroid.

Using Eq. (9.17), the primary and secondary leakage inductances can be calculated as:

$$l_1 = \Lambda_\sigma n_\text{p}^2 \text{ and } l_2 = \Lambda_\sigma n_\text{s}^2, \tag{9.18}$$

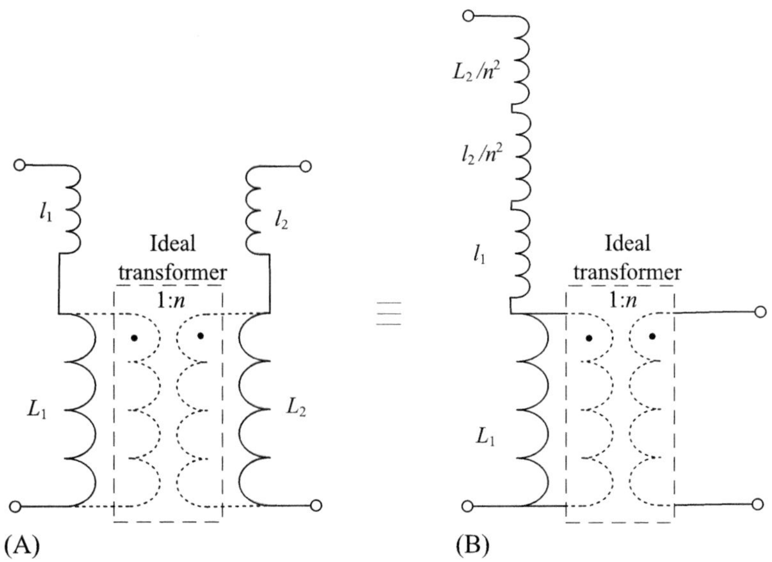

FIG. 9.9 Equivalent circuit of the SCASA coupled inductor with (A) primary and secondary leakage inductances, and (B) secondary referred to the primary. Both these equivalent circuits have omitted the winding resistance and the core loss resistance for simplicity.

where n_p and n_s are primary and secondary number of turns. The magnetizing inductance values can be calculated using the core permeance (inductance per square turn), Λ_m, as:

$$L_1 = \Lambda_m n_p^2 \text{ and } L_2 = \Lambda_m n_s^2. \tag{9.19}$$

The required effective primary inductor value, L_p, based on design specifications according to Eq. (9.5) can be equated to the equivalent circuit inductor value:

$$L_p = \frac{L_2}{n^2} + \frac{l_2}{n^2} + L_1 + l_1 = \frac{\Lambda_m n_s^2}{n_s^2/n_p^2} + \frac{\Lambda_\sigma n_s^2}{n_s^2/n_p^2} + \Lambda_\sigma n_p^2 + \Lambda_m n_p^2 = 2n_p^2(\Lambda_m + \Lambda_\sigma). \tag{9.20}$$

When the surge passes through the primary and secondary will have an induced voltage. A part of the magnetic core flux created by the primary coil will be lost due to the uncoupled secondary leakage inductance, l_2. Hence, the effective induced voltage is of:

$$L_s = L_2 - l_2 = \Lambda_m n_s^2 - \Lambda_\sigma n_s^2 = n_s^2(\Lambda_m - \Lambda_\sigma). \tag{9.21}$$

The bracketed terms $(\Lambda_m + \Lambda_\sigma)$ and $(\Lambda_m - \Lambda_\sigma)$ are dependant on the selected core materials and geometry. Hence the two following limiting conditions were used to select an appropriate core.

Magnetic wire insulation breakdown voltage should be greater than 2000 V AC. Because the coupled inductor is directly subjected to the incoming surge to avoid insulation breakdown between two consecutive turns the worst-case voltage per turn should be less than the allowed maximum insulation breakdown voltage [8] as described in Eq. (9.22):

$$V_{\text{supply}}^{\max} + V_{\text{surge}}^{\max} - V_{\text{var}} < 2\text{kV}. \tag{9.22}$$

To minimize the potential difference between windings, both windings should fit into a single layer on the toroid. Hence, the following inequality should be satisfied, where d_{wire} is the diameter of the magnetic wire. This can be chosen based on the rms load current.

$$\pi d_{\text{int}} > (n_p + n_s) d_{\text{wire}}. \tag{9.23}$$

9.7 Coupled Inductor Design Calculation

The design procedure starts by selecting the appropriate wire gauge for the rms current of the application. Then the smallest possible core size is selected using inequalities (9.22), (9.23). This requires calculating L_p and n using Eqs (9.5), (9.8); 17 SWG enamel wire can handle continuous 10 A rms current of a 2.5-kW load. Hence, $d_{\text{wire}} = 1.422$ mm.

Select Magnetic Inc. core 0077071A7 with internal diameter, $d_{\text{int}} = 19.4$ mm, effective core cross-sectional area, $A_c = 65.6\ \text{mm}^2$, $\Lambda_m = 61 \pm 8\%$ nH/turn^2.

Based on specifications [9] $L_p = 14.97\mu H, L_s = 17.14\mu H$, and $n = 1.07$. Inserting the specifications into inequality (9.22).

$$\begin{aligned} n_p &> \frac{325 + 6000 - 710V}{2000V} = 2.8\text{turns} \\ \Rightarrow n_p &> 3\text{turns}. \end{aligned}$$

From inequality (9.23),

$$\begin{aligned} n_p + n_s &< \frac{\pi \times 19.4\text{mm}}{1.422\text{mm}} = 42.8 \\ \Rightarrow \quad n_p + n_s &< 42\text{turns}. \end{aligned}$$

From Eq. (9.20),

$$\begin{aligned} n_p^2 &= \frac{14.97\mu H}{2(61\text{nH/turn}^2 + 2300\text{nH/m} \times 0.0194\text{m})} \\ \Rightarrow \quad n_p &= 8.4 \approx 8\text{turns}. \end{aligned}$$

From Eq. (9.21),

$$\begin{aligned} n_s^2 &= \frac{17.147\mu H}{61\text{nH/turn}^2 - 2300\text{nH/m} \times 0.0194\text{m}} \\ \Rightarrow \quad n_s &= 32\text{turns}. \end{aligned}$$

These two values satisfy both the inequalities. Choosing a smaller core will not satisfy all conditions, hence this core is the optimum size. The resulting transformer's primary and secondary self-inductance values will be

$$\begin{aligned} L_{self-p} &= 61\text{nH/turn}^2 \times 8^2 \quad = 3.9\mu H, \\ L_{self-s} &= 61\text{nH/turn}^2 \times 32^2 = 62.5\mu H. \end{aligned}$$

These are the values expected when the primary and secondary coils are measured separately using an RLC metre. Using the chosen core and 17 SWG enamel wire, a toroid transformer was built with 8 and 32 turns in the primary and the secondary. A summary of the measured inductance values of the transformer built is listed in Table 9.1.

Based on series and inverse series connected inductor measurements as shown in Fig. 9.10, the transformer-coupling coefficient can be found as described follows:

$$\begin{aligned} L_a &= L_1 + L_2 + 2M, \\ L_b &= L_1 + L_2 - 2M, \end{aligned}$$

where M is the mutual inductance.

$$\begin{aligned} M &= \frac{L_a - L_b}{4} \\ \text{and } k &= \frac{M}{\sqrt{L_1 L_2}}. \end{aligned}$$

TABLE 9.1 Theoretical and Experimental Self-Inductance Values Comparison for the Prototype Transformer

Description	Primary (μH)	Secondary (μH)
Theoretical	3.9	62.5
Experimental	3.6	56.4
% Mismatch w.r.t. theoretical value	−7.7	−9.7

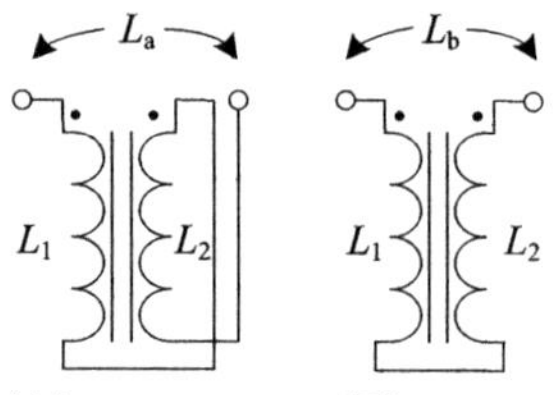

FIG. 9.10 Measuring inductances L_a and L_b of a coupled inductor (A) series connection, L_a; (B) inverse series connection, L_b.

Based on experimental measurements, $L_a = 100\mu H$ and $L_b = 45\mu H$, hence

$$M = \frac{100-45}{4} = 13.7\mu H,$$

$$k = \frac{13.7}{\sqrt{5.6 \times 67.3}} = 0.7.$$

The toroidal-core flux, Φ, when surge current, I, passes through a winding with inductance, L, can be expressed as follows:

$$\Phi = \frac{L}{NT}\int_0^T I\,dt, \tag{9.24}$$

where N is the number of turns and T is the duration of the surge current I. If the mathematical representation of the surge current is given by:

$$I(t) = A_I I_p t^3 e^{-t/\tau}$$

by substituting the surge current I into Eq. (9.24), the toroidal-core flux can be written as follows:

$$\Phi = \frac{LA_I I_p \tau}{NT}\left[6\tau^3 - (T^3 + 3\tau T^2 + 6\tau^2 T + 6\tau^3)e^{-T/\tau}\right].$$

Because the flux density, $B = \Phi/A_c$, where A_c is the cross-section of the toroid.

$$\therefore B = \frac{LA_I I_p \tau}{NTA_c}\left[6\tau^3 - (T^3 + 3\tau T^2 + 6\tau^2 T + 6\tau^3)e^{-T/\tau}\right]. \tag{9.25}$$

Substituting the parameters of combination current waveform, primary winding and toroid; $I_p = 3000$ A, $A_I = 1.243 \times 10^{-2}\mu s^{-3}$, $\tau = 3.911\mu s$, $T = 100\mu s$, $N = 5$, $L = 3.9\mu H$, and $A_c = 65$ mm^2 in Eq. (9.25)

$$B = 5798.5G.$$

This value is well below the maximum flux density of the kool mu material—10,500 G—of the selected core.

9.8 LTSpice Simulations

LTSpice was used to simulate the behaviour of the SCASA. These simulations were used to verify the theoretical arguments made during the circuit synthesis process. The simulation used the equivalent circuit model for the V20E275 varistor provided by Littelfuse, Inc. The transformer was modelled with the experimental primary and secondary inductance as the self-inductances of the primary and secondary. Experimentally found coupling coefficient of 0.7 was used to model the transformer.

The complete LTSpice simulation circuit used to generate IEEE Standard combination waveform is given in Fig. 9.11. The wave shaping circuit and the models are available in [10]. Fig. 9.12 shows the complete SCASA arrangement including the oscilloscope probes. The 1-F SC is modelled using the *RLC* model found in Chapter 4, with a surge resistance of 52.8 mΩ and inductance of 69.3 nH as given in Table 4.4.

Fig. 9.13 illustrates the effect of the transformer turns ratio omitting the SC subcircuit as shown in Fig. 9.4. A value higher than the calculated-theoretical turns ratio would result a voltage undershoot and an overshoot for a lower turns

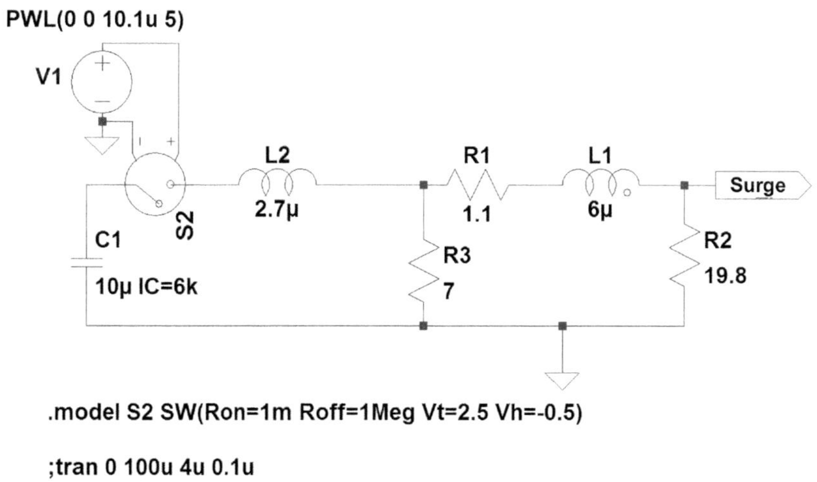

FIG. 9.11 LTSpice simulation circuit of the lightning surge simulator (Noise laboratory Co. Ltd, Japan).

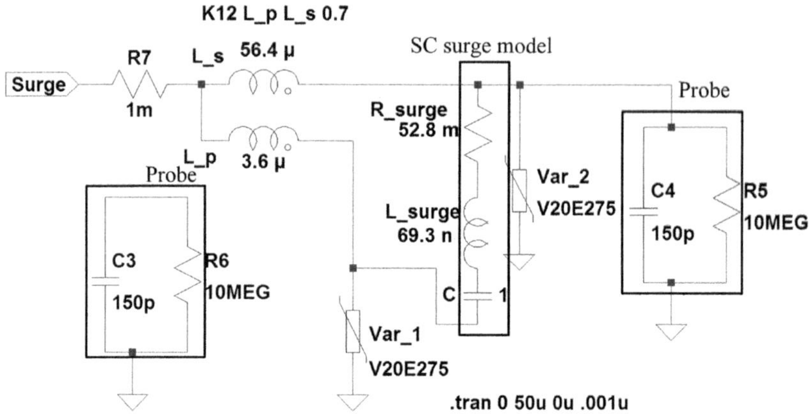

FIG. 9.12 LTSpice simulation circuit for the prototype supercapacitor-assisted surge absorber.

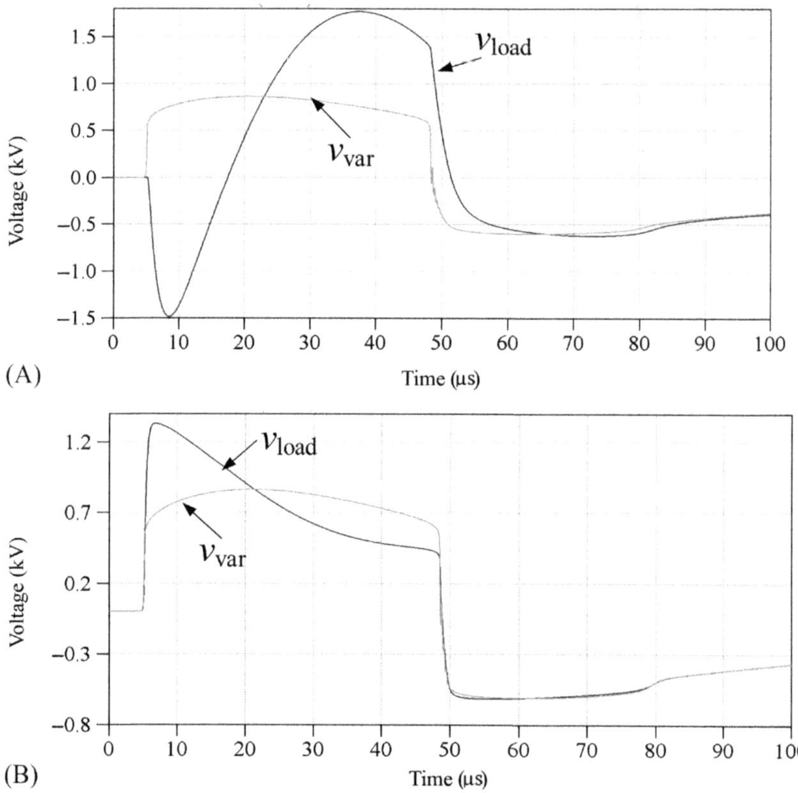

FIG. 9.13 LTSpice simulation results for SCASA without the capacitor subcircuit: (A) a transformer turns ratio greater than the calculated value causing an initial voltage undershoot; (B) a turns ratio less than the calculated value causing an initial voltage overshoot.

ratio. Varistor clamped voltage will stay constant irrespective of the transformer configuration. Load voltage and varistor voltage become equal at the point when the surge current maximizes, hence $L\frac{di}{dt}=0$.

Fig. 9.14 shows experimental results from the prototype design. Surges with increasing voltage were applied to the SCASA circuit with high winding ratio without the SC subcircuit. Experimental results with the nonoptimal turns ratio, results in load voltage values beyond the safe load voltage range.

Fig. 9.15 gives a comparison of simulation and experimental results using a capacitor of 47 nF for the subcircuit capacitor. This simulation and experiment were carried out to show the importance of using the optimum capacitor size and validate simulation circuit. Based on the graph resonant frequency can be calculated. This value should match with the calculated resonant frequency value based on Eq. (9.9). Table 9.2 lists the experimental results. The percentage mismatch is within the 10% tolerance range of the capacitor. This confirms that the analysis carried out on the magnetic circuit is correct.

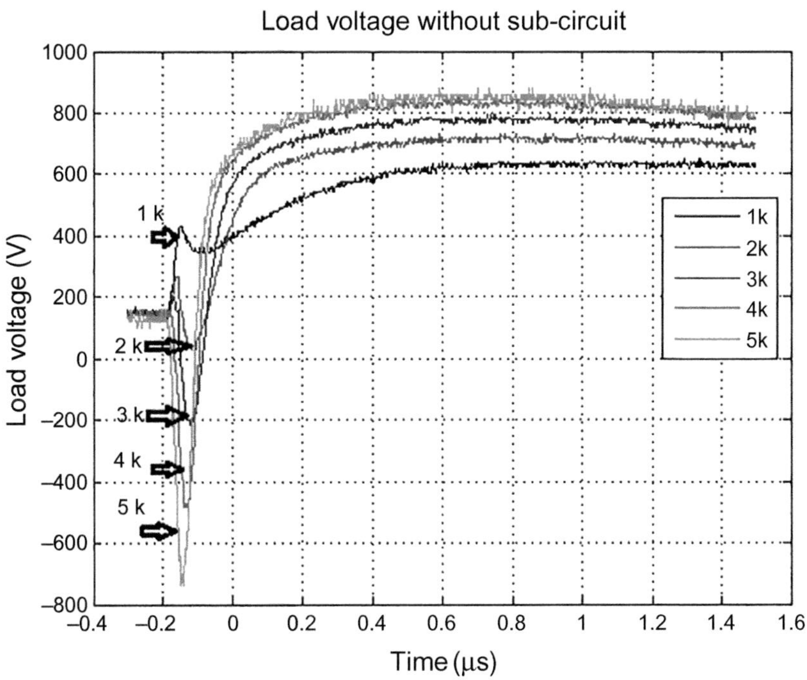

FIG. 9.14 Effect of the secondary winding: the induced-reverse voltage in the secondary helps to reduce the load voltage lower than the varistor voltage. Results are for the prototype SCASA system without the SC subcircuit capacitor. Turns ratio is set at the nonoptimal value (1:10).

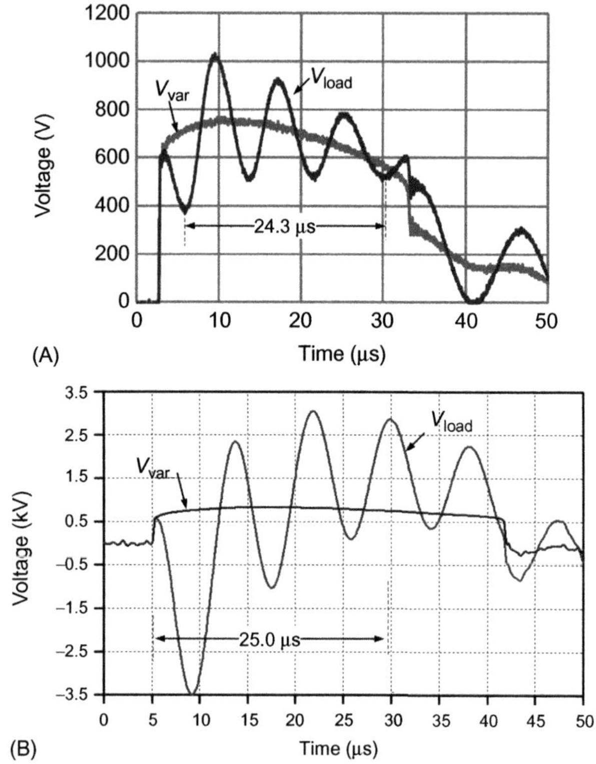

FIG. 9.15 Effect of subcircuit capacitor on load voltage during surge (A) experimental results for 47 nF capacitor and (B) simulation results for 47 nF capacitor.

TABLE 9.2 Effect on Subcircuit Capacitor Size on Resonant Frequency

	Experimental	Simulated
Resonant frequency	$\frac{1}{24.3\mu s/3}$	$\frac{1}{25.0\mu s/3}$
Based on Fig. 9.15	124 kHz	120 kHz
Resonant frequency	$\frac{1}{2\sqrt{(3.6+56.4+2.7\sqrt{3.6\times 56.4})\mu H\times 47nF}}$	$\frac{1}{2\sqrt{(3.9+62.5+2.7\sqrt{3.9\times 62.5})\mu H\times 47nF}}$
Based on Eq. (9.9)	116 kHz	109 kHz
% Error	−6.4	−9.1

9.9 Final Circuit

Fig. 9.16 shows the final experimental circuit built. This used a second varistor to further reduce the load voltage and reduce the probability of surge passing on to the load as a way of increasing reliability. The present version of SCASA was developed [11] only for differential-mode surge protection. Fig. 9.17 gives the simulation and experimental results of the final circuit. The figure clearly shows that SCASA reduces the load voltage less than the varistor voltage during the peak current. The discrepancy between the experimental and simulated surge wave shapes is due to the possible simplifications within the manufacturer-supplied signal shaping circuits and the corresponding simulation of the LSS to produce the exact surge current waveform. LSS has a complex circuit, which is not addressed in this thesis. However, an analysis on LSS circuitry is available in [12]. It can be seen that even with the use of a single varistor the load voltage is maintained lower than the varistor voltage for most of the surge duration. The ability of SCASA to reduce the load voltage lower than the varistor voltage is shown in Fig. 9.18. When applying surges from 1 to 6.6 kV, SCASA is capable of reducing the load voltage up to 200 V than the varistor voltage. This reduction is further measured by the use of a second varistor. In the next section, SCASA achievement is compared with commercially available surge protectors, and it is seen that SCASA can reduce load voltage by about 150–200 V than the varistor at 6.6 kV surge voltage (Fig. 9.19).

9.10 Comparison With Other Commercial Products

Smart TViQ built based on SCASA is shown in Fig. 9.20 compares two commercially available surge protectors with Smart TViQ for differential-mode surge protection. The high-end products would use more varistors for reliable operation and low-end product uses one MOV and a capacitor as seen in Table 9.3. In total, a high-end product has 11 components and an inexpensive one will need only 2 components. Whereas SCASA will sit in the middle with four components. On the other hand based on the output load voltages as shown in Fig. 9.20 it can be seen that SCASA reduces the load voltage by

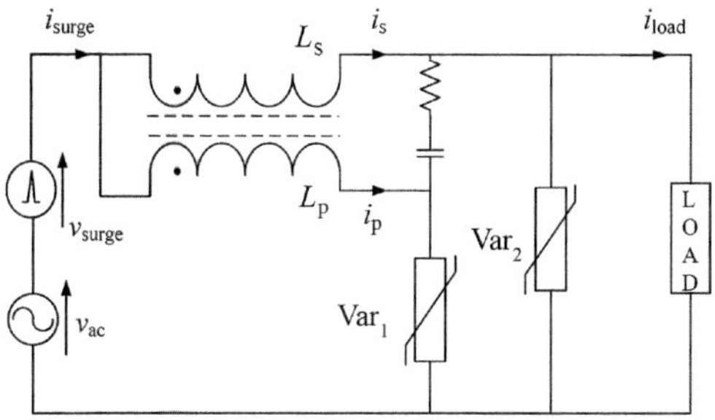

FIG. 9.16 The SCASA technique with an additional second varistor for further protection.

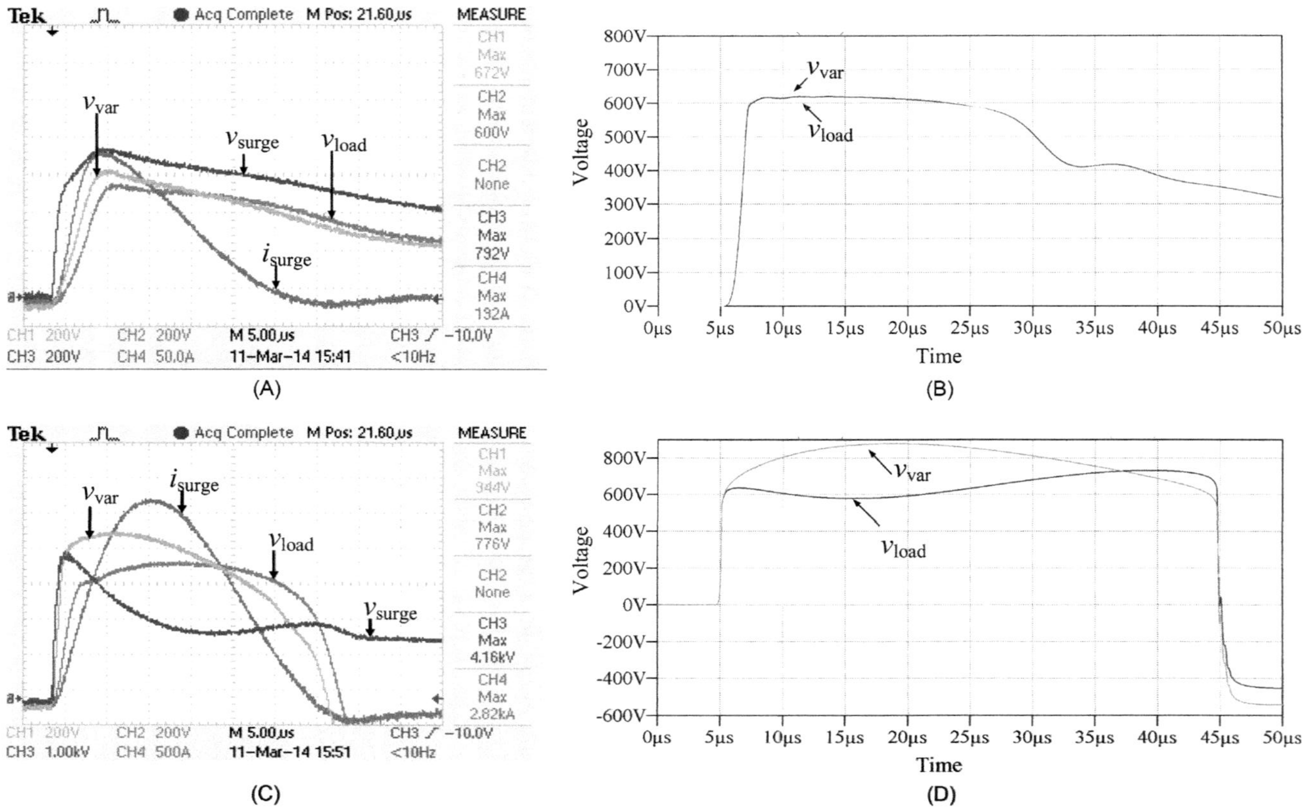

FIG. 9.17 Experimental and simulation results for different surge voltage inputs: (A) oscillogram for 1 kV surge; (B) simulated results for 1 kV surge; (C) oscillogram for 6.6 kV surge; (D) simulated results for 6.6 kV surge. [Courtesy of Thor Technologies, Perth, Australia]

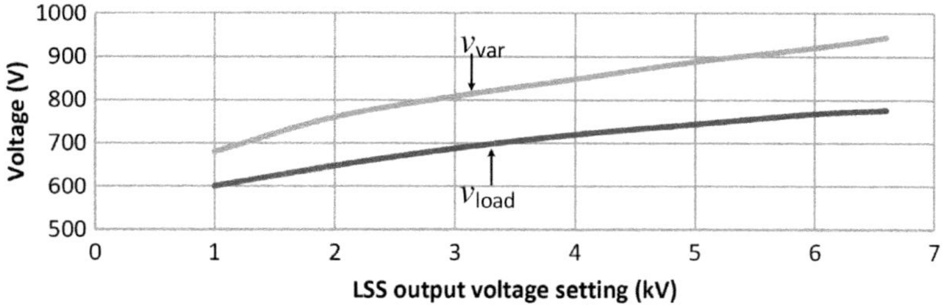

FIG. 9.18 SCASA technique reduces the load voltage lower than the varistor voltage.

FIG. 9.19 The commercial product, which was launched in the Australian market in 2014.

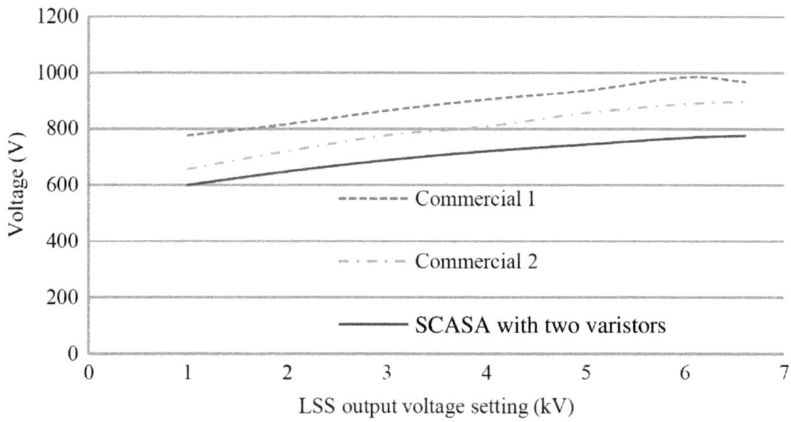

FIG. 9.20 Performance comparison of SCASA with two commercial surge protectors.

TABLE 9.3 Component Count Comparison for Differential-Mode Section in Commercial Surge Protectors vs SCASA

Component Type	High End	Low End	SCASA
MOVs	4	1	2
Inductors	2	0	1
X-type capacitors	5	1	0
Supercapacitors	0	0	1

approximately 200 V irrespective of the lightning surge voltage. This is achieved by the coupled inductor set up ahead of the varistor to induce a reverse voltage on the load path to reduce the surge voltage. This will block the incoming surge, from propagating into the load side.

Table 9.4 presents details of a destructive test carried on varistors to show another important feature of SCASA. By the introduction of the coupled inductor—SC subcircuit, SCASA diverts part of the surge energy dissipated in the varistor to the subcircuit and increase the lifetime of the varistor. The test carried out used the two varistors in the Smart TViQ. Continuous 6 kV surges were applied to each stand-alone varistor until failed. Then surges were applied to an SCASA with the same two varistors. Even up to one hundred surges none of the components did fail. As shown in Table 9.4 none of the varistor could withstand even 40 surges. This illustrates the ability of SCASA to improve the lifetime of varistor.

TABLE 9.4 Destructive Varistor Test

Varistor	Rated Energy (J) 10/1000 μs	Rated Voltage (V)	Test Results
Varistor 1 (431KD14)	66	275	Sample 1: failed after 7 surges
			Sample 2: failed after 10 surges
Varistor 2 (431KD10)	132	275	Sample 1: failed after 34 surges
			Sample 2: failed after 27 surges

9.11 A Final Remark About the SCASA Technique

SCASA is a technique currently under development into various advance versions at the University of Waikato, Hamilton, New Zealand. The design details and approaches shown in this chapter indicate that despite the very low DC or AC voltage rating of commercial supercapacitors, they can be combined with other commonly used components in SPDs.

If a supercapacitor-resistor combination is fed by a short duration high-voltage transient of several kilovolts, lasting only a few microseconds, resistive parts of the loop tend to dissipate a huge share of the transient energy, whilst the supercapacitor develops an overall voltage of only a small fraction of its DC voltage rating.

SCASA technique used in the commercial product example in this chapter should be considered as a case study to show new ways to develop SPDs based on SC energy recirculation technique.

References

[1] J. Fernando, N. Kularatna, A supercapacitor based enhancement technique for stand-alone surge protection circuits, in: 2013 IEEE International Symposium on Industrial Electronics (ISIE), 2013, pp. 1–6.

[2] J. Fernando, N. Kularatna, Supercapacitor assisted surge absorber (SCASA) technique: selection of supercapacitor and magnetic components, in: Energy Conversion Congress and Exposition (ECCE), IEEE, 2014, pp. 1992–1996.

[3] J. Fernando, N. Kularatna, H. Round, S. Talele, Implementation of the supercapacitor-assisted surge absorber (SCASA) technique in a practical surge protector, in: IECON 2014: 40th Annual Conference of the IEEE on Industrial Electronics Society, 2014, pp. 5191–5195.

[4] L. Fernando, Supercapacitor-Assisted Surge Absorber (SCASA) and Supercapacitor Surge Modelling (Doctoral thesis), School of Engineering, The University of Waikato, 2016.

[5] Inc Littelfuse, Metal-oxide varistor (MOV), Littelfuse, Inc., 2015. Tech. Rep.

[6] Underwriters Laboratories, Transient Voltage Surge Suppressors, Standard, Underwriters Laboratories, 2014.

[7] V.C. Valchev, A.V. den Bossche, Inductors and Transformers for Power Electronics, CRC Press, 2005.

[8] H. Pops, J. Walker, Origins of high voltage continuity failures in enamel coated copper magnet wire, in: Electrical Electronics Insulation Conference and Electrical Manufacturing and Coil Winding Conference, 1993. Proceedings, Chicago '93 EEIC/ICWA Exposition, 1993, pp. 457–484.

[9] Magnetics Inc, 2014. Tech. Rep. 0077071a7.

[10] S. James, N. Kularatna, A. Steyn-Ross, R. Künnemeyer, Estimation of transient surge energy transferred with associated time delays for individual components of surge protector circuits, IET Power Electron. 8 (5) (2015) 685–692.

[11] N.A.D.V. Kularatna, L.H.J.D.K. Fernando, Power and Telecommunications Surge Protection Apparatus, WO2014007647, 2014.

[12] S. James, Investigation of Surge Propagation in Transient Voltage Surge Suppressors and Experimental Verification (Doctoral thesis), School of Engineering, The University of Waikato, 2014.

Chapter 10

Application of Surge Protection Systems

10.1 Introduction

As we discussed in some detail in the first chapter, modern electronic products and systems are designed using very delicate silicon and other compound semiconductor-based high transistor density integrated circuits. Today's complex electronic systems carry multiple processor-based digital components with millions to billions of nanometre feature size transistors, with a huge amount of engineering systems knowledge stored in semiconductor memories in the form of long strings of software to carry out many specialized tasks, while these systems today communicate with the rest of the world via international communication platforms residing in all areas of the five continents.

In protecting these complex and interconnected systems, against transient high energy signals, an SPD acts as the "*security post*' at the entrance of a transient-wise vulnerable operational complex. These security posts are not expected to block or retard the flow of wanted electrical signals such as the power and communication, but it should block the flow of unwanted transient signals which could harm the hardware components and/or disturb the running of complex software routines.

This chapter summarizes the common applications of SPD systems, with a brief comparison of different standards for SPDs which was reasonably accurate around 2016.

10.2 Transient Over Voltages and Their Occurrences

Surges are extremely high transient voltages that could damage or completely destroy insulation. This could hence impair or completely disrupt the function of the electrical circuit, irrespective of whether it is simple or complex. Every electrical component generally carry some insulation to isolate the contact with unwanted electrical signals or other circuit components. The insulation strength is dependent on the rated voltage and the type of electrical component, as stipulated by the relevant standards such as IEC/VDE/NEMA or similar national or

Design of Transient Protection Systems. https://doi.org/10.1016/B978-0-12-811664-7.00010-0

international standards applicable to the subject area. It is tested by applying the prescribed voltage for a defined period of time. If the test voltage is exceeded beyond the specified limits of the relevant standard the safety effect of the insulation is not guaranteed and the component could be damaged partly or completely ruined.

Surges are voltage occurrences that are higher than the test voltages, and these could have a detrimental effect on an electrical component and hence on the overall system. These transient surges could either affect the insulation of a component or it could impact on the rated voltage which can be safely handled by a given component.

Fig. 10.1 depicts rise times of a transient high voltage such as a lightning-related signal or an inductive dump on an input circuit, compared with the 50 Hz power supply waveform. This Figure shows the rising edge of the transient to be very much faster than the quarter cycle time of the 50 Hz power supply waveform, which is only 5 ms, whereas the transient could have a rise time of one μs, which is usually nonrepetitive. The shape, peak amplitude, and the timing of transient is highly unpredictable, but they are usually considered nonrepetitive and in a worst case if that repeats, it will be only few repeats which may be usually few seconds to fractional seconds apart.

Transients are primarily caused by (i) lightning due to atmospheric discharges, (ii) transient switching operations, (iii) faulty switching operations, and (iv) electrostatic discharges.

In selection of SPDs suitable for the previous transients, we can group the first three types into one subset which could be covered by the international

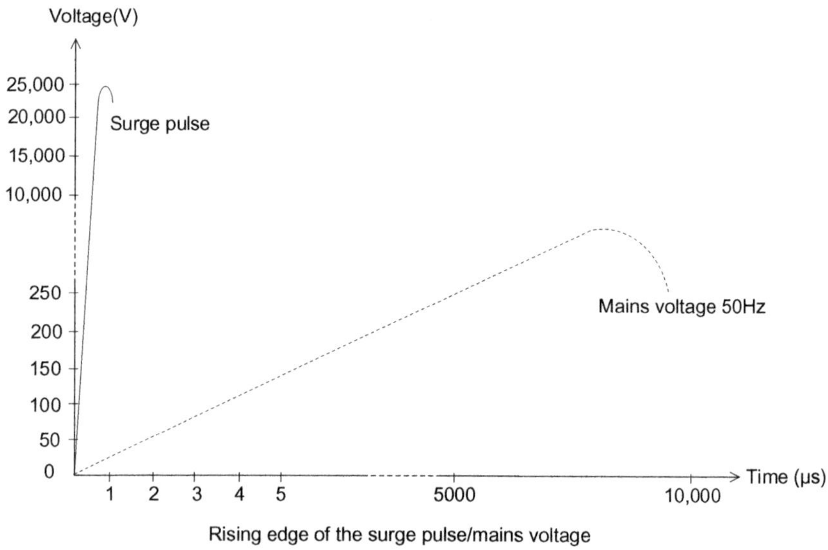

FIG. 10.1 Rise time (edge) behaviour of a transient surge versus the 50 Hz sine waveform.

standards compared in Section 10.3. Electrostatic discharges are to be treated according to different standards.

10.3 A Review of International Standards and SPDs

With the earlier review of transient voltages, including the most important case of lightning transmitted through the power supply systems, we can now compare the common standards used in SPDs. Reader should bear in mind that the standards such as IEEE C62 series, UL 1449 series, and the IEC 61643 series may evolve with time, and latest version should be consulted in case of any designing or selection of commercial SPDs are envisaged. Fig. 10.2 depicts the relationships between the IEEE C 62 series categories and the UL1449—Edition 3 categories.

One of the important revisions was the 3rd Edition of the UL1449 standard, which was made effective on the 29th September 2009. Prior to this standard, the surge protectors were referred to as Transient Voltage Surge Suppressors (TVSS), and this name was changed to SPD as a result of the 3rd Edition of the UL 1449. According to UL 1449 Edition 3, SPDs are divided into 4 major categories, namely: (i) Type 1, (ii) Type 2, (iii) Type 3, and (iv) Type 4. UL 1449 Type 4 means SPDs which are considered an integral part of the equipment, where it is not a stand-alone external unit. For this reason Type 4 of UL could match any of the IEEE categories of standards.

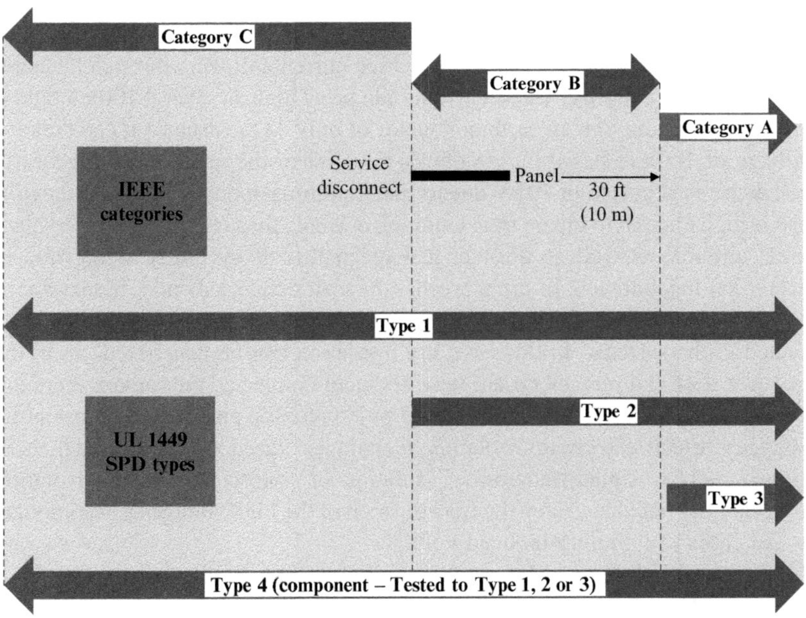

FIG. 10.2 Comparison between the IEEE C62 categories and the UL 1449 Types 1 to 4.

IEC also divides them into types: (i) Class I (or Type 1), (ii) Class II (or Type 2), (iii) Class III (or Type 3). Ratings of the commercial products depend on the intended use and the corresponding magnitude of the surge suppression capabilities.

IEEE also specified location categories. Category C is outdoor and the service entrance on the incoming power line side. Category B is the Service entrance and the branch panels, and the Category A is equipment level. Table 10.1 presents the relationship between IEEE categories and the UL 1449 categories, with the specified test waveforms.

10.4 Lightning Protection—As a Need in a Building Wiring System for Reliable Protection Against Transient Overvoltages

Lightning is an atmospheric discharge, which cannot be easily predicted for its statistical occurrence or the level of amplitude. When lightning strike occurs on a building it can cause a direct hit, which causes a kilo-amperes order current to flow through the lightning conductor, creating an Ohm's law related very high voltage at the top of the conductor. In addition to this directly conducted current, induced or radiated voltages also can be superimposed on the AC mains waveforms, or the signal lines such as data or telecom signals into the wiring complexes.

10.4.1 Lightning-Related Overvoltages and Protection Systems

Bolts of lightning can create extremely high currents flowing through the earth conductor of a building. These currents can be as high as several 100 kA, and if 100 kA passes through an earth conductor of only 1 Ω resistance it can create a voltage of 100 kV, based on the ohm's law. When the upper end of the earth conductor is at even 5 or 10 kV due to such lightning-induced current, that voltage is high enough to create transient loop currents through the connected electrical circuits, as well as through leakage paths, or by direct arcing due to insulation breakdowns. In other words, this can create galvanic, inductive, or capacitive coupling of surge voltages within the circuits of equipment connected to the systems. In this case any insulation can be penetrated. So in the real world of complex electrical networks and connected equipment there are no electrical isolation methods which provide reliable protection against surge voltages. Electrical circuit isolation techniques based on well-known techniques such as signal-transformer isolation or opto-coupler-based isolation are generally not able to stop the transmission of the high voltage transients such as the cases of lightning-induced voltages.

A natural lightning strike usually consists of a main discharge and a time-shifted post-discharge. The strength of this postdischarge is usually far below the energy level of the main discharge. Both discharges, however, have enough

TABLE 10.1 Comparison of Different Standards for SPDs

UL or IEC Category/Type	Service Entrance: *Line Side of Mains Disconnect*	Service Entrance: *Load Side of Mains Disconnect*	Branch Panel	Equipment Level	Test Requirements
	IEEE Category C Locations	*IEEE Category B Locations*	*IEEE Category B and A Locations*	*IEEE Category A Locations*	
1449 Type 1	Similar				15 pulses of 8/20 μs waveform
1449 Type 2		Similar	Similar		15 pulses of 8/20 μs waveform
1449 Type 3				Similar	Combination wave 6 kV, 3 kA
1449 Type 4	Similar	Similar	Similar	Similar	Based on application
IEC Type 1/Class 1	Similar				10/350 μs waveform
IEC Type 2/Class II		Similar	Similar		8/20 μs waveform
IEC Type 3/Class III				Similar	Hybrid waveform—open circuit voltage 1.2/50 μs; short circuit current of 8/20 μs

transient energy to cause significant damage. For a detailed discussion on lightning, Ref. [1] is suggested.

Fig. 10.3 depicts the case of lightning scenario where there could be direct strike with several 100 kA transient current can be pushed into the earth conductor on the left-hand side of the figure, while the lightning bolt between a set of clouds could generate an induced voltage and current through the right side of the figure. As a rule, 50% of the current is discharged via the lightning protection system and the remaining 50% is coupled into the conductors and conductive parts of the building and distributed uniformly. In a typical case, left-hand side of Fig. 10.3 could be equivalent to the telecom or data circuits coming into the building, while the right-hand end is same as the case of power supply. For these reasons, two ground conductors could have different resistances R_{Telecom} and R_{Power}. Due to the flow of lightning-related current, these ground conductors will develop high voltages based on the Ohm's law.

10.4.2 Fundamental Concepts Applied to Protection Against Lightning

One important aspect of surge protection is the area of AC power supply and the distribution systems. The procedure is based on the systematic subdivision prescribed by the protective zones we discussed in Chapter 1 and the corresponding coordination of surge arrestors with different surge suppression capabilities. Protection of the power supply lines forms the basis for protecting all electrical and electronic equipment right down to the smallest and the most sensitive components.

Fig. 10.4 shows the case of a 3-phase power supply system, where the conductor to earth voltage is lower than 300 V, according to the standards such as IEC 60664-1. In this figure the whole distribution system is divided into four different zones. Just before the electricity Watt-hour meter, a lightning arrestor of the Type 1 is required and this is expected to withstand transient surge voltages up to 6 kV up to the main distribution board of the incoming power supply.

Next lower level protector, which is in the area of subdistribution board should be a Type II arrestor, which can protect up to 4 kV surges. In between the subdistribution board to individual equipment area, an arrester suitable for Type III protection of up to 2.5 kV is required. At the final equipment end a lower 1.5 kV surge protector will have to be used.

The requirements placed on surge protection and the necessary tests for surge components are stipulated by national and international standards and a SPD can only be considered safe after the product has been fully tested. For rated voltages up to 1000 VAC, the standards are available for the manufacturers and the installers of SPDs within a facility. Insulation coordination for electrical equipment in low voltage systems to relevant standards such as IEC 60664-1 is critical for the design of a SPD system. These specify different dielectric strengths within the electrical systems. Based on this, individual

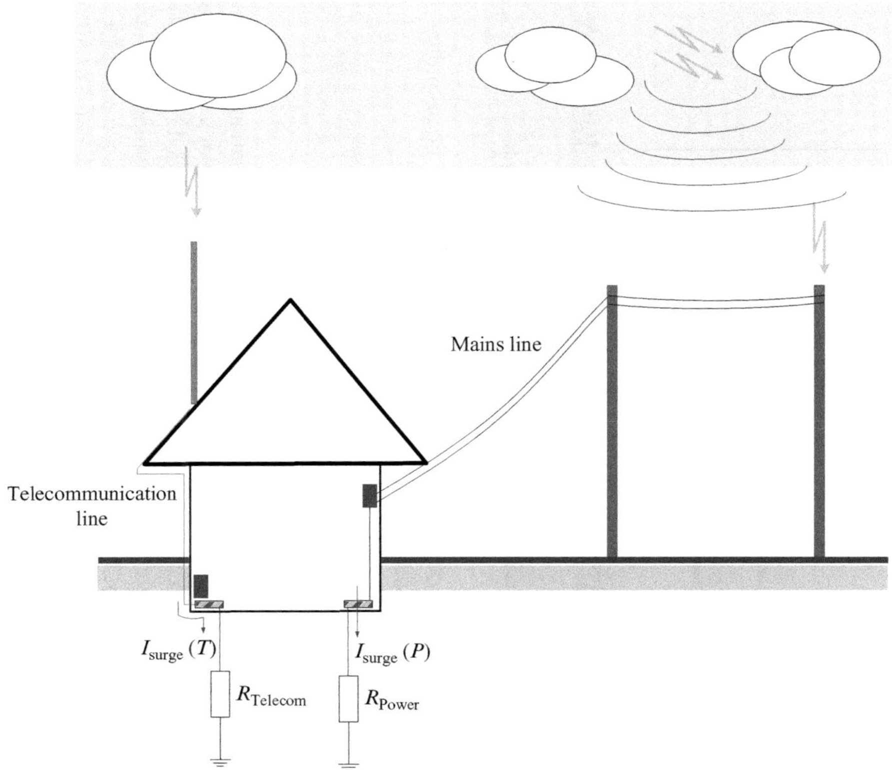

FIG. 10.3 Indication of lightning-related currents possible through the earth conductors.

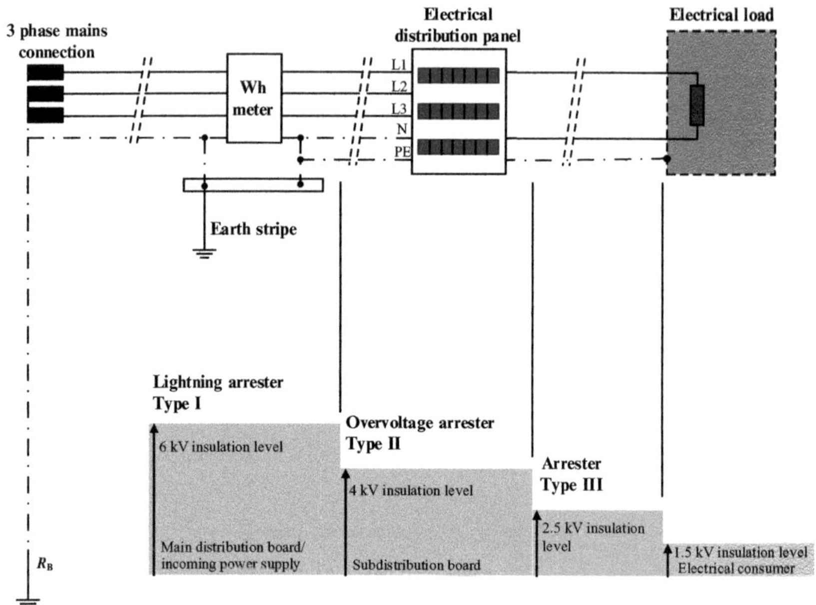

FIG. 10.4 Principle of selecting arrestors according to international standards.

lightning protection zones (discussed in Chapter 1) can be set up according to IEC 62305-3.

A detailed account of this subject is available in Ref. [2].

10.4.3 Lightning Protection Levels

Surge protection for electrical systems cannot be treated in isolation with lightning protection requirements related to buildings. If a building or a complex of buildings are not protected against lightning, protection provided by the SPDs may be compromised, since direct lightning on to building, lightning within the vicinity of building, or the induction effects of lightning could generate severe transient surges, which may easily exceed the maximum protection capability of SPDs. Lightning protection of buildings is a separate subject beyond the scope of this book. However, the Refs [2, 3] can provide some useful overview on lightning protection systems covered by standards such as BS/EN/IEC 62305. Following summary paragraph on lightning protection levels (LPL) are based on this standard.

LPL refers to the case of maximum surge current, based on the 10/350 μs pulse current waveform, which is applied to the case of Type I protectors used in power entry level into a building. Fig. 10.5 depicts the three different levels of protection, based on the maximum possible peak surge current through the lightning arrestor and the earth wire.

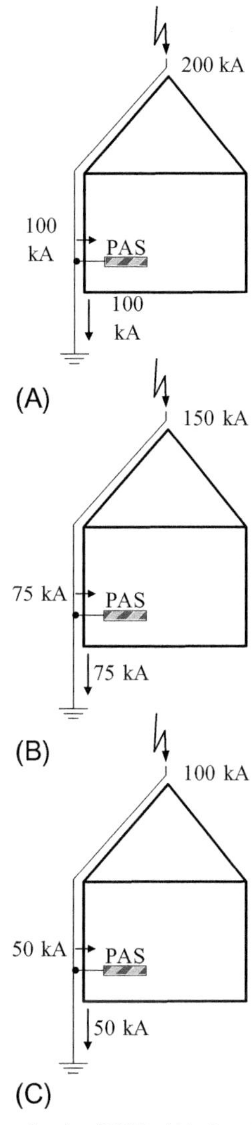

FIG. 10.5 Lightning protection levels (LPL) (A) Level I-200kA (B) Level II-150kA (C) Level III-100kA.

LPL-I covers a pulse of 200kA as a worst-case scenario of a direct strike of a lightning and pertains to external lightning protection facilities as shown in Fig. 10.5A. Half of this current is expected to conduct into earth, and in a 3-phase-4 wire system a current of 25kA is expected to flow through each of the four wires. In the case of LPL-II, the pulse peak will be at 150kA, and with half of this peak (75kA) will flow into the earth connection, while each of the four wires is expected to carry around 17kA (Fig. 10.5B). In the case of LPL-III in Fig. 10.5C.

The Level I protection applies to large areas, such as petrochemical facilities and explosive material complexes. Level II covers multiple areas like hospitals, warehouses, and telecom towers, and so on. Level III/IV covers 80% of all applications such as houses, office complexes, industrial plants, and so on.

More details on these areas are available in Refs [2, 3].

10.5 Levels of Surge Protection

In the application of SPD systems, general protection levels may be divided into three categories, namely: (i) primary protection, (ii) secondary protection, and (iii) board-level protection.

10.5.1 Primary Protection

Primary protection applies to power lines and data lines exposed to an outdoor environment, service entry, and AC distribution panels. Transient currents can range from tens to hundreds of kiloamps at these sources as discussed before. This has to be coordinated with the lightning protection systems, associated with the building and the IEEE Category C/UL Type 1/IEC Class 1 systems to be installed at the input side of the service entrance.

10.5.2 Secondary Protection

Secondary protection is for equipment inputs, including power from long-branch circuits, internal data lines, PBX, wall sockets, and lines that have primary protection at a significant distance from the equipment. Transient voltages can exceed several kilovolts with transient currents ranging from several hundred to several thousand amps. These should be covered by IEEE Category B/A, UL Type II, or IEC Class 2/3 standards.

10.5.3 Board-Level Protection

Board-level protection is usually internal to the equipment; it is for protection against residual transient from earlier stages of protection, system generated transients, and Electro Static Discharge (ESD). Transients at this level range from tens of volts to several thousand volts with currents usually in the tens of amps.

10.6 Circuit Concepts Used for Surge Protection

In general, practical circuits developed for surge protection are based on few simple concepts: (i) protection circuit is usually designed separately as an add-on block to the base circuit, (ii) attenuate the incoming transient using passive series impedances or bypass the surge currents via passive low impedance circuits [they act as filters for high frequency components of the surge], (iii) use nonlinear devices such as GDTs, MOVs, or TVS to divert the surge currents and absorb the transient energy. Fig. 10.6 depicts three different kinds of TVSS circuits.

10.6.1 SPD Systems for Power Line Input End

In the case of power line protection, usually GDT devices are not used in the SPDs, since they are acting as crowbar devices when fired creating a temporary short circuit across the line. Fig. 10.6A depicts a typically available packaged SPD which can be tested according to the requirements under relevant standards.

In Fig. 10.6A, which is designed to protect against both common and differential mode transients, M_1 to M_3 are MOVs which will enter their firing or conduction mode when the transient exceeds the threshold voltage limit. Inductors L_1 and L_2 act as high impedances for the transient signal, since its spectrum contains many higher frequency components, compared to the impedance of the inductors at the line frequency. Similarly the capacitors indicated as C_{nx} and C_{ny} act as low impedance paths to the transient signal, which helps diverting the transient energy from reaching the critical load end. Overall effect of these circuitry is to minimize the transfer of the transient voltage towards the critical load side. However due to clamping voltages of the MOVs, some share of the transient energy reaches the load side. The TVS diodes (or thyristors) are used to absorb the remaining transient energy to minimize the possible damage.

10.6.2 Telecom and Data Line Protection

An SPD designed for a power line is expected to carry the 50 or 60 Hz line frequency power into the protected electrical systems or the sensitive electronic product. In this case the series equivalent impedance of the SPD at the line frequency has to be pretty low, so that the AC voltage is not dropped across the series inductance.

Compared to a power line input which can be subjected to severe levels of external and internal surges, a telecom, signal input, or the data connections do not transfer watts to kilowatt order signal powers into the system to be protected. For example, a telephone line supplies 48 V DC which normally carries only few milliamps order currents into the telephony or the data connections into the home or the office building. Given this case of milliamps or microamps order current a signal line carried into the system to be protected, the series equivalent impedance of the SPD could be quite high. For this reason, in such SPDs few ohms to few 100 Ω order series resistance could replace the series inductors required in the SPD for lowering the cost. Fig. 10.6B depicts such a generic circuit.

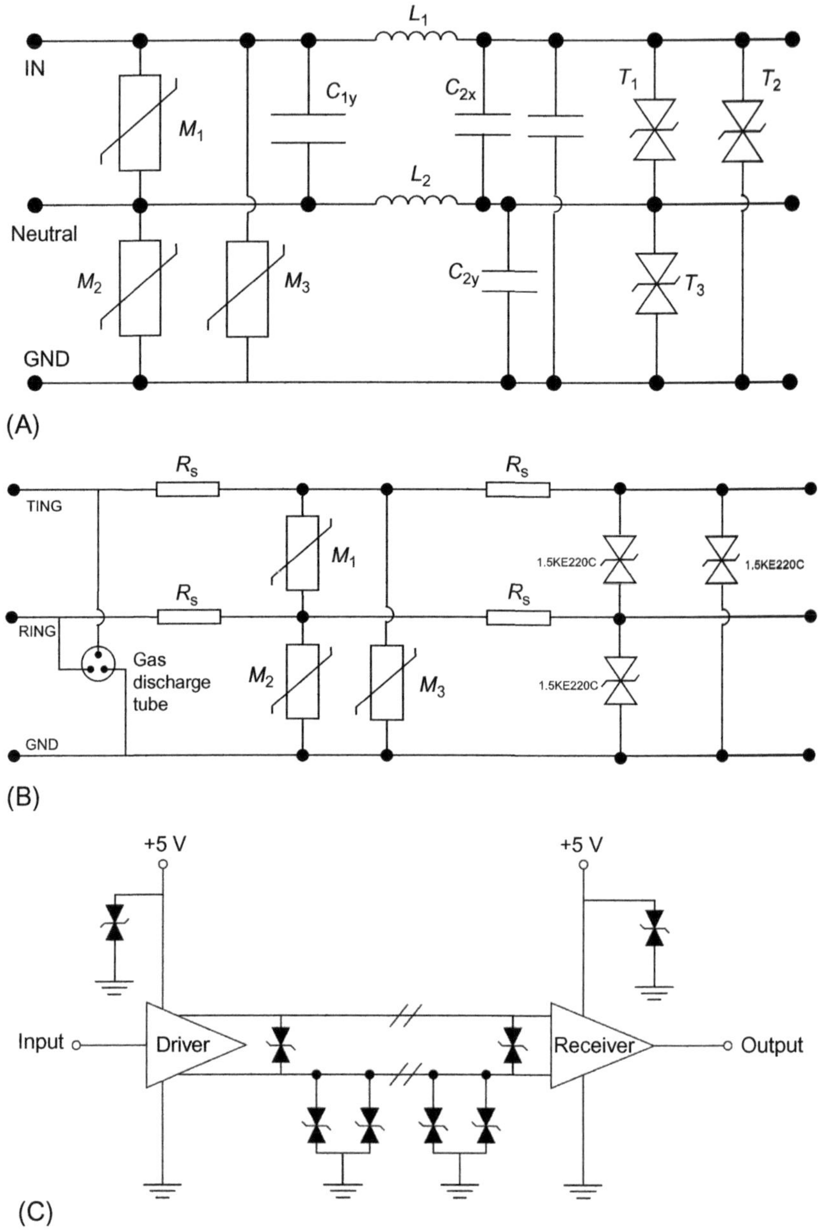

FIG. 10.6 Different types of surge protection circuits (A) Multistage surge suppressor (SS) using a combination of devices (B) communication or signal line protection using multistage SS devices including a GDT (C) line driver-receiver protection using back-to-back thyristor elements.

10.6.3 Line Driver-Receiver Protection

Fig. 10.6C is an example of line driver-receiver protection. As indicated in Fig. 10.6C a line driver-receiver pair is protected by BBD-type semiconductor devices for transients which could creep into the driver or the receiver part of the circuit, and there are many practical BBD devices specially developed for these applications.

10.7 Electrostatic Discharge and Circuit Protection

Electrostatic discharge (ESD) is a serious issue in assembly and testing of systems containing semiconductor devices. There are various standards developed for testing the ESD capability of semiconductor products. These standards have been generated with regard to a specific need related to the electromagnetic compatibility of the system environment. They include the Human Body Model (HBM), Machine Model (MM), and the Charged Device Model (CDM). Each such standard relates to the nature of electrostatic discharge generated within a system application and the potential for damage to the IC. Among the better known standards are:

- Human Body Model using the MIL-STD-883, Method 3015.7
- Machine Model using EIAJIC121
- Human Body model using the IEC 1000-4-2 standard

Testing for ESD immunity is more broadly defined to include a device, equipment, or system. Both direct contact and air discharge methods of testing are used with four discrete steps in the severity level ranging up to 8 and 15 kV, respectively. In its simplest form, Fig. 10.7A test circuit provides a means of charging the 150 pF capacitor, R_C, through the charge switch and discharging ESD pulses through the 330 Ω resistor, R_D, and discharge switch to the Equipment or Device Under Test. The test equipment for the IEC 1000-4-2 standard is constructed to provide the equivalent of an actual human body ESD discharge and has the waveform shown in Fig. 10.7B.

10.8 Protection of Printed Circuit Board Components

As we discussed earlier, any remaining components of a transient or an ESD can ultimately end up on a PCB, physically destroying its components or temporarily corrupting any stored data. In order to safeguard physical components in PCBs, semiconductor manufacturers such as Littelfuse and ON semiconductor have introduced components suitable for board-level protection. Two basic component solutions exist. They are TVS diode arrays and SCR-diode arrays, which come in common IC packages. These devices cater for ESD protection requirements in general.

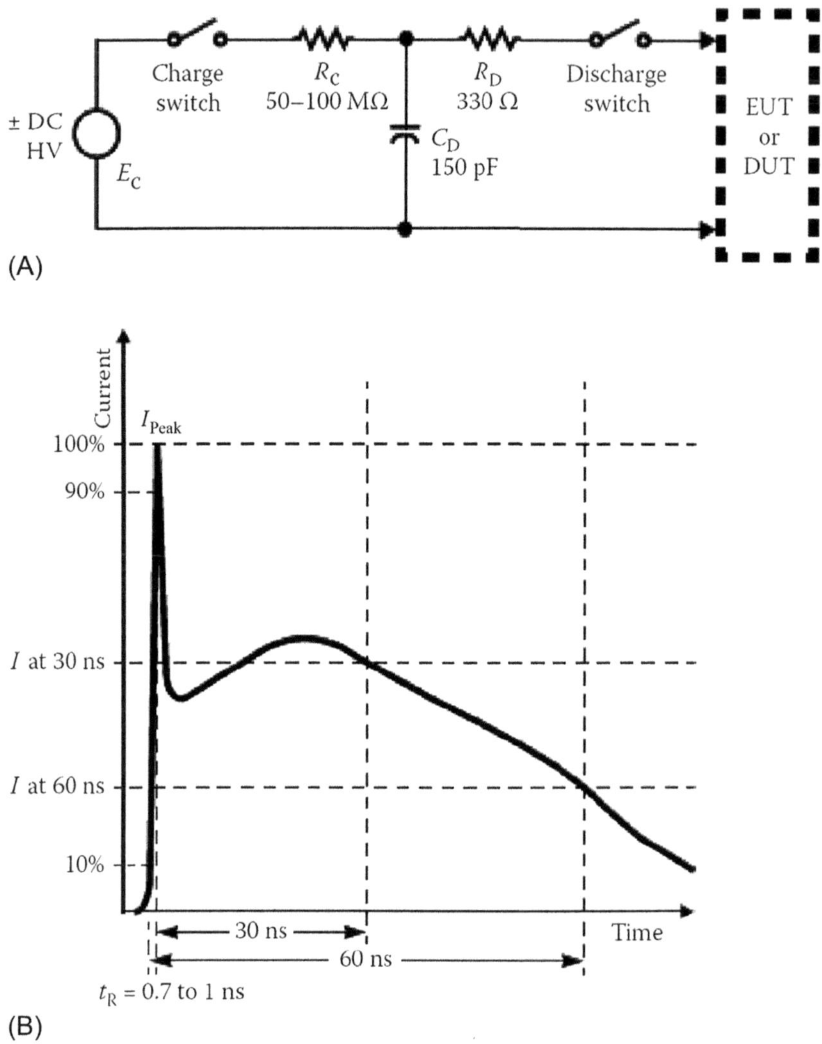

FIG. 10.7 ESD test energy source and output current of an ESD generator. (A) Test system and (B) waveform.

10.8.1 Electronic Components for PCB-Level Protection

The SP720, SP721, and SP723 are protection ICs from Littelfuse with an array of SCR-diode bipolar structures for ESD and overvoltage protection of sensitive input circuits. They have two protection SCR-diode device structures per input. See Fig. 10.8A. The SCR structures are designed for fast triggering at a threshold of one $+V_{BE}$ diode threshold above $+V$ (positive supply terminal) or a $-V_{BE}$ diode threshold below $V-$ (negative or ground). A clamp to $V+$ is activated at each protection input if a transient pulse causes the input to be increased to a voltage level greater than one V_{BE} above $V+$. A similar clamp to $V-$ is activated

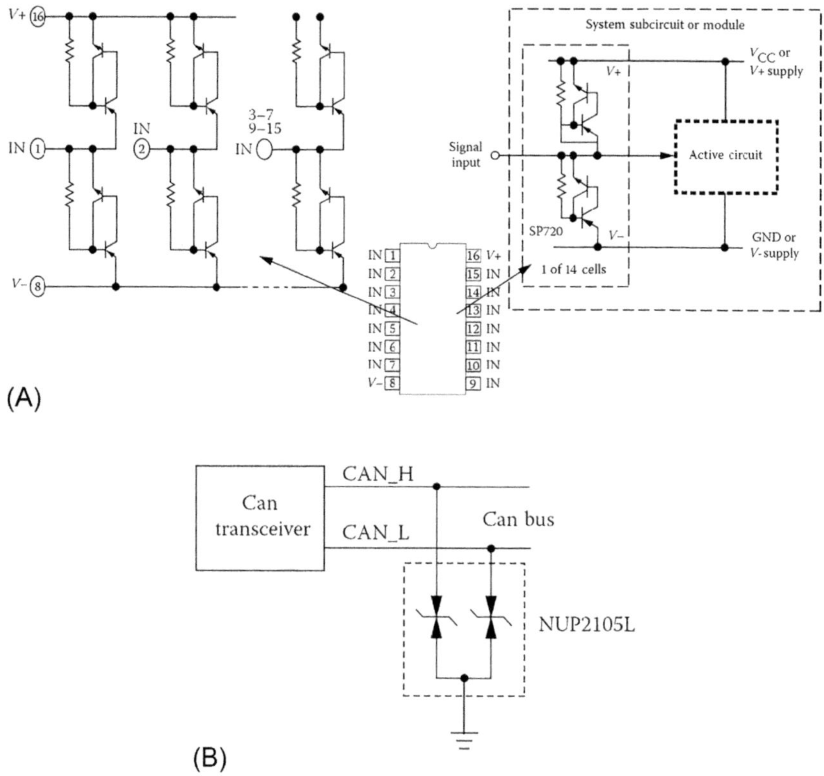

FIG. 10.8 ESD protection and techniques (A) use of SCR type device arrays (B) use of TVS diodes. *(Sources: ON Semiconductor and Littelfuse; Adapted from W. Austin, IEC 1000-4-2 ESD immunity and transient capability for the SPX72X series protection arrays, Littelfuse Application Note AN 9612.2, 1999, 6 pp.; On Semiconductor Data sheet, NUP2105 Dual Line CAN Bus Protector, 2005, July, Rev 4.)*

if a negative pulse, one V_{BE} less than $V-$, is applied to an input. For further details Refs [4, 5] is suggested.

Similar to the case of SCR-type devices, TVS diode is also an effective low cost option to protect sensitive devices from ESD and EMI. In these cases the devices are connected between signal lines and a ground connection or a ground plane on a PCB. Typical examples are NUP 2105 from ON Semiconductor [6], which is typically used in automotive CAN bus environments, and so on. Usually the device connection is typically as shown in Fig. 10.8B, where the diodes provide common mode paths between the signal lines and the ground.

10.8.2 Practical Design Considerations

Having discussed a background on surge protection devices, and practices, it is also important to discuss the PCB layout and assembly issues, since these could enhance or degrade the level of protection provided by the devices, particularly in the case of board-level protection. In developing board-level protection stages, we have to keep in mind that capacitive and inductive parasitic elements in a PCB can compromise the level of protection offered by the device such as TVS diodes, or TVS thyristor structures, which comes in compact IC packages, or as individual elements.

In general, TVS devices should be used on all data and power lines that enter or exit a PCB at the I/O connector. Locating the TVS devices as close as possible to the noise source ensures that a surge voltage will be clamped before the pulse can be coupled into any adjacent PCB traces. Fig. 10.9A depicts this scenario, indicating the good and bad practices. In general, TVS device should be very close to the *I/O* connector and minimizing the length of the TVS traces. In the figure on the left-hand side, due to inductances of the traces, part of the energy in the surge pulse can be entering the IC. Unless this component can be absorbed by the ESD protection stages inside the IC, it can have disastrous consequences. Figure on the right-hand side indicates a multilayer PCB, where a separate GND plane is used where return end of the TVS diode is connected, to minimize any inductive impedance. It is simply important to remember that a short trace length equates to a low impedance, which ensures that the surge energy will be dissipated by the TVS device instead of the IC's internal ESD protection stage. Locating sensitive traces in the centre rather than near the edge of the PCB is a simple method to protect against any ESD issues during handling [7].

It is important to consider that when a transient is diverted by a TVS device, it can cause a significant current to flow through the device into ground connection or the ground plane. This high current can cause a momentary voltage spike on the ground paths, which can have complex connections within the PCB. In case the signal and chassis grounds are separately laid out, if the surge transient-related currents are returned to chassis ground, it may not create a significant transient voltage rise on the signal path ground. Fig. 10.9B depicts the good and bad practices, noting the two independent GND connections in the right-hand figure. In the left-hand figure, shunting the surge voltage directly to IC's signal can cause ground bounce.

As shown in Fig. 10.9C, another possible issue is the radio frequency susceptibility issue due to loop area formed by high-speed signal lines and the ground return paths. One effective method to minimize this loop problem is to incorporate a separate ground plane. In Fig. 10.9C, we can see that maximizing the separation distance from the TVS device in IC provides isolation, however, at the expense of forming a larger loop area.

Fig. 10.9D indicates a case of a diode set such as NUP 2201 from ON Semiconductor, is used with a semiconductor chip where ESD protection is built in. In such cases, placing a series resistor, such as *R* in the Fig. 9.17D, ensures that majority of surge energy is absorbed by the external TVS diode. More details on device selection guidelines and placement options are discussed in Ref. [7]. ESD failure modes and mechanisms are discussed in Ref. [8].

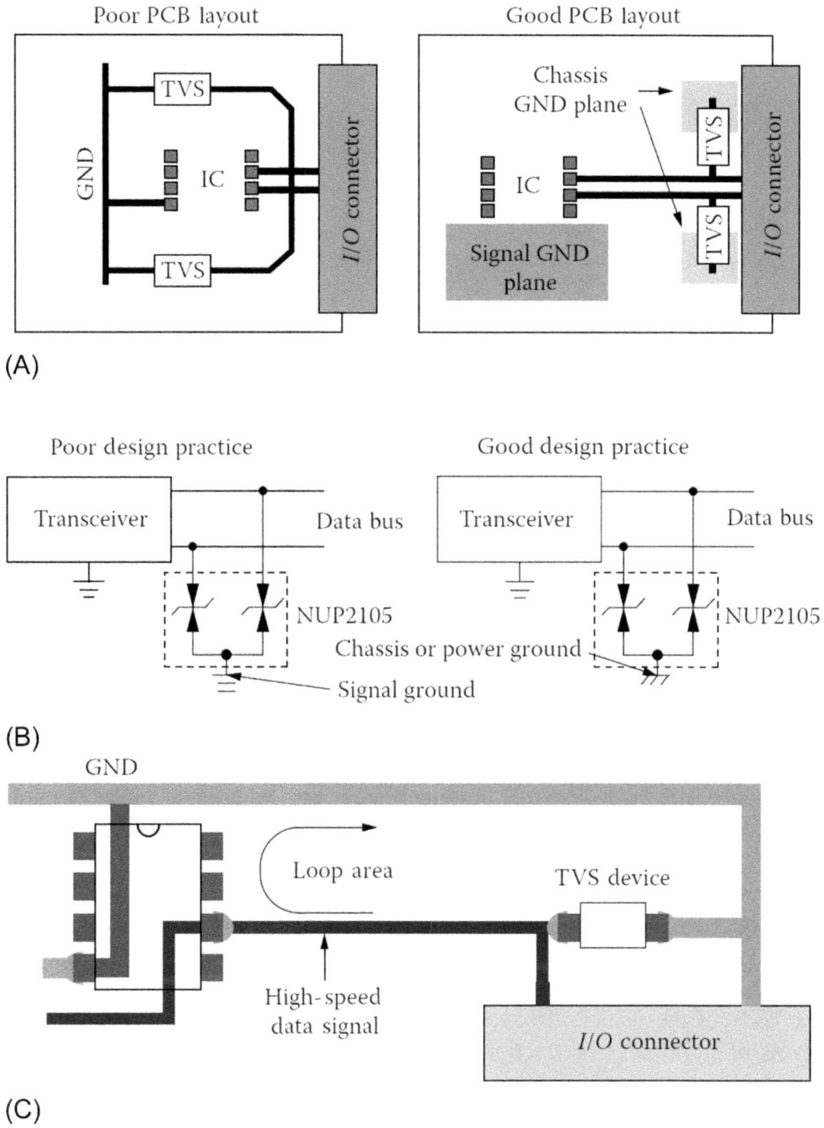

FIG. 10.9 PCB Layout issues: (A) good and the bad practices in placing the TVS device closer to *I/O* connector with minimum inductive traces

Continued

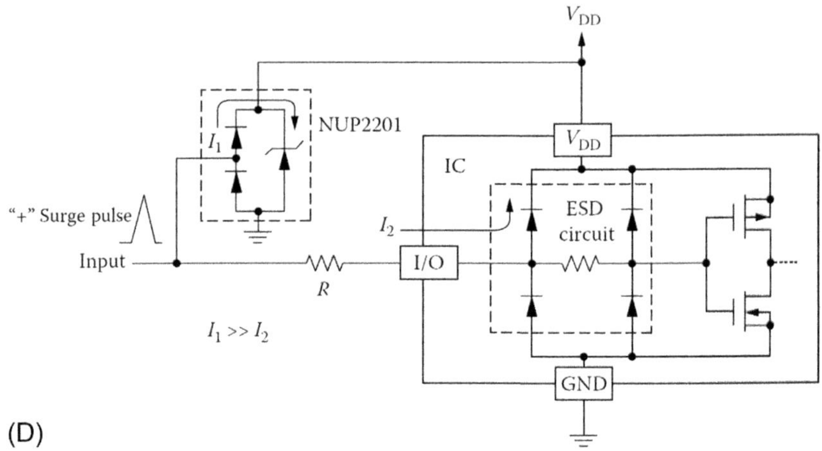

(D)

FIG. 10.9—CONT'D (B) good and the bad practices in using the correct ground return, when separate signal and chassis grounds exist (C) data and ground traces forming a loop that could act as an antenna with increased RF susceptibility (D) use of resistor to absorb the majority of surge energy by the external diodes. *(Adapted from J. Lepkowski, Designers maximize TVS diode performance, Power Electron. Technol. (2006) 48–49.)*

10.9 Automotive Electronics and Surge Protection

All electronic equipment should reliably work in its electromagnetic environment without unduly influencing its environment. This requirement is generally known as the electromagnetic compatibility [9], especially important in all automotive electrical systems. In modern cars with significant safety features (airbags, electronic stability programme, driving assistance, etc.), comfort systems (air conditioning and memory seats, for example), infotainment systems (Bluetooth phone connections, GPS, traffic information), and engine management systems, the modern automotive systems are based on huge amount of electronic sensors (may be over 200 sensors in a midsize car), 10 to 100 processor boards, and vehicle-wide CAN bus-based signalling systems within the body of the vehicle. In such automotive systems, transient surge energy levels of millijoules to fractional joules are sufficient to disturb or destroy the electronic systems within a car, which may be accounting for about 30% of the price we pay for a car today.

SPD component manufacturers such as EPCOS-TDK, Littelfuse, and so on have developed a wide range of special MOV and semiconductor devices to the particular demands encountered in automotive power supplies. Automotive range of components to be used as SPDs, generally come with the following properties [10]:

- Extra high energy absorption (load dump)
- Effective limiting of transients
- Low leakage current

- Jump start capability (no device damage at twice the battery voltage)
- Insensitive to reverse polarity
- Wide operational temperature range
- High resistance to cyclic temperature stress

ISO 7637 (DIN 40839) standard details the EMC in automotive electrical systems. The toughest test here is the requirement for suppression is the simulation of the load dump, which is called Pulse 5. This critical fault could occur when the battery is disconnected from the generator while the engine is running, for example due to broken cable. Under this situation peak voltages of up to 200 V with energy levels of up to 100 J can occur. EPCOS-TDK has developed a device range for these called SIOV-Auto varistors [10].

In an automotive, components and electronic modules are often far apart, and EMC requirements cannot be satisfied by a single suppressor module. Instead the strategy is to provide extra-fine protection at individual module levels, using large numbers of MOVs, and so on. Fig. 10.10 depicts a representative automotive electrical system in a case with an integrated starter generator (ISG) with a supercapacitor bank within the engine management system, safety systems package, car comfort package, and so on interconnected via a CAN bus system. In such ISG-based systems 12 and 42 V supply systems could be present. Fig. 10.10 depicts this system with many surge protector devices, individually placed at each vulnerable electrical system location.

10.9.1 Tests for EMC Requirements in Automotive Applications

Adherence to EMC requirements could be checked with conventional test generators, as shown in Fig. 10.11. Fig. 10.11A shows the case of a load dump generator, with the case of a battery connected in parallel and the case of a battery connected in series in Fig. 10.11B. In these figures the varistors under test are shown at the right-hand side of the figures.

Fig. 10.12A depicts a typical test waveform (Pulse 5 example with a peak voltage of up to 200 V) generated in a case such as in Fig. 10.11A. Fig. 10.12B shows the result of the applied test pulse on the DUT and when the DUT is removed. Fig. 10.12C depicts the current through the DUT, and the energy dumped over a time period of several 100 ms. A detailed discussion on this is available in Ref. [10].

10.10 LED Lighting and Surge Protection

LED lighting is increasingly replacing the traditional light sources like tungsten filament, mercury vapour, metal halide, and sodium vapour in both indoor and outdoor applications, due to higher lumen efficiency of the new LEDs and other

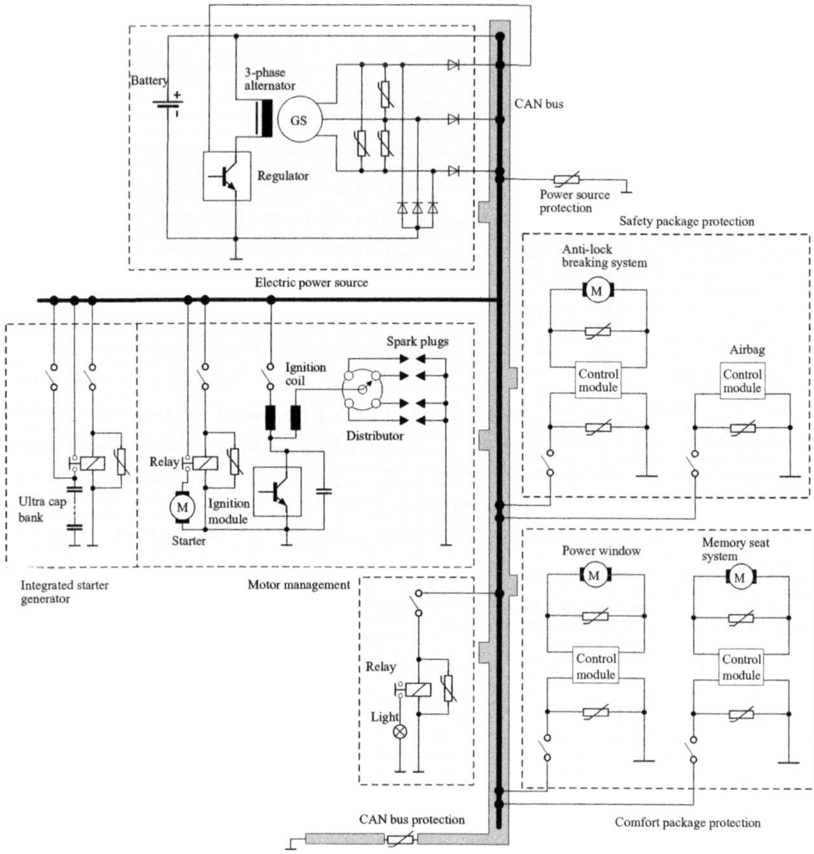

FIG. 10.10 Use of varistors in a representative automotive electrical system to satisfy EMC requirements. *(Courtesy of EPCOS AG; From TDK-EPCOS, SIOV Metal Oxide Varistors, App note January, 17 pp. 2018.)*

technical reasons. While any indoor LED lighting could be safeguarded by suitable SPD s installed in relevant areas of the AC power distribution systems, outdoor LED lighting may need special attention, due to their vulnerability to transients in the outdoor situations like sports stadiums, container yards, and jetty in ports.

In general, most outdoor lighting units are in the power range of 100 W to several 100 W, and they are usually powered using AC/DC converters built into the LED modules. In selecting and installing SPD modules for outdoor LED luminaires, end-users have to consider the possible acts of the god situations in the outdoor areas, due to atmospheric situations like lightning and thunderstorms. Given these situations, particularly in countries like equatorial belt and

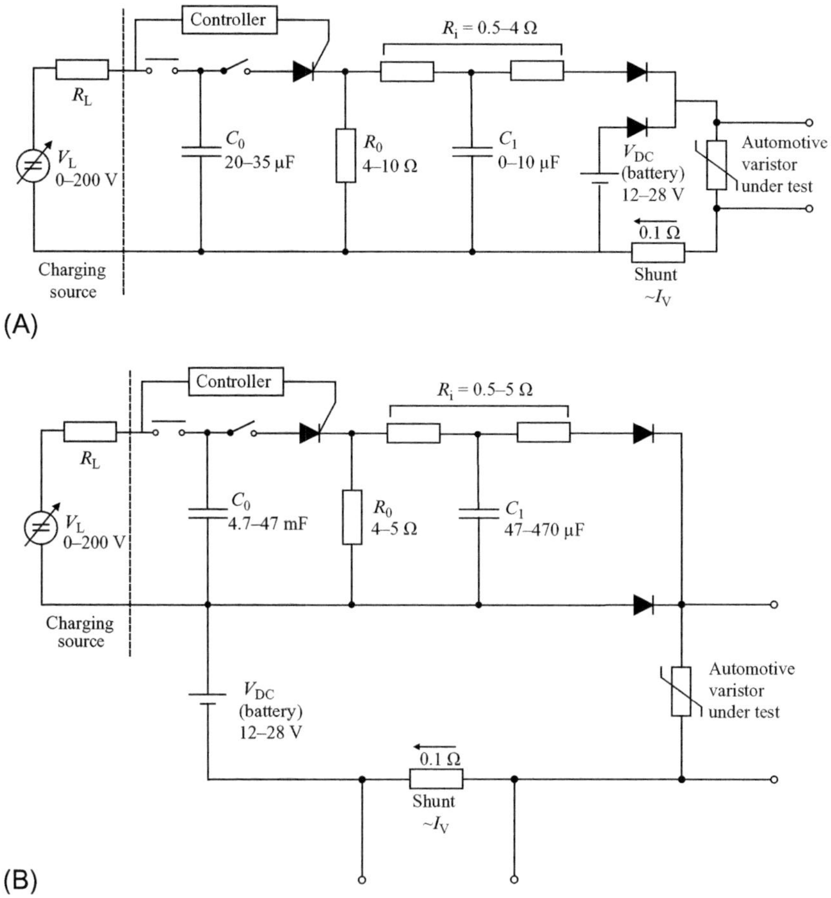

FIG. 10.11 Test generators for load dump test (A) with battery in parallel case (B) battery in series. *(Courtesy of EPCOS AG; From TDK-EPCOS, SIOV Metal Oxide Varistors, App note January, 17 pp. 2018.)*

similar areas the users have to consider the frequent transients possible. In these situations, MOVs in SPDs could quickly get worn-off and the physical SPD connected to the outdoor LED systems may not be an effective one. Given these cases, SPDs designed with thermally protected (by a built in fuse) MOVs are frequently used, and they are expected to have a simple LED indicator lamp to show its effective operation (by a green indication) versus a noneffective SPD based on a red indicator.

In these situations, two different systems are possible, namely the parallel SPD units and the series SPD units. More details on these are available in the Ref. [11].

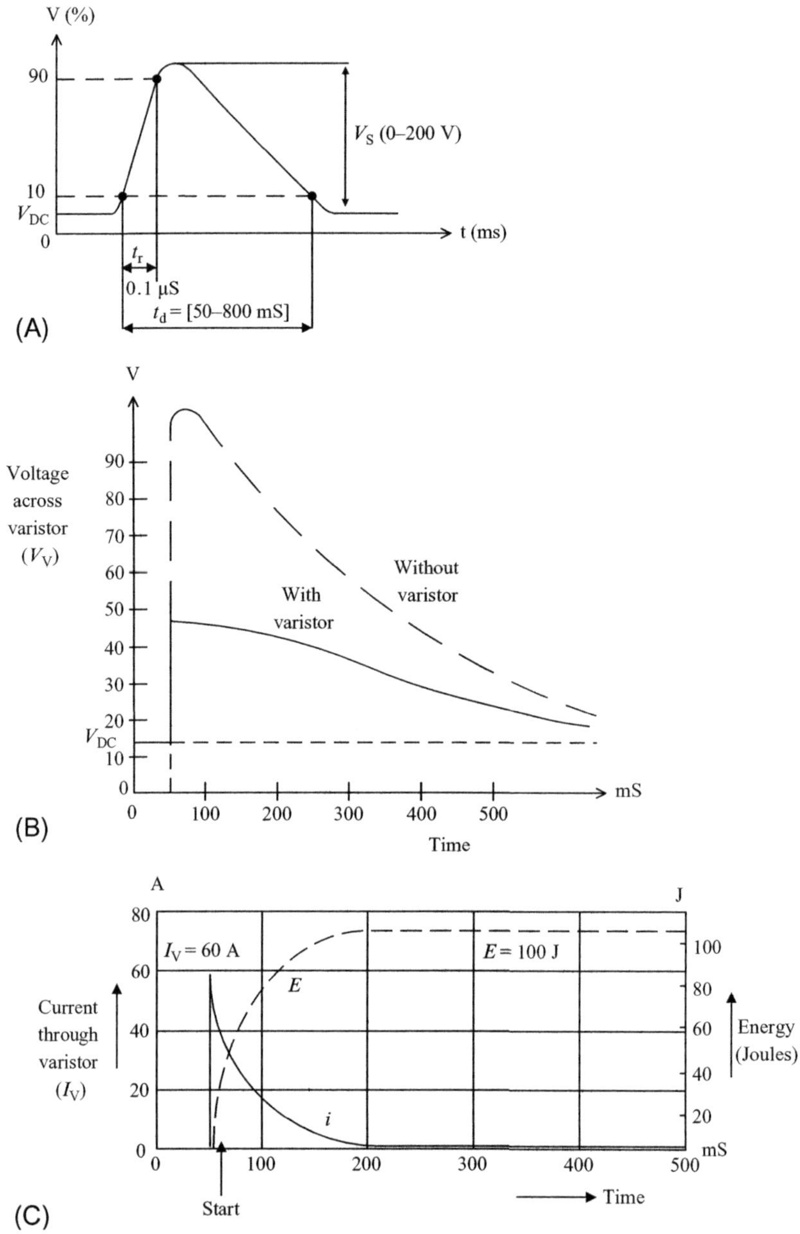

FIG. 10.12 EMC test Pulse 5 and typical test results (A) shape of the Pulse 5 (DIN 40839) (B) effect of the MOV as the DUT, compared with the case of MOV removed (C) current through the MOV and the energy absorbed by the MOV over time. *(Courtesy of EPCOS AG; From TDK-EPCOS, SIOV Metal Oxide Varistors, App note January, 17 pp. 2018.)*

References

[1] V. Cooray, The Lightning Flash, second ed., IET, London, 2014. 928 pp.
[2] DEHN–SÕHNE, Lightning Protection Guide-3rd Updated Edition, December 2014.
[3] Furse Inc., n.d. Guide to BS EN/IEC 62305.
[4] W. Austin, R. Sheatler, Transients vs. electronic circuits: survival of the fittest, the need for transient protection (part II)—ESD immunity and transient current capability for electronic protection array circuits, in: PCIM, May, 1996, pp. 46–53.
[5] W. Austin, IEC 1000-4-2 ESD immunity and transient capability for the SPX72X series protection arrays, Littelfuse Application Note AN 9612.2, 1999. 6 pp.
[6] On Semiconductor Data sheet, NUP2105 Dual Line CAN Bus Protector, 2005. July, Rev 4 .
[7] J. Lepkowski, Designers maximize TVS diode performance, Power Electron. Technol. (2006) 48–49.
[8] A. Amerasekera, W.V. Ableen, L. van Roozendal, M. Hannemann, P. Schofield, ESD failure modes: charateristics, mechanisms, and process influences, IEEE Trans. Electron Devices 39 (2) (1992) 430–436.
[9] D. Morgan, A handbook for EMC testing and measurement, IEEE Elect. Meas. Ser. 8 (1996). 290 pp.
[10] TDK-EPCOS, SIOV Metal Oxide Varistors, App note January, 17 pp, 2018.
[11] Littelfuse Inc., LED Lightning Surge Protection Modules–Design and Installation Guide, (2014). 23 pp.

Further Reading

[12] N. Kularatna, Power Electronics Design Handbook, Elsevier-Newnes, 1998. 300 pp. (Chapter 9) .
[13] N. Kularatna, Digital and Analogue Instrumentation: Testing and Measurement, (2003)690 pp. (Chapter 2) .
[14] S. Duety, et al., The key for processor performance beyond a GHz is clean power, in: Proc. of PCIM, 2000, pp. 68–74.
[15] N. Kularatna, Power Electronics Design Handbook, Elsevier-Newnes, 1998. 300 pp. (Chapter 4) .
[16] B. van Beneden, Varistors: ideal solution for surge protection, Power Electron. Technol. Mag. (2003) 26–32.
[17] Littelfuse, The ABCs of MOVs, Application note, AN 9311-6, July, 3 pp, 1999.
[18] Littelfuse, The ABCs of Littelfuse multilayer suppressors, Application note, AN 9671.2, July, 3 pp, 1999.
[19] ON Semiconductor Datasheet, 1N6267A Series, 1500 Watt Morsorb Zener Transient Voltage Surge Suppressors.
[20] ON semiconductor datasheet, MMT08B310T3 thyristor surge protectors, August, Rev. 3, 6 pp. 2005.
[21] J. Lepkowski, Evaluating TVS protection circuits with SPICE, Power Electron. Technol. Mag. (2006) 44–49.
[22] Harris/Littelfuse Inc., Littlefuse varistors- Basic properties, terminology and theory, Application Note-AN 9767.1, July, 13 pp, 1999.
[23] B. Walaszczyk, Multiple protection devices guard against transients, Power Electron. Technol. (2002) 55–64.
[24] D. Dunlap, Thermally protected MOVs resist overvoltage failures, Power Electron. Technol. (2001) 64–68.

[25] M. Clark, Expand the data sheet on silicon transient voltage protection devices, Power Electron. Technol. (2000) 42–73.
[26] O.M. Clark, Transient Voltage Suppressor Types and Applications, in: PCIM, November, 1990, pp. 19–26.
[27] W. Russell, Transients versus electronic circuits: survival of the fittest- part I-the need for transient protection, in: PCIM, April, 1996, pp. 66–71.
[28] R. Winters, Transients in Power Supply Systems, Solid State Power Conversion, September/October, 1976, pp. 10–15.
[29] IEEE Std C62.41.2-2002, IEEE Recommended practice on characterization of surges in low-voltage (1000 V and less) AC power circuits, IEEE Power Engineering Society, April, 2003.
[30] P. Lewis, Transient voltage surge suppression response time, Power Quality Assurance, September/October, 1995 pp. 56–63.
[31] Eaton Corporation, UL® 1449 3rd Edition, Key Changes, May, 2 pp, 2008.
[32] Siemens Industry Inc., UL 1449 Third Edition: SPD/TVSS Changes Effective September 29, (White Paper by Siemens), 2 pp, 2009.
[33] L.M. Levinson, H.R. Philip, The physics of metal oxide varistors, J. Appl. Phys. 46 (3) (1975) 1332–1341.
[34] G.D. Mahan, L.M. Levinson, H.R. Philip, Theory of conduction of ZnO varistors, J. Appl. Phys. 50 (4) (1979) 2799–2812.

Appendix A

Surge Capability Testing of Supercapacitor Families Using a Lightning Surge Simulator

*Details of this work was published as a paper in IEEE Transactions on Industrial Electronics, Vol. 58, No. 10, October 2011, pp. 4942–4949. All figures in this documents are with the permission of IEEE.

A.1 Introduction

Super capacitor families based on electric double layer effect have entered into a mature stage of usage in many application areas such as electric vehicles, renewable energy sources, UPS systems, battery-supercapacitor hybrids, cellular phones, and back up/ride thorough requirements in electronic systems [1–14]. It is interesting to see from the websites of supercapacitor manufacturers that their suggested commercial uses of these are now in many consumer electronics and industrial applications. For example, the Australian manufacturer Cap-XX aims their thin profile supercapacitors ranging from 0.1 F to over 1 F in applications such as high brightness cellular phone or digital camera flash systems, instantaneous power requirements in GPRS, and transient power requirements in class D audio amps [15]. Maxwell Technologies (United States) who supplies single cell elements up to about 4000 F (2.3 V) and modules up to about 100 F (voltage rated up to 125 V) aim their products at uninterruptible power supplies, telecom network systems, wind turbine pitch systems, peak power for drive systems and actuators, peak shaving and graceful power-down of robotic systems, augmenting the primary energy source for portable devices such as power tools, renewable energy storage sources, and high power pulse forming in power generators[16]. Some special applications under development are improving the end-to-end efficiency of linear regulator techniques using low-frequency supercapacitor circulation, and, the attempts to use supercapacitors in surge-free energy transfer in UPS systems [17–19].

During the last decade there have been many attempts to measure the commonly used technical parameters to characterize the supercapacitor behaviour [20–27]. However, none of these publications have attempted to characterize supercapacitors for their transient and surge voltage endurance, since continuous DC ratings of single cell supercapacitors are between 1 and 5 V only.

In a simple preliminary test conducted by the authors where several supercapacitors were subjected to a single-shot high voltage surge as well as multiple surges of identical shape from a lightning surge simulator indicated that these devices don't get destroyed by a few 100 μs duration high voltage transient surges. The waveforms used were as prescribed in standards such as IEEE C62-41 and IEC 61400-4-5. Fig. A.1 indicates how three different supercapacitors, develop their terminal voltage as we repeatedly surge them with identical high voltage transient waveforms with peak values from 1.5 to 6.5 kV. Devices used were (i) 120 F, 2.3 V device from Maxwell Technologies, (ii) 90 F 2.7 V device from Nesscap, Korea, and (iii) 0.4 F from Cap-XX, Australia.

Given this observation where the devices under test (DUT) do not develop adequate DC terminal voltage even after 20 pulses that would exceed the rated DC voltage, it was decided to develop an automatic test interface for repeated

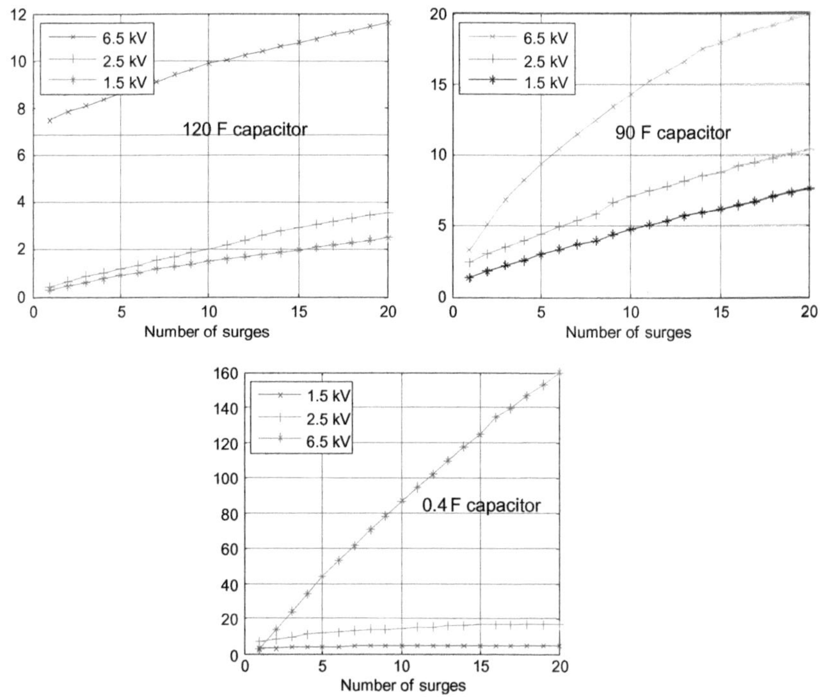

FIG. A.1 Terminal voltage development versus number of surges at different peak values of a combined waveform as per IEC61400-4-5.

application of standard surge waveforms to the DUT in order to determine their surge endurance. The purpose of this was to eliminate the tedious and time-consuming nature of repeated tests, allowing a lightning surge simulator (LSS) to recharge within about 20 s before each pulse was applied.

This paper presents details of the above measurement system to test the surge endurance capability of supercapacitor families. Surge withstanding capability of three different commercial supercapacitor families is presented, with encouraging results. Given many possible applications for supercapacitor families currently available [10, 15, 16] and the new technology developments envisaged for supercapacitor families [28–34], the test results could be helpful in developing the reliability estimates of electronic products using supercapacitors, as well as using them in novel applications, particularly in the power conversion, power conditioning, and surge protection systems.

A.2 Supercapacitor Models and Their Energy Storage Capabilities in Relation to Test Waveforms Specified in International Standards

The double-layer capacitor (DLC) is a complex device in which charge is stored in a double layer formed at the interface between a large surface area material such as activated carbon and a liquid electrolyte. In literature there are many models and equivalent circuits proposed for this device [35]. Two of the common and useful models are depicted in Fig. A.2 [36]. The classical equivalent circuit shown in Fig. A.2A is the most useful in circuit design environments and represents the data sheet parameters in a very applicable manner. The ladder circuit model shown in Fig. A.2B is useful in analysing slow discharge and pulse load applications [36].

Given the relationship for energy storage (E) versus the capacitance value (C) and the terminal voltage (V),

$$E=\frac{1}{2}CV^2 \tag{A.1}$$

a supercapacitor of 10 F with a terminal DC voltage rating of 2 V gives an energy storage capability of 20 J. A typical computer power supply's smoothing capacitor within the rectifier stage of value 220 μF with a voltage rating of 400 V will have a maximum energy storage capacity of 17.6 J.

As per the IEEE C62.41Std series Table A.1 provides a summary of various location categories, frequency of surge occurrences and test waveforms, and so on, as applicable to the testing of surge protectors [37, 38]. Comparing the energy and current ratings for various test waveforms recommended by this ANSI/IEEE standard and other international standards, one can easily see that the energy storage capacities of supercapacitors are well within these recommended values, despite their very low DC voltage ratings.

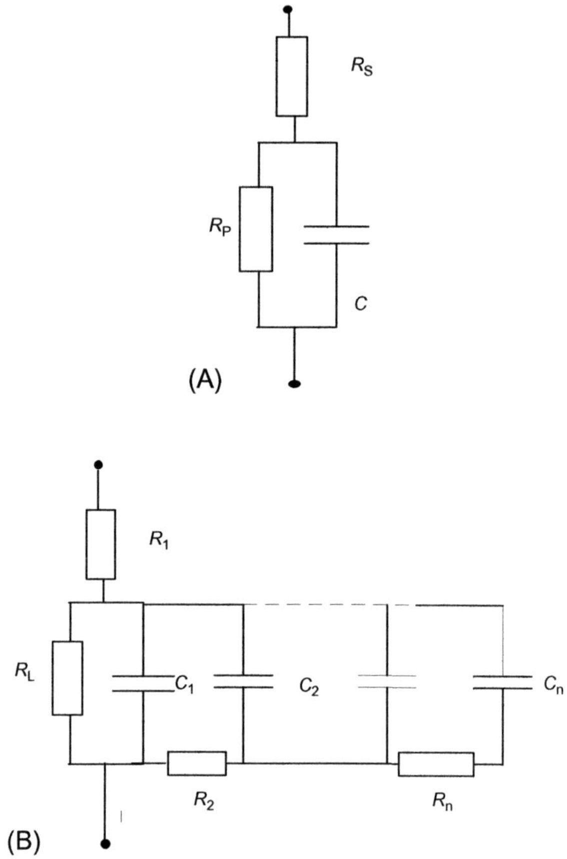

FIG. A.2 Supercapacitor equivalent circuits (A) classical equivalent circuit (B) ladder circuit.

TABLE A.1 IEEE C62.41Location Categories, Surge Waveform Details, and Frequency of Occurrences as Applicable to Testing of Surge Protectors

		Ring Type Test Waveform Details		
IEEE Location Category	Level of Exposure	*2kV, 70A, 0.63J*	*4kV, 130A, 2.34J*	*6kV, 200A, 5.4J*
A1	Low	0	0	0
A2	Medium	50	5	1
A3	High	1000	300	90

TABLE A.1 IEEE C62.41Location Categories, Surge Waveform Details, and Frequency of Occurrences as Applicable to Testing of Surge Protectors—cont'd

		Combination Type Test Waveform Details		
IEEE Location Category	Level of Exposure	*2 kV, 1 kA, 9 J*	*4 kV, 2 kA, 36 J*	*6 kV, 3 kA, 81 J*
B1	Low	0	0	0
B2	Medium	50	5	1
B3	High	1000	300	90

A.3 Justification and the Basis for Testing Surge Endurance Capability of Commercial Supercapacitors

Fig. A.3 depicts some examples of practical waveforms recommended for testing the surge endurance capability of electronic systems as per international standards. Time parameters defining these surge test waveforms are all within fractional microseconds to over 100 μs. Voltage and current values presented in Table A.1 are the peak voltages applicable to Fig. A.3A and the peak current values applicable to Fig. A.3B, respectively. Current values presented in Table A.1 also refer to the short-circuit current capabilities of test systems such as a LSS used to test industrial products. Fig. A.4 depicts the case of an open-circuit voltage waveform indicated in Fig. A.3A applied to a supercapacitor, simplified by the model shown in Fig. A.1. A surge waveform as depicted in Fig. A.3A can be represented mathematically as,

$$V_{SG}(t) = \frac{\alpha\beta}{\alpha-\beta}\left(e^{-\alpha t} - e^{-\beta t}\right) \tag{A.2}$$

where, α represents the inverted value of the time constant related to the decaying part of the exponential waveform and β is the same for the rising part of the exponential waveform. By differentiation and estimating the maxima of the time varying surge, we can obtain the peak value of the surge waveform, V_p as,

$$V_\mathrm{p} = \frac{\alpha\beta}{\alpha-\beta}\left[\left(\frac{\alpha}{\beta}\right)^{\frac{\alpha}{\beta-\alpha}} - \left(\frac{\alpha}{\beta}\right)^{\frac{\beta}{\beta-\alpha}}\right] \tag{A.3}$$

From Eqs. (A.2), (A.3), we can arrive at a normalized open-circuit voltage as similar to Fig. A.3A. By suitable mathematical manipulations and

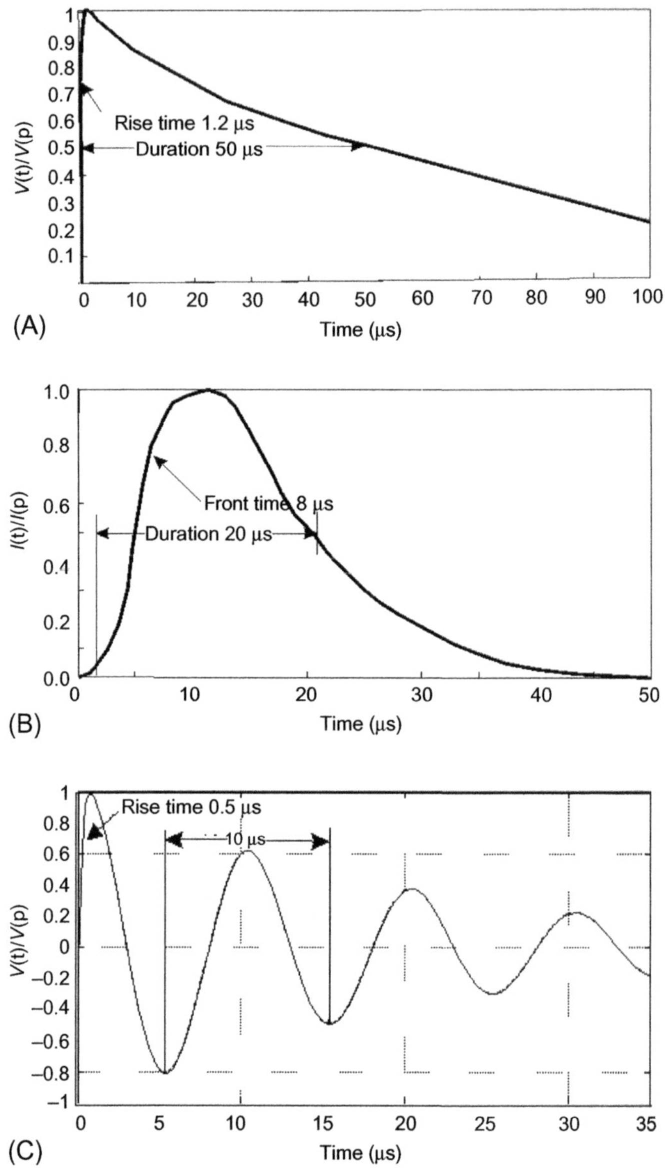

FIG. A.3 Few examples of wave shapes defined in surge test standards (A) open-circuit voltage waveform (B) short-circuit current waveform (C) ring wave.

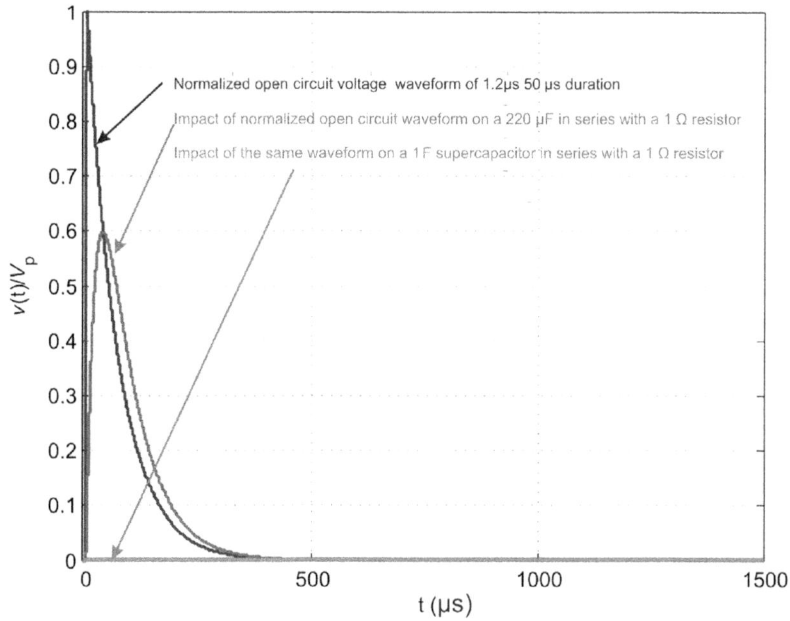

FIG. A.4 MATLAB simulation to study the impact of applying an IEC61400-4-5 combined surge waveform to 220 μF and 1 F capacitors in series with a 1 Ω resistor.

approximations [39] we can get the following relationship to approximate the normalized open-circuit voltage waveform as a function of time in microseconds,

$$V_{SG,\mathrm{nor}}(t) = 1.02032\left(e^{-4.16t} - e^{-0.0139t}\right) \tag{A.4}$$

Fig. A.4 indicates the MATLAB generated waveform, which closely matches the shape of the normalized open-circuit voltage waveform together with its impact on an RC circuit comprising of 220 μF and 1 F capacitors in series with a 1 Ω resistor. Fig. A.4 clearly indicates that due to very large time constants in supercapacitors, such a high voltage waveform which has a total duration of about 100 μs will not develop an adequate voltage across the capacitor. For example, if the peak of the surge waveform is 1 kV, a 220 μF capacitor in series with a 1 Ω resistor could develop an approximate peak voltage of 600 V, while a 1 F capacitor will develop only an extremely minimal voltage.

Given the above situation, we simply predicted that a supercapacitor of Farads order value, with 10–100 mΩ ESR may withstand high voltage transients of durations less than few 100 μs. However, no publication indicated if this is the case, in the context of any other possible failure phenomenon which can occur due to a high voltage pulse.

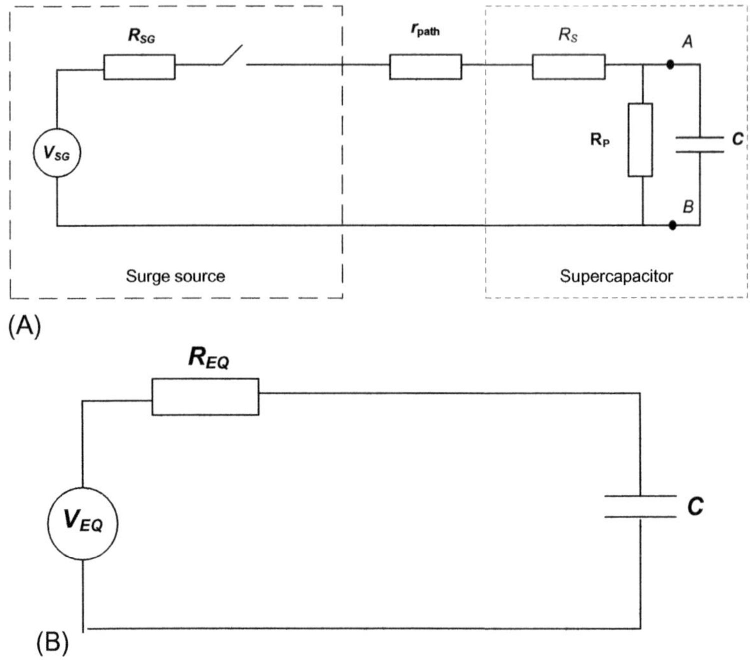

FIG. A.5 Surge source connected to a supercapacitor (A) case with simplified equivalent circuit of supercapacitor (B) simplified Thevenin equivalent circuit.

In analysing the effect of a surge or a transient on a supercapacitor, similar to the case of a high voltage LSS output applied on the DUT, we can use the circuits in Fig. A.5. In this case we have simplified the supercapacitor using the equivalent circuit in Fig. A.2A. Based on the Thevenin equivalent circuit seen by the two points A and B in Fig. A.5A, the case can be simplified to the circuit as shown in Fig. A.5B. In this case,

$$V_{EQ} = V_{SG}\left(\frac{1}{1+\left[\left(R_{SG}+R_s+r_{\text{path}}\right)/R_p\right]}\right) \tag{A.5}$$

$$R_{EQ} = R_{\text{Thevenin}}\left(\frac{R_s+R_{SG}+r_{\text{path}}}{1+\left[\dfrac{R_s+R_{SG}+r_{\text{path}}}{R_p}\right]}\right) \tag{A.6}$$

where R_{SG} and r_{path} are the internal resistance of the surge source and the path resistance of interconnections between the supercapacitor and the surge source.

It can be easily seen that the time constants applicable to the combined effect will be always significantly higher than the case for $\tau = R_S C$ which will be the lowest possible time constant for the supercapacitor charging. This is based on the assumption that the value of R_P is significantly higher than other series path resistances. Given the above scenario, if a capacitor of value C is subjected to a step voltage of a duration T with maximum amplitude of V_{max}, the worst case voltage developed across the capacitor at the end of the step voltage will be

$$V_{C,\max} = V_{\max}\left(1 - e^{-T/R_s C}\right) \tag{A.7}$$

This could be further simplified to,

$$V_{C,\max} = V_{\max}\left[\frac{T}{\tau} - \frac{1}{2}\left(\frac{T}{\tau}\right)^2 + \frac{1}{3!}\left(\frac{T}{\tau}\right)^3 - \ldots\right] \tag{A.8}$$

This gives an approximate value of $\frac{T}{\tau}V_{max}$ and for a supercapacitor with $10\,\mathrm{m}\Omega$ ESR and a capacitance value of 100 F, the capacitor will be charged only up to $10^{-4}V_{max}$. This comfortably predicts that a supercapacitor will not acquire enough charge to exceed its DC voltage ratings during the period of a single-shot application of a combination surge waveform as in Fig. A.3A. It was also identified that novel applications developments such as in [9, 19] require that we have some data to guarantee that a (low voltage DC rated) supercapacitor has adequate endurance to withstand single-shot or multiple high voltage surges.

A.4 Testing Supercapacitors and Practical Issues

The test set-up discussed here allows the device under test (DUT) to be applied with repeated high voltage surges from a LSS with or without a precharged DC voltage. This greatly reduces the time required in repeatedly triggering the LSS to apply its output to the DUT, and in turn saves time required in testing the surge endurance of supercapacitors significantly. In developing the test system, following important considerations had to be met:

1. Allow the LSS internal circuits to precharge after each hit on to a supercapacitor. The instrument used in the experiment, Noiseken LSS 6110, required a minimum of 20 s in between surges.
2. To keep the processor subsystem isolated from high voltage leakages through circuit paths at all times.
3. Immediately after the surge application to the DUT the ADC stage of the processor subsystem to be coupled to the capacitor for measurement.
4. Noiseken LSS 6110 system to be remotely triggered at selected intervals, either on a single shot basis or repeated basis at different time intervals, subjected to a minimum interval of 20 s.

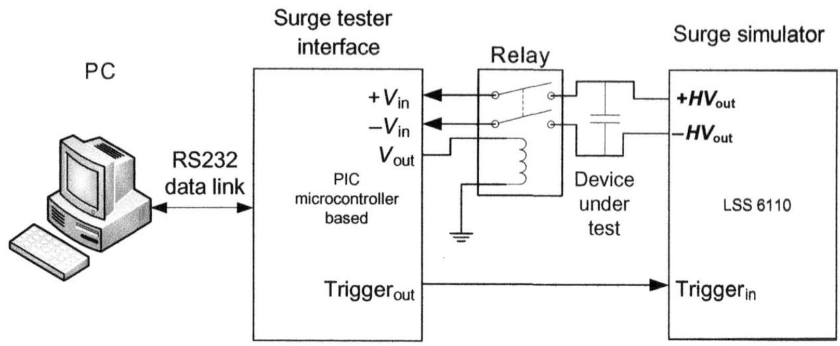

FIG. A.6 Test set-up with interconnections.

Fig. A.6 indicates the overall test set-up and the simplified details of the automatic test interface. The surge tester interface in Fig. A.6 allows the user to trigger the LSS automatically at specified intervals with a minimum value of 20 s in between the surges. After applying the surge to the device under test (DUT), it completely isolates the LSS from the DUT and connects the DUT to the measurement system comprising the ADC and the required probing systems for the oscilloscope. This is to eliminate the risk of low voltage measuring systems and processor interfaces getting damaged due to any remaining charge or any possible high voltage leakages from the LSS.

Fig. A.7A depicts the flow chart applicable to the PIC processor based surge tester interface and Fig. A.7B depicts the flow chart applicable to the PC software which drives the RS 232 interface.

A.5 Supercapacitor Test Results

Using the LSS and the automatic connection and trigger interface developed, three different families of supercapacitors as per details in Table A.2 were tested. Three different tests were conducted:

1. Uncharged supercapacitors with repeated combined surge waveforms applied
2. Partly charged supercapacitors with repeated combined surge waveforms applied
3. Charging and discharging of supercapacitors subjected to repeated surges

Fig. A.8 indicates a sample test result applicable to the 0.18 F supercapacitor from Cap-XX indicating that the devices could easily withstand over 200 hits of high voltage surges before a failure. Multiple samples were further tested without any precharging as well as with precharging up to different voltages within their specified DC ratings. It was noted that at higher values of precharge voltages (over and above 1.8 V for 0.18 F) the capacitor keeps discharging during the tests, while lower pre-charge voltages did not show the same effect.

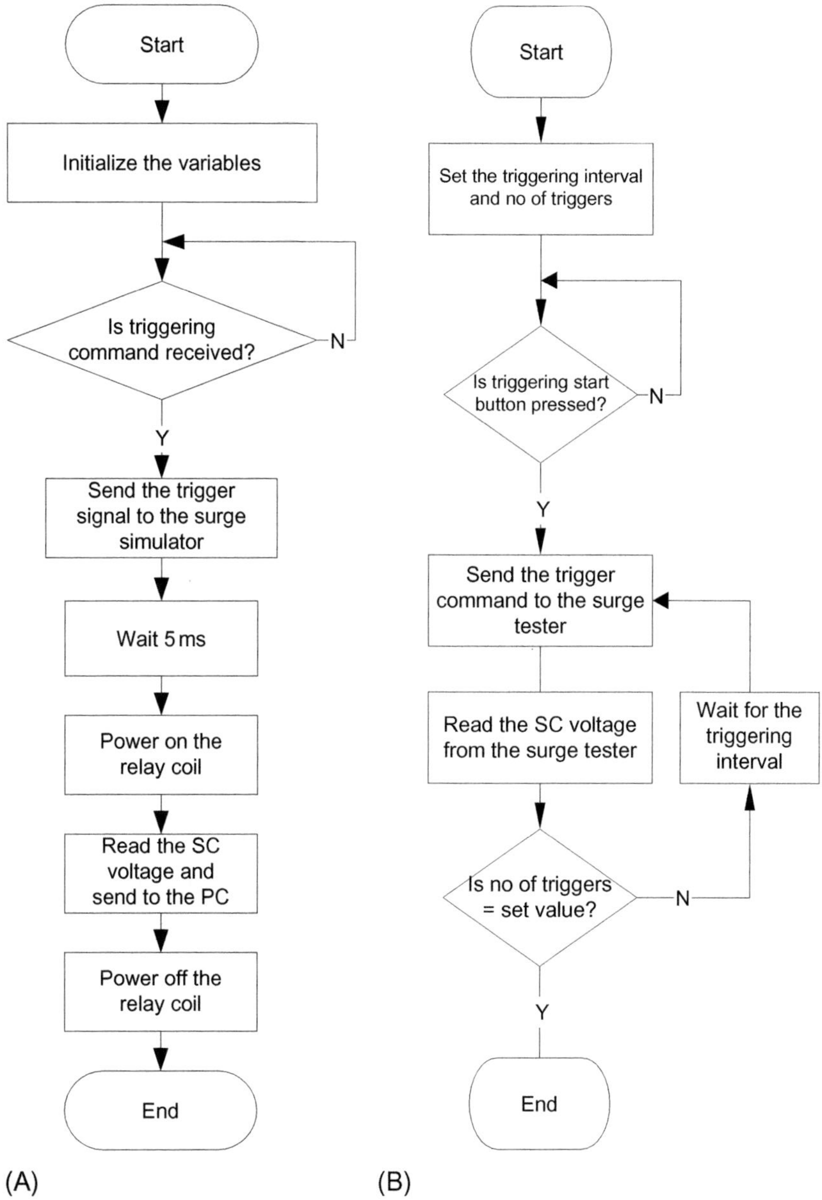

FIG. A.7 Flow charts applicable to test system (A) PIC processor based tester interface (B) PC interface software.

TABLE A.2 Supercapacitor Data Summary and Summary of Test Results

Parameters as Per Data Sheets	Capacitor Family		
	Cap-XX	*Nesscap*	*Maxwell Boostcap*
Capacitance	0.18 F	90 F	230 F
Rated DC voltage	2.3 V	2.7 V	2.5
ESR	45 mΩ	0.33 mΩ	2.2 mΩ
Temperature range	−40°C to 70°C	−40°C to 65°C	−40°C to 70°C
Leakage current	3.5 μA	0.75 mA	0.45 mA

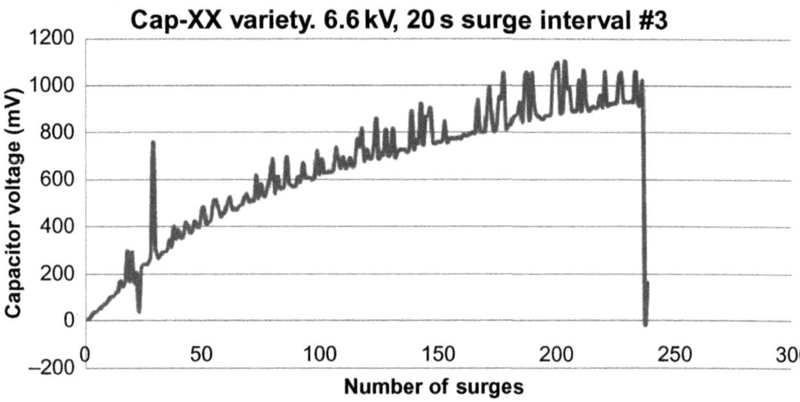

(A)

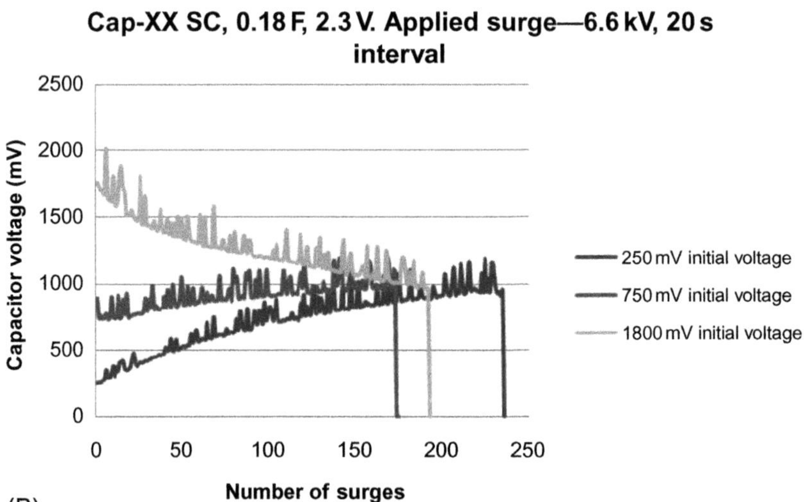

(B)

FIG. A.8 Waveforms show voltage across Cap-XX SC after repeated 6.6 kV surges. Failure is indicated by the sudden decline in voltage (A) with no initial charge on capacitor (B) with initial charge on capacitor varying from 200 mV to 1.8 V.

Photographs in Fig. A.9 indicate the destructive stages of these fractional Farad value capacitors after about 200–230 surges.

It was useful to observe that larger value supercapacitors were very much less prone to any failure due to repeated surges. Figs A.10 and A.11 indicate the results of some sample test cases for Maxwell and Nesscap brands of supercapacitors supporting this condition, where no failure is seen even after 700 hits.

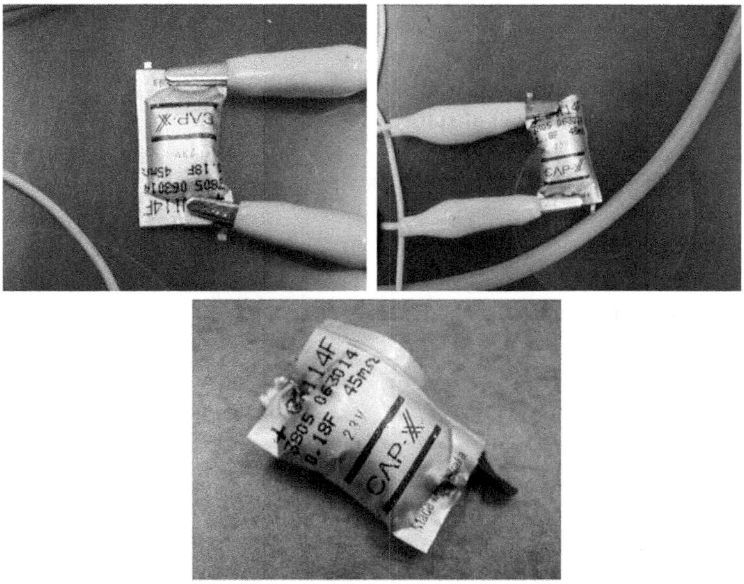

FIG. A.9 Progressive destruction of sub 1 F supercapacitors from Cap-XX.

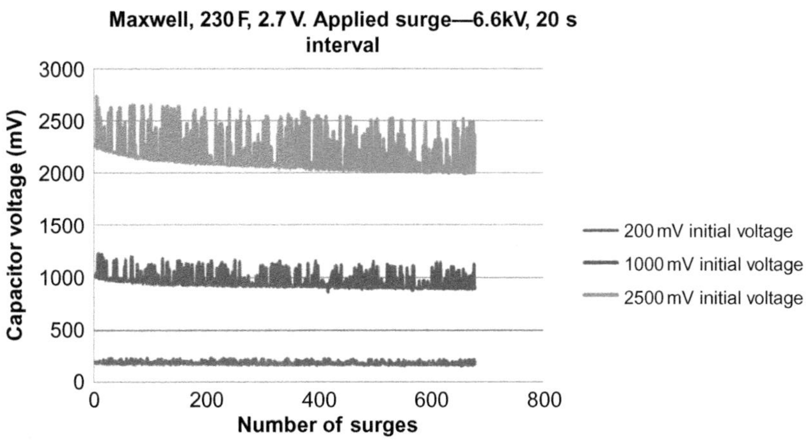

FIG. A.10 Waveforms show voltage across Maxwell SC after repeated 6.6 kV surges. Note: none of the test samples failed after approximately 600 repeated surges.

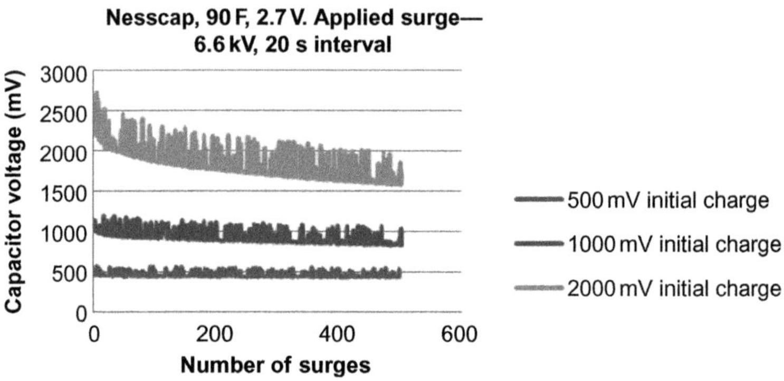

FIG. A.11 Waveforms show voltage across Nesscap SC after repeated 6.6 kV surges. Note: none of the test samples failed after approximately 600 repeated surges.

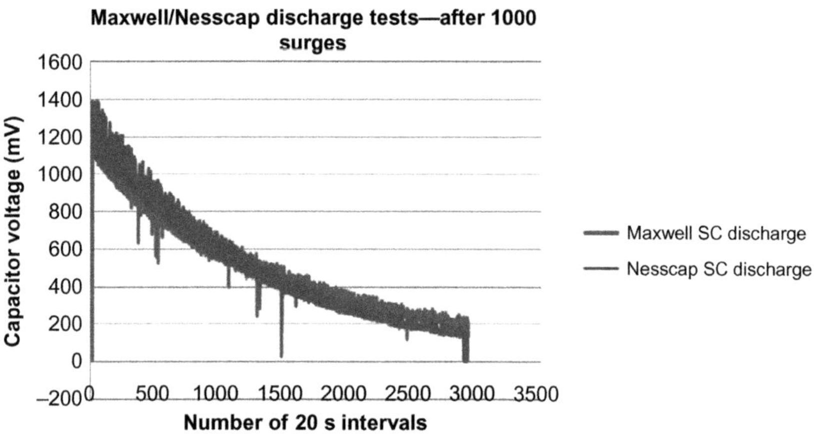

FIG. A.12 Discharge of Maxwell/Nesscap SC's after 1000 repeated surges—Capacitor was placed across a 200 Ω resistor. Waveforms indicate that capacitor integrity remains after surge application.

In order to make sure that the DUTs were having no fundamental failures after repeated surging we connected them to a resistor of 200 Ω for discharge after surge application. Using the same test set-up with the automatic tester interface, this discharge performance of the DUT was plotted against the time axis (using 20 s surge intervals as the unit of measurement). As depicted in Figs A.12 and A.13 all the sample devices behaved as a supercapacitor of the same approximate value. In Fig. A.12 for both samples (Maxwell and Nesscap), initial voltage in the graph was due to the charge accumulation from repeated surges, and the subsequent discharge curve is based on the time constant of

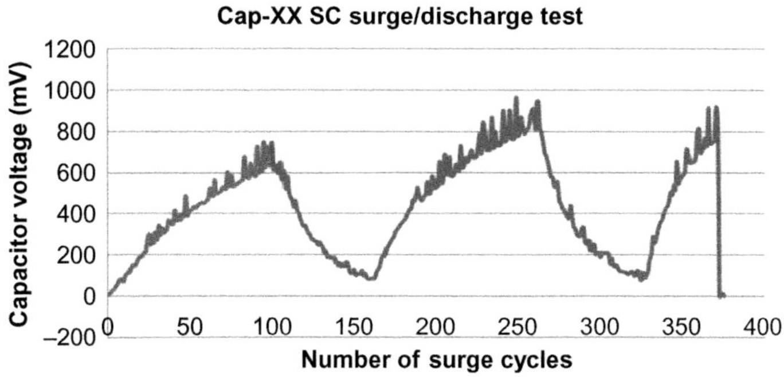

FIG. A.13 Surges are applied to charge the Cap-XX branded SC, after 100 surge applications SC discharges across a 200 Ω resistor. This is repeated until failure

the capacitor with the 200 Ω series resistor. Fig. A.13 indicates the situation of subjecting a Cap-XX 0.18 F device to 100 repeated surges and discharging them across a 200 Ω resistor. These tests confirmed that none of the device families show any performance degradation after large number of repeated surges with peak voltage values up to 6.6 kV (open-circuit voltages applicable to combined waveform generated from the LSS).

Table A.3 provides a summary of test results applicable to the three supercapacitor families used in the experiment, which indicates the behaviour of these devices when subjected to random, repeated surges.

A.6 Observations on Supercapacitor Surge Performance

The work reported in this paper describes the efforts of a research team to develop an automatic test system to repeatedly surge commercial supercapacitor families to characterize the surge behaviour of supercapacitors. Based on testing up to ten samples of each type, the results demonstrate the fact that a limited number of high voltage transients, up to 100 μs duration, do not destroy most commercial supercapacitor families. This useful observation confirms that a limited number of repeated high voltage surges can be safely tolerated by the current commercial supercapacitors, and it leads us to consider more statistical type tests to develop more detailed surge endurance test data for these devices. However, further testing was limited by the available funding.

In a repeated application of surges with a maximum open-circuit voltage of 6.6 kV, the team was unable to see any capacitor failure within the first 150 surges for their sub-1 F supercapacitors (Cap-XX family), and within the first 600–1000 surges for larger value supercapacitors [90 F (from Nesscap) to 230 F (from Maxwell)]. These results (on multiple samples tested) provide the basis and confidence to consider them useful in power electronic product design environments, where nonrepetitive and random surges could occur.

TABLE A.3 A Summary of Test Results Applicable to the Three Supercapacitor Families Used in the Experiment

Capacitor Data	LSS Output Peak Voltage	Number of Surges to Destroy the Device			
		No Precharge	*Precharge Voltage = 250 mV*	*Precharge Voltage = 500 mV*	*Precharge Voltage = 1.8 V*
Cap-XX 0.18 F, 2.3 V	6.6 kV	220 to over 250	Over 220	Over 180	Over 160
Maxwell 230 F, 2.5 V	6.6 kV	No precharge	Precharge voltage = 200 mV	Precharge voltage = 1.0 V	Pre-charge voltage = 2.5 V
		Did not fail after 1000	Did not fail after 600; Charge accumulation was not observed	Did not fail after 600; Over the period of 700 repeated surges 0.1 V discharge was observed	Did not fail after 600; Over the period of 700 surges a discharge of 0.2 V was observed
Nesscap 90 F, 2.7 V	6.6 kV	No precharge	Precharge voltage = 500 mV	Precharge voltage = 1.0 V	Precharge voltage = 2.0 V
		Did not fail after 600	Did not fail after 600; No charge accumulation observed	Did not fail after 600; Slight discharge of 0.15 V observed during 700 surges	Did not fail after 600; Discharge of approx 350 mV was observed over 700 surges

Typical examples are power conversion interfaces where random cases of lightning or inductive power dumps may occur.

A.7 Conclusion

A low-cost PIC processor-based tester interface can help automating the surge endurance testing of supercapacitors. Significant amount of testing done on three different commercial brands of supercapacitor families indicated clearly that (a) most supercapacitors under 0.5 F do not get destroyed by a single voltage surge up to 6.6 kV as per IEEE C62.41/IEC 61400-4-5 specifications, (b) most supercapacitor families over 0.5 F value can withstand over 100 repeated surges, (c) much larger (over 90 F) tank-type supercapacitors do not get destroyed due to less than 500 repeated surges with open-circuit voltages up to 6.6 kV.

These test results have allowed the authors to confidently expand their work initiated in a series of proof-of-concept-type research work to prove the applicability of supercapacitors for special power electronic applications. Furthermore, this experimental work encouraged the research team to go into another phase of testing to estimate if any of the design parameters such as the ESR value and the leakage current could be affected by repeated short duration high voltage surges. More discussion and feedback are welcome from the readership.

Acknowledgment

Authors are grateful to WaikatoLink Ltd., The University of Waikato for their Tech Jump Start grant, at the preliminary stage of the test system development, and to the Research Office of the University of Waikato for the summer research grant to complete the preliminary stage of testing of three commercial device families. Mr. Stewart Finlay is gratefully acknowledged for all assistance in setting up the systems. Authors are very thankful to the Dept. of Electrical and Computer Engineering of the University of Auckland for lending them the Lightning Surge Simulator.

References

[1] Y.-P. Yang, et al., An electric gearshift with ultracapacitors for the power train of an electric vehicle with a directly driven wheel motor, IEEE Trans. Veh. Technol. 56 (2007) 2421–2431.

[2] L. Shuai, et al., A unique ultracapacitor direct integration scheme in multilevel motor drives for large vehicle propulsion, IEEE Trans. Veh. Technol. 56 (2007) 1506–1515.

[3] A.I. Belyakov, D.A. Sojref, High power supercapacitor's solutions for reliable power supply, in: Presented at the International Conference on Power Engineering, Energy and Electrical Drives, 2009, pp. 348–352.

[4] C. Abbey, G. Joos, Supercapacitor Energy Storage for Wind Energy Applications, IEEE Trans. Ind. Appl. 43 (2007) 769–776.

[5] N. Kakimoto, et al., Ramp-Rate Control Of Photovoltaic Generator With Electric Double-Layer Capacitor, IEEE Trans. Energy Convers. 24 (2009) 465–473.

[6] C. Attaianese, et al., High performances supercapacitor recovery system including power factor correction (PFC) for elevators, in: European Conference on Power Electronics and Applications, 2007, pp. 1–10.

[7] C. Attaianese, et al., High performances supercapacitor recovery system for industrial drive applications, in: Nineteenth Annual IEEE Applied Power Electronics Conference and Exposition, APEC '04, vol. 3, 2004, pp. 1635–1641.

[8] The-Infoshop, n.d. Ultracapacitors: Emerging Technologies for High-Power Energy Storage. http://www.the-infoshop.com/report/tn115412-hi-power-energy-stor.html.

[9] M. Zarghami, et al., A novel approach to interarea oscillation damping by unified power flow controllers utilizing ultracapacitors, IEEE Trans. Power Syst. 25 (2010) 404–412.

[10] Mat Dirjish, Ultracapacitors branch out into wider market, (2008)Electronic Design November 17th, *Electronic Design Online ID #20034.*

[11] S.M. Lukic, J. Cao, R.C. Bansal, F. Rodriguez, A. Emadi, Energy storage systems for automotive applications, IEEE Trans. Ind. Electron. 55 (6) (2008) 2258–2267.

[12] F. Khoucha, S.M. Lagoun, K. Marouani, A. Kheloui, M.E.H. Benbouzid, Hybrid cascaded H-bridge multilevel-inverter induction-motor-drive direct control for automotive applications, IEEE Trans. Ind. Electron. 57 (3) (2010) 892–899.

[13] H. Yoo, S.K. Sul, Y. Park, J. Jeong, System integration and power-flow management for a series hybrid electric vehicle using supercapacitors and batteries, IEEE Trans. Ind. Appl. 44 (1) (2008) 108–114.

[14] M.B. Camara, H. Gualous, F. Gustin, A. Berthon, Design and new control of DC/DC converters to share energy between supercapacitors and batteries in hybrid vehicles, IEEE Trans. Veh. Technol. 57 (5) (2008) 2721–2735.

[15] http://www.cap-xx.com/resources/reviews/advantage.htm#reviewsXx.

[16] http://www.maxwell.com/ultracapacitors/applications/industrial.aspXx.

[17] N. Kularatna, J. Fernando, A supercapacitor technique for efficiency improvement in linear regulators, in: Proc of 35th Annual Conf on Industrial Electronics, IECON 09, 2009, pp. 132–135.

[18] US patent application number 627542001000- High current voltage regulator, filing date Dec, (2008).

[19] U.K. Madawala, D.J. Thrimawithana, N. Kularatna, An ICPT-supercapacitor hybrid system for surge-free power transfer, IEEE Trans. Ind. Electron. 54 (6) (2007) 3287–3297.

[20] S. Buller, et al, n.d. Modeling the dynamic behavior of supercapacitors using impedance spectroscopy, IEEE Trans. Ind. Appl., vol. 38, 1622 – 1626

[21] E.-H. El Brouji, O. Briat, J.-M. Vinassa, N. Bertrand, E. Woirgard, Impact of calendar life and cycling ageing on supercapacitor performance, IEEE Trans. Veh. Technol. 58 (8) (2009) 3917–3929.

[22] D. Linzen, S. Buller, E. Karden, R.-W. De Doncker, Analysis and evaluation of charge-balancing circuits on performance, reliability, and lifetime of supercapacitor systems, IEEE Trans. Ind. Appl. 41 (5) (2005) 1135–1141.

[23] H. Gualous, H. Louahlia-Gualous, R. Gallay, A. Miraoui, Supercapacitor thermal modeling and characterization in transient state for industrial applications, IEEE Trans. Ind. Appl. 45 (3) (2009) 1035–1044.

[24] J.-N. Marie-Francoise, H. Gualous, A. Berthon, Supercapacitor thermal- and electrical-behaviour modelling using ANN, Proc. Electr. Power Appl. 153 (2) (2006) 255–262.

[25] D.L. Smith, M.E. Savage, G.R. Ziska, R.L. Starbird, ZR Marx capacitor vendor evaluation and lifetime test results, IEEE Trans. Plasma Sci. 33 (4) (2005) 1273–1281.

[26] S. Buller, Impedance-based simulation models of supercapacitors and Li-ion batteries for power electronic applications, IEEE Trans. Ind. Appl. 41 (3) (2005) 742–747.

[27] Y. Cheng, Assessments of Energy Capacity and Energy Losses of Supercapacitors in Fast Charging-Discharging Cycles, IEEE Trans. Energy Convers. 25 (1) (2010) 253–261.

[28] M. Kaempgen, J. Ma, G. Gruner, G. Wee, S.G. Mhaisalkar, Bifunctional carbon nanotube networks for supercapacitors, Appl. Phys. Lett. 90 (2007)2 pp.

[29] L. Ma, Y. Yang, Solid-state supercapacitors for electronic device applications, Appl. Phys. Lett. 87 (2005)3 pp.

[30] W. Li, G. Joos, J. Belangor, Real_time simulation of a wind turbine generator coupled with a battery supercapacitor energy storage system, IEEE Trans. Ind. Electron. 57 (4) (2010) 1137–1145.

[31] A. Xu, S. Xie, X. Liu, Dynamic voltage equalization for series-connected ultracapacitors in EV/HEV applications, IEEE Trans. Veh. Technol. 58 (8) (2009) 3981–3987.

[32] F.I. Simjee, P.H. Chou, Efficient Charging of Supercapacitors for Extended Lifetime of Wireless Sensor Nodes, IEEE Trans. Power Electron. 23 (3) (2008) 1526–1536.

[33] Z. Amjadi, S.S. Williamson, Power-electronics-based solutions for plug-in hybrid electric vehicle energy storage and management systems, IEEE Trans. Ind. Electron. 57 (2) (2010) 608–616.

[34] M.B. Camara, H. Gualous, F. Gustin, A. Berthon, B. Dakyo, DC/DC converter design for supercapacitor and battery power management in hybrid vehicle applications—polynomial control strategy, IEEE Trans. Ind. Electron. 57 (2) (2010) 587–597.

[35] R.L. Spyker, R.M. Nelms, Classical equivalent circuit parameters for a double-layer capacitor, IEEE Trans. Aerosp. Electron. Syst. 30 (3) (2000) 829–836.

[36] R.M. Nelms, D.R. Cahela, B.J. Tatarchuk, Modeling double-layer capacitor behavior using ladder circuits, IEEE Trans. Aerosp. Electron. Syst. 39 (2) (2003) 430–438.

[37] IEEE C62.41, IEEE Recommended Practice on Surge Voltages in Low-Voltage AC Power Circuits, (1991).

[38] James, S., Kularatna, N. and Steyn-Ross, A. A Mathematical Approach to Studying Transient Propagation in Power Electronic Interfaces Using Wavelets and MATLAB, n.d. (to be published)

[39] James, S. and Kularatna, N and Pandey, A., Testing of Computer Power Supplies for Surge Withstand Capability Using a Lightning Surge Simulator to Generate a Database for a Surge Propagation Study, n.d. (to be published)

Further Reading

[40] N. Kularatna, Power Electronics Design Handbook-Low Power Components and Applications, Elsevier-Newnes(1998)300 pp. (Chapter 6).

Appendix B1

Explanatory Notes on the Test System Used for UL1449 3rd Ed. Related Tests Done at Waikato Uni HV Labs

B1.1 General

The document below is developed as an aid to explain the test set-up used at the UOW HV-Test-Lab while testing a commercial SPD.

B1.2 Noiseken LSS and Its Monitoring Outputs

It is important to note at the first instance that the Noiseken LSS 6130/6230 [6.6 kV/3.3 kA testers], whose internal circuit diagram is depicted in Fig. B1.6, has two monitoring outputs for the following:

1. *Voltage output*—It is via a voltage divider of three resistors, where the middle resistor voltage is supplied to the oscilloscope via an isolation transformer unit.
2. *Current output*—This is via a current transformer which is to monitor the overall loop current through the DUT. However, the voltage measurement channel under Item 1 above is considered part of the loop current. We have measured the loop current via this current transformer, with the DUT port kept open, and it reads about 138 A (peak) and this reading is directly related to the current through the voltage monitoring channel.

All commercial LSS systems have a setting based on the highlighted area in brown dotted lines, as in Fig. B1.6 where a set of resistors and capacitors are injected or removed from the loop for respective loop current and VPR measurements.

As per our observations, the peak voltage reading at the DUT, fed into the oscilloscope, reads a value less than the set voltage on the front panel of the LSS. For example, as depicted in Fig. B1.2A which indicates the voltage at the 0.5 ohm resistor (wire-wound version of 10 W capability) [VLR] in yellow, voltage at the monitoring channel [LSV] in purple plotted against the injection voltage setting of the LSS. For example, when the LSS output is set to 6.6 kV (UL requirement of voltage), monitoring output shows only 3.76 kV with a loop current of 2760 A. This simply indicates the relationship between the LSS set voltage versus the voltage measured at the voltage monitoring output of the LSS.

Fig. B1.2C shows the loop current through the test loop in green when the DUT connected is a 0.5 ohms resistor. For example, when the voltage setting of the LSS is at 6.6 kV, loop current read is 2760 A. Brown and blue graphs show the cases with short-circuited DUT and a DUT of 1.0 ohms.

As shown in Fig. B1.4, when the LSS output put is kept open, [i.e., DUT not connected], shows a loop current 138 A at the set voltage of 6.6 kV. This is due to the effect of the voltage monitoring output of the LSS loaded by the oscilloscope probe capacitance, and so on.

This measurement shows the case of how the measurements and the waveforms on the scope relates to the calibrated voltage output set (UL 1449 Ed 3 requirement) on the LSS and the worst case current through the loop (typically of 3 kA). As we see from Fig. B1.2C measured loop current depends on the value of the DUT, since the total loop impedance is composed of multiple components such as:

- Internal resistance of the high voltage source of the LSS
- Wave-shaping circuit of the LSS
- Coupling circuit at the LSS output (which is composed of a combination of 10 ohms resistor, and two 9 μF capacitors)
- Path resistances and inductances created by the connections

In addition to all above the input impedance of the DUT will get added and the equivalent circuit of a fired MOVs has a parallel combination of a capacitance and a resistance (of few ohms). In Fig. B1.1, Z_p *will* include the effect of the coupling circuit as well.

It is important to note that, when a fully assembled SPD is tested, the loop current will depend on the effective input impedance of the SPD under fired condition, where we record the VPR value at the output of the SPD. So this value will be slightly different to the reading from the current monitoring output of the LSS.

B1.3 Equivalent Simplified Circuit for LSS

Fig. B1.1 is a simplified case of a LSS internal circuit diagram, showing the effect of various internal impedances which affects the total loop resistance

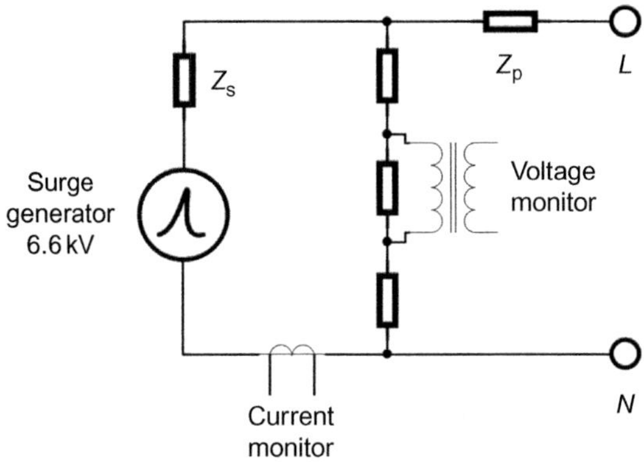

FIG. B1.1 LSS internal simplified circuit diagram (Z_s—internal series impedance, Z_P—path series impedance).

during a measurement, and its voltage monitoring and current monitoring outputs, providing electrical isolation for the oscilloscope input channels.

The maximum current which can be drawn when the *L* and *N* terminals of the LSS are short circuited is 3.42 kA (higher than the UL value of 3000 A) and the monitored voltage through LSS monitor is 3.8 kV (for a front panel setting of 6.6 kV). If you open the short circuit, LSS voltage monitoring channel output rises to 7.6 kV.

B1.3.1 Calibration Measurements: Based on a Constant 10 W Wire Wound Resistors Used as the DUT (Load)

Voltage across the load resistor—VLR
Voltage at the output of the LSS (*LSS Monitor Output*)—LSV
Loop Current (*LSS Monitor Output*)—LSLC (Fig. B1.2)

B1.4 Additional Measurements Done to Confirm the Validity of Above Data

For our own understanding and to make sure that simplifications used for Fig. B1.1 are valid, we have done some additional measurements to establish the variation of the total loop resistances with applied surge voltage, as well as the effect of vertical channel setting on the leakage current measurement.

B1.4.1 Loop Resistance in the Closed Circuit Loop

Characteristics of the 0.5 ohms closed loop circuit have been measured to investigate about the resistivity of the loop circuit. Fig. B1.1 shows the simplified case of the measurement loop, showing the simplified model of the surge generator, connection path resistances, and the relative locations of the monitoring

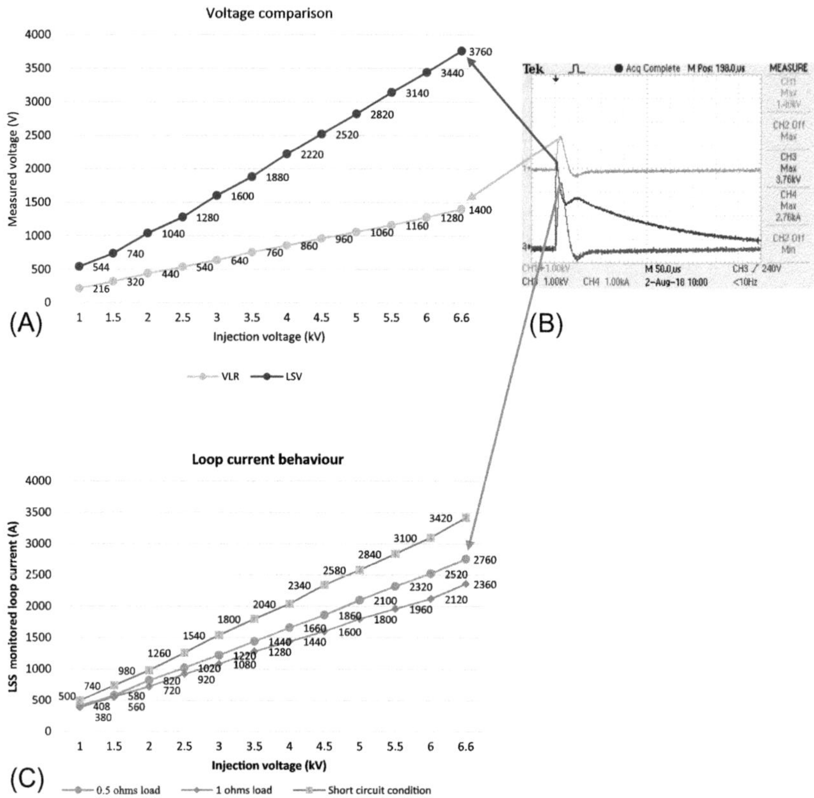

FIG. B1.2 (A) Measured voltages against surge injection. (B) Oscillographs at 6.6kV Surge to a 0.5 ohms resistive load. (C) Monitored current in at different loads (DUTs of 0.5 ohms, 1.0 ohms and a short circuit).

points within the measurement systems. As we see from the graphs of Fig. B1.3, there is a small variation of the total loop resistance with the applied surge voltage. However, this variation is due to simplification that all impedances within the loop are considered resistive and linear.

B1.4.2 Leakage Current Investigation

In the open loop condition at the DUT, the results below show that some leakage current in the LSS can be directly monitored through the LSS Current sensor. This is mainly due to the case that the current monitoring channel always monitors current drawn by the voltage monitoring path of the LSS. Fig. B1.4 shows the effect of vertical channel setting of the Tektronix TPS 2014 versus the leakage current reading, which leads to the conclusion that approximately 138 A value shown in the Fig. B1.4 is a reasonable estimate. This is due to multiple

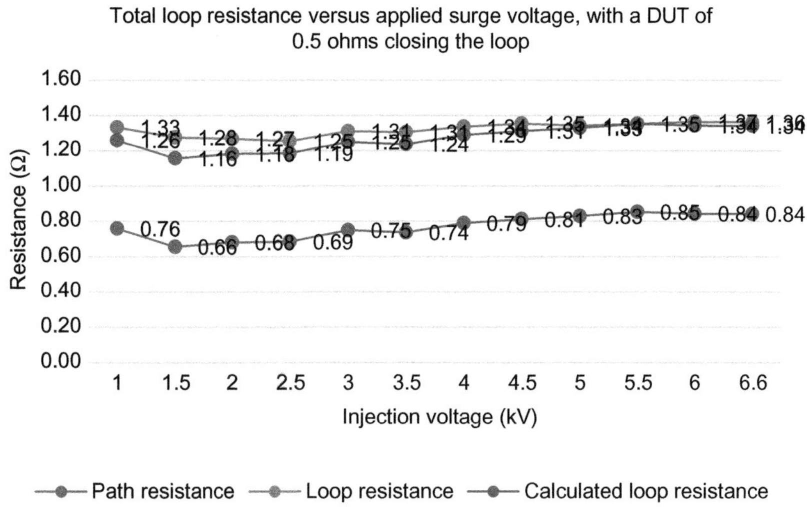

FIG. B1.3 Loop resistance investigation.

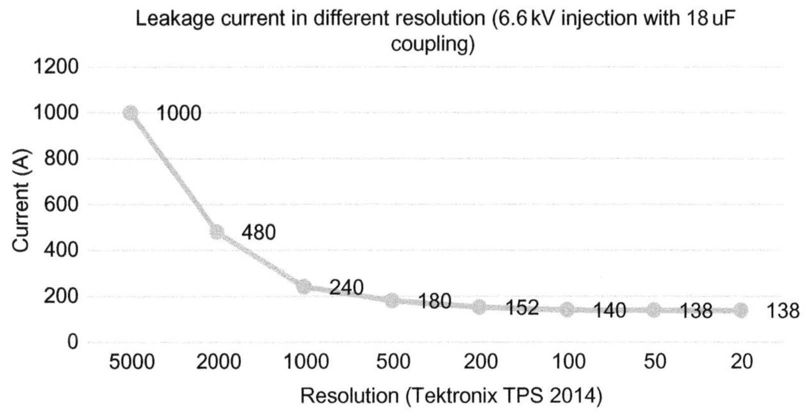

FIG. B1.4 Leakage current in different resolutions.

effects of the 8-bit resolution of the vertical channels combined with the probe effects at high voltage inputs.

B1.4.2.1 Test Set-Up in the Lab

(See Fig. B1.5.)

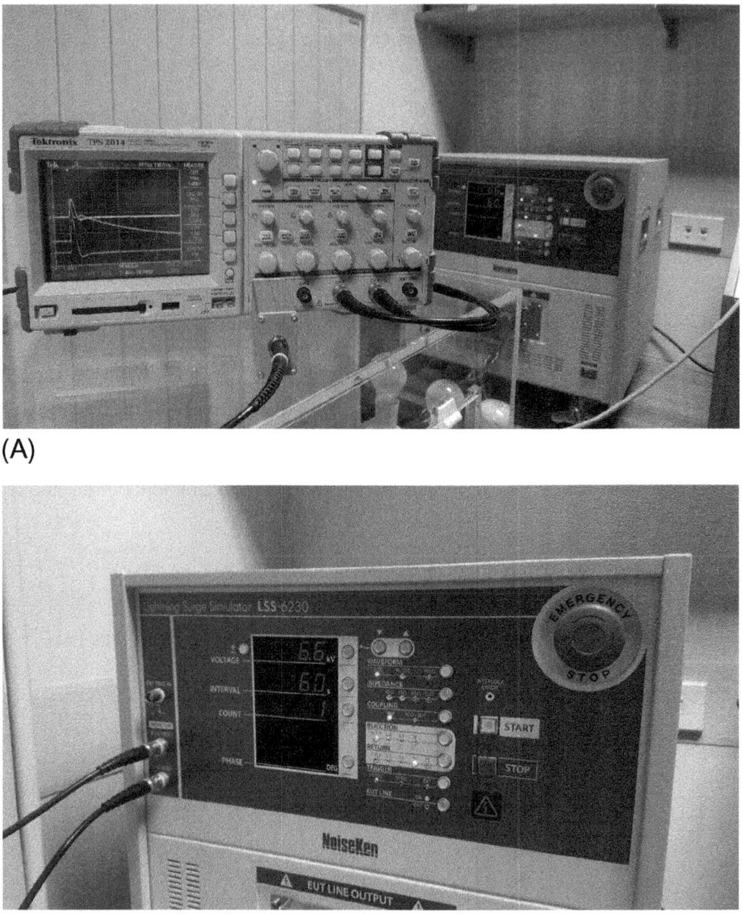

(B)

FIG. B1.5 (A) Test set-up (Tektronics TPS2014, LSS 6230). (B) LSS-6230 (set to 6.6 kV Injection and 18 uF coupling).

B1.4.2.2 Internal Circuits of the LSS-6130 Indicating Its Voltage and Current Outputs

(See Fig. B1.6.)

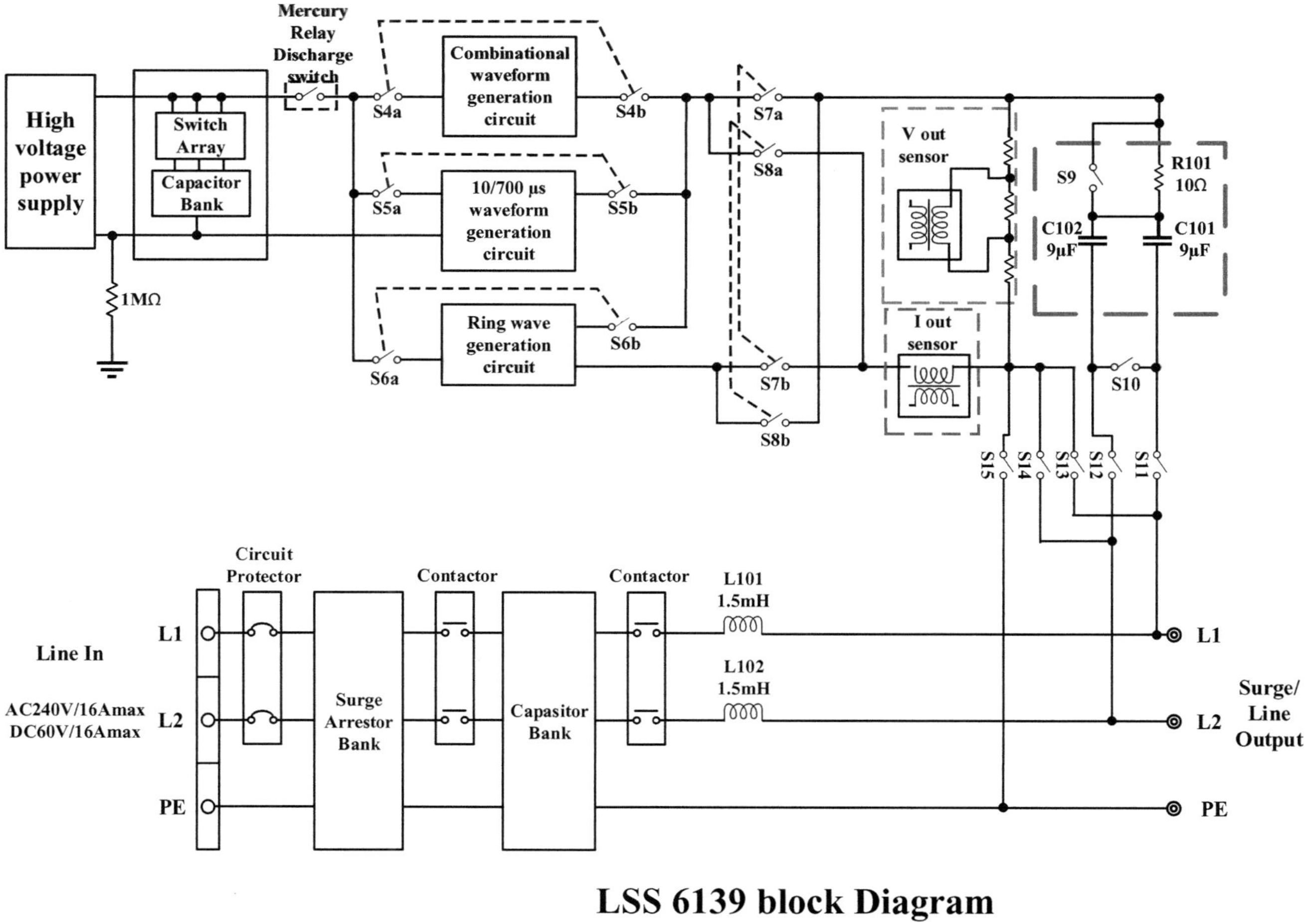

FIG. B1.6 LSS internal circuit diagram.

B1.5 Conclusion

It can be observed that the open circuit waveform peak is 7.6 kV and the short-circuit current is 3.42 kA for the setting of 6.6 kV on the front panel of the LSS. This means effective impedance is 2.21 ohms which satisfies the UL1449 effective impedance requirement. The test results of the Smart TVIQ^2 show the current as 3 kA as the value through the device. Therefore, the injected voltage to the DUT is 6.63 kV.

Appendix B2

Supercapacitor specifications – Selected devices usable in surge protectors

Manufacturer	Part no.	Capacitance *(F)*	Rated voltage *(V)*	ESR *(mΩ)*	Rated Max Current[1] *(A)*	Temperature range	Leakage current *(mA)*	Typical mass *(g)*
Maxwell	BCAP0001	1	2.7	700	0.1	-40°C to 65°C	0.006	1.1
	BCAP003	3.3	2.7	290	0.33	-40°C to 65°C	0.012	1.7
	BCAP005	5	2.7	170	0.5	-40°C to 65°C	0.015	2.3
	BCAP0010	10	2.7	75	1	-40°C to 65°C	0.03	3.5
Vina Tech	VEC 2R7 105 QG	1	2.7	195	1	-40°C to 65°C	0.002	1.1
	VEC 2R7 105 QG	3.3	2.7	95	3	-40°C to 65°C	0.007	1.5
	VEC 2R7 505 QG	5	2.7	60	5	-40°C to 65°C	0.01	1.8
	VEC 2R7 106 QG	10	2.7	35	10	-40°C to 65°C	0.02	3.2
	VEC 3R0 105 QG	1	3.0	220	1	-40°C to 65°C	0.003	1.1
	VEC 3R0 335 QG	3.3	3.0	105	3.5	-40°C to 65°C	0.01	1.5
	VEC 3R0 505 QD	5	3.0	70	5.5	-40°C to 65°C	0.015	1.8
	VEC 3R0 106 QG	10	3.0	40	9	-40°C to 65°C	0.03	3.2
NESSCAP	ESHSR-0003C0-002R7	3	2.7	70	2.9	-40°C to 65°C	0.005	1.4
	ESHSR-0005C0-002R7	5	2.7	45	3.8	-40°C to 65°C	0.008	2.1
	ESHSR-0010C0-002R7	10	2.7	30	5.6	-40°C to 65°C	0.023	3.1

1 It should be noted that in the supercapacitor based surge protector technique discussed in the relevant chapter this maximum current specification in manufacturer data sheets do not relate to microseconds order transient currents occurring within the surge duration.

Appendix C

State Equations and MATLAB Code for Chapter 6

C.1 Laplace Solution for the Open-Circuit Voltage of the LSS Validated With a SPICE Simulation

MATLAB code used to produce Fig. 6.3

```
function LSS_open_circuit_voltage_analysis
% LSS equivalent circuit
% Laplace solution for the open-circuit voltage of the LSS
% validated with a SPICE simulation (SPICE data included)
% Capacitor initial charge = 100V

L1 = 2.7e-6;
C  = 10e-6;
R1 = 7;
R2 = 1.1;
L2 = 6e-6;
R3 = 1e6;

A = R1+R2+R3 ;
B = R1^2 ;

format short e
a0 = 1000*C*R1
b3 = L1*L2*C
b2 = C*R1*L2 + A*L1*C
b1 = A*C*R1 + L2 - C*B
b0 = A
```

```
p = [b3 b2 b1 b0]
r = roots(p)
t = linspace (0,200e-6,10001);
f = @(a) a0*exp(a*t)/(3*b3*a^2+2*b2*a+b1);
% setup current I in output loop
I = zeros(size(t));

for j = 1:3
    I = I + f(r(j));
    Voc = I*1e6;
end

% SPICE data is read from a comma-separated values (CSV) file

M = csvread('LSS_OCV_2loops_spice.csv', 1,1); %reads data from the
% csv formatted file starting at row 1 and column 1

time_spice = M(:,2);    % takes all rows of column three
Voc_spice = M(:,3);       % takes all rows of column four

figure(1);clf;
    plot (t*1e6,Voc,'linewidth', 2);hold on;
    plot(time_spice*1e6, Voc_spice, 'xr','linewidth', 1);
    legend('Laplace','Spice');
    axis ([-10 200 0 1000])

    xlabel('Time (us)'); ylabel('Voltage (V)');
end
```

C.2 SPICE Input File to Trace the Open-Circuit Voltage of the LSS

(The results obtained were included in the MATLAB code to validate the Laplace solution—Appendix C.1)

```
** LSS - Open-circuit Voltage **

** Circuit Description **
C1  1  0  10UF  IC=1000V      ;Capacitance of 10μF with initial
                              voltage of 1000 volts
L1 1 2 2.7UH                  ; Inductor of 2.7 micro-henries
R1 2 0 7                      ; Resistance of 7 ohms
R2 2 3 1.1                    ; Resistance of 1.1 ohms
L2 3 4 6UH                    ; Inductor of 6 micro-henries
R3 4 0 1MEG                   ; Resistance of 1 meg-ohm

** Output Requests **
* Transient analysis from 0 to 100 us with a 1 us time increment
and using initial conditions (UIC)
.TRAN 1US 100US UIC
```

```
* Plot the result of the transient analysis with the voltage at
node 4
.PLOT TRAN V(4)
.PROBE V(4)
.END     ; End of circuit file
```

C.3 Laplace Solution for the Open-Circuit Voltage of the LSS Validated With Numerical Simulations Based on Euler and RK4 Method

MATLAB code LSS_OCV_Euler_RK4 used to produce Fig. 6.3

```
%--------------------------------------------------------------
function LSS_OCV__Euler_and_RK4
% Solve for open-circuit voltage of the two loop LSS
% equivalent circuit using Euler and RK4 methods.
% Assume the capacitor is fully charged at time zero

R1 = 7;
C = 10e-6;
L1 = 2.7e-6;
R2 = 1.1;
L2 = 6e-6;
R3 = 1e6;
Vc0 = 1000;

a = R1/L1; b = 1/(L1*C);
c = R1/L2; d = (R1+R2+R3)/L2;
tic
A = R1+R2+R3 ;
B = R1^2 ;

format short e

a0 = 1000*C*R1
b3 = L1*L2*C
b2 = C*R1*L2 + A*L1*C
b1 = A*C*R1 + L2 - C*B
b0 = A

p = [b3 b2 b1 b0]

r = roots(p)

t = linspace (0,100e-6,10001);

f = @(h) a0*exp(h*t)/(3*b3*h^2+2*b2*h+b1);

% setup current I in output loop
```

```
I = zeros(size(t));

for j = 1:3
    I = I + f(r(j));
    Voc = I*1e6;
end
%
fignum = 3;
   figure(fignum); clf;
   plot(t*1e6, Voc, 'linewidth', 1);
   zoom on; hold on;
toc
%———————————————————————————————
% DE definition for LSS circuit
function xdot = RLC_dot(t, x, flag, params)
    [x1 x2 x3] = deal(x(1), x(2), x(3));
    x1_dot = x2;
    x2_dot = -a*x2 - (b-a*c)*x1 - (a*d)*x3;
    x3_dot = c*x1-d*x3;
    xdot = [x1_dot; x2_dot; x3_dot]; % return column vector
end

% Euler solution of DEs: I = current; J = dI/dt
Nsteps = 1e6;
[I1, J, I2] = deal(zeros(Nsteps,1));
dt = 1e-11;
time = [0: Nsteps-1]*dt;

% choose integration method
% 1 = Euler; 2 = improved Euler
METHOD = 2;
tic
switch METHOD

% Euler with fixed timesteps
   case 1
       I1(1) = 0;
       J(1) = Vc0/L1; % integrate delta-function to give initial
                      % condition
       I2(1) = 0;
       for n = 1 : Nsteps-1
        I1(n+1) = I1(n) + J(n)*dt;
        J(n+1) = J(n) + (-a*J(n) - (b-a*c)*I1(n) - a*d*I2(n)) * dt;
        I2(n+1) = I2(n)+(c*I1(n)-d*I2(n))*dt;
        end
```

```
% RK4 with fixed timesteps
    case 2
        X0(:,1) = [0; Vc0/L1; 0]; flag = ''; params = [];
        [t_out, X_out] = rk4fixed(@RLC_dot, time, X0, flag, params);
        time=t_out;
        I1 = X_out(:,1); J = X_out(:,2);I2 = X_out(:,3);
end

figure(fignum);
    plot(time*1e6, I2*1e6, '--r', 'linewidth', 2);
    axis ([-2 100 0 1000]);
    legend('Laplace solution','Simulation with rk4fixed solver');
    xlabel('Time (us)'); ylabel('Voltage (V)');
toc
end
```

C.4 Development of State Equations (6.13)–(6.17) for the MOV "Off" State Driven by the LSS

In order to solve this 3-loop circuit we will assume that the loop currents are i_1, i_2, and i_3. The Laplace transformed network representation of Fig. C.1 is given in Fig. C.2.

The loop equations for the circuit shown in Fig. C.2 are given by

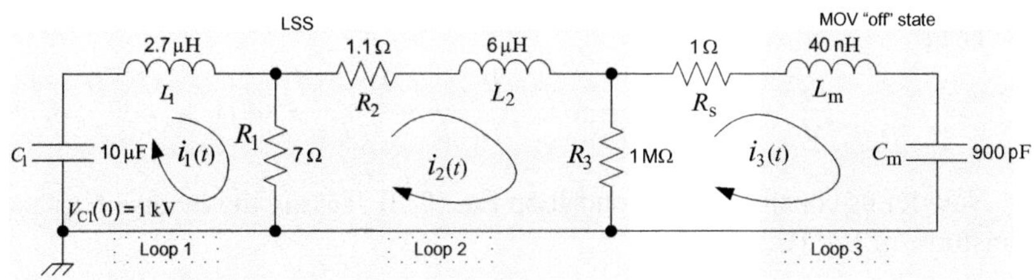

FIG. C.1 MOV "off" state driven by the LSS (Combined equivalent circuit).

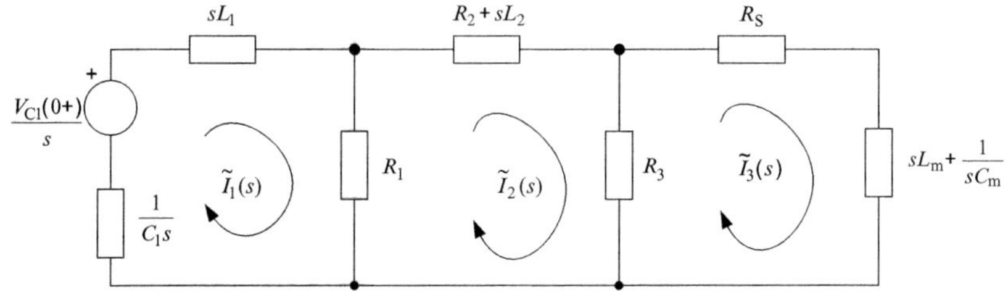

FIG. C.2 Transformed network for the circuit in Fig. C.1.

$$\left(sL_1+R_1+\frac{1}{sC_1}\right)\tilde{I}_1(s)-R_1\tilde{I}_2(s)=-\frac{V_{c1}(0+)}{s} \tag{C.1}$$

$$-R_1\tilde{I}_1(s)+(R_1+R_2+R_3+sL_1)\tilde{I}_2(s)-R_3\tilde{I}_3(s)=0 \tag{C.2}$$

$$\left(R_3+R_S+sL_m+\frac{1}{sC_m}\right)\tilde{I}_3(s)-R_3\tilde{I}_2(s)=0 \tag{C.3}$$

Let us first consider the first loop Eq. (C.1). Multiplying Eq. (C.1) by sC_1

$$(s^2L_1C_1+sR_1C_1+1)\tilde{I}_1(s)-(sR_1C_1)\tilde{I}_2(s)=C_1V_{c1}(0+) \tag{C.4}$$

Taking the inverse Laplace transform of Eq. (C.4)

$$\frac{d^2i_1}{dt^2}L_1C_1+\frac{di_1}{dt}R_1C_1+i_1-R_1C_1\frac{di_2}{dt}=C_1V_{c1}(0+)\cdot\delta(t) \tag{C.5}$$

Dividing Eq. (C.5) by L_1C_1, we get

$$\frac{d^2i_1}{dt^2}+\frac{R_1}{L_1}\frac{di_1}{dt}+\frac{1}{L_1C_1}i_1-\frac{R_1}{L_1}\frac{di_2}{dt}=\frac{V_{c1}(0+)}{L_1}\cdot\delta(t) \tag{C.6}$$

In order to solve this second-order DE numerically, we recast it into normal form: a system with two interacting first-order DEs. Let

$$\frac{di_1}{dt}=j_1 \tag{C.7}$$

Then

$$\frac{dj_1}{dt}=-\frac{R_1}{L_1}j_1-\frac{1}{L_1C_1}i_1-\frac{R_1}{L_1}\frac{di_2}{dt}+\frac{V_{c1}(0+)}{L_1}\cdot\delta(t) \tag{C.8}$$

Now let us consider the second-loop Eq. (C.2). Taking the inverse Laplace transform of Eq. (C.2), we get

$$i_2(R_1+R_2+R_3)+L_2\frac{di_2}{dt}-i_1R_1-i_3R_3=0 \tag{C.9}$$

Dividing by L_2 and rearranging

$$\frac{di_2}{dt}=\frac{R_1}{L_2}i_1+\frac{R_3}{L_2}i_3-\frac{R_1+R_2+R_3}{L_2}i_2 \tag{C.10}$$

Substituting Eq. (C.10) in Eq. (C.8), we get

$$\frac{dj_1}{dt}=-\frac{R_1}{L_1}j_1-\frac{1}{L_1C_1}i_1-\frac{R_1}{L_1}\left(\frac{R_1}{L_2}i_1+\frac{R_3}{L_2}i_3-\frac{R_1+R_2+R_3}{L_2}i_2\right)+\frac{V_{c1}(0+)}{L_1}\cdot\delta(t) \tag{C.11}$$

Rearranging Eq. (C.11)

$$\frac{dj_1}{dt} = -\frac{R_1}{L_1}j_1 - \left(\frac{1}{L_1C_1} - \frac{R_1^2}{L_1L_2}\right)i_1 - \frac{R_1(R_1+R_2+R_3)}{L_1L_2}i_2 - \frac{R_1R_3}{L_1L_2}i_3 + \frac{V_{c1}(0+)}{L_1} \cdot \delta(t) \tag{C.12}$$

Now let us consider the third-loop Eq. (C.3). Multiplying Eq. (C.3) by sC_m

$$\tilde{I}_3(s)\left[1+s(R_3+R_S)C_m+s^2L_mC_m\right] - \tilde{I}_2(s)sR_3C_m = 0 \tag{C.13}$$

Taking the inverse Laplace transform of Eq. (C.13), we get

$$\frac{d^2i_3}{dt^2}L_mC_m + \frac{di_3}{dt}(R_3+R_S)C_m + i_3 - R_3C_m\frac{di_2}{dt} = 0 \tag{C.14}$$

Dividing Eq. (C.14) by L_mC_m

$$\frac{d^2i_3}{dt^2} + \frac{(R_3+R_S)}{L_m}\frac{di_3}{dt} + \frac{1}{L_mC_m}i_3 - \frac{R_3}{L_m}\frac{di_2}{dt} = 0 \tag{C.15}$$

In order to solve the second-order DE Eq. (C.15) numerically, we recast it into normal form: Let

$$\frac{di_3}{dt} = j_3 \tag{C.16}$$

Then Eq. (C.15) can be written as

$$\frac{dj_3}{dt} = -\frac{(R_3+R_S)}{L_m}j_3 - \frac{1}{L_mC_m}i_3 + \frac{R_3}{L_m}\left(\frac{R_1}{L_2}i_1 + \frac{R_3}{L_2}i_3 - \frac{R_1+R_2+R_3}{L_2}i_2\right) \tag{C.17}$$

Rearranging Eq. (C.17) gives

$$\frac{dj_3}{dt} = -\frac{(R_3+R_S)}{L_m}j_3 + \frac{R_1R_3}{L_2L_m}i_1 - \frac{R_3(R_1+R_2+R_3)}{L_2L_m}i_2 + \left(\frac{R_3^2}{L_2L_m} - \frac{1}{L_mC_m}\right)i_3 \tag{C.18}$$

Now Eqs (C.7), (C.10), (C.12), (C.16), and (C.18) make up the set of state equations required to simulate Fig. C.1 (same as Fig. 6.9). These five equations can be written as state equations (6.13)–(6.17) provided we make the following substitutions:

$a=R_1/L_1, b=1/L_1C, c=R_1/L_2, d=(R_1+R_2+R_3)/L_2, e=R_3/L_2, f=(R_3+R_S)/L_m$, $g=R_3/L_m$, and $h=1/L_mC_m$.

The impulse response that results from the delta function input in Eq. (C.12) can be taken care of by using the initial condition that arises from the impulse input, and making the input itself equal to zero. Hence the initial conditions would be $i_1(0)=0$, $j_1(0)=V_{C1}(0+)/L_1$, $i_2(0)=i_3(0)=j_3(0)=0$.

C.5 Numerical Simulation of the "Off" State Current in a MOV Validated With a SPICE-Based Simulation

MATLAB code used to produce Fig. 6.10

```
%-------------------------------------------------------------
function LSS_MOV_three_loops_ODE15S
% Solve for third loop current thru the "off" state of the MOV.

R1 = 7;
C = 10e-6;
L1 = 2.7e-6;
R2 = 1.1;
L2 = 6e-6;
R3 = 1e6
Rs = 1;
Lm = 40e-9;
Cm = 900e-12;
Vc0 = 100;

a = R1/L1, b = 1/(L1*C)
c = R1/L2, d = (R1+R2+R3)/L2
e = R3/L2, f = (R3+Rs)/Lm
g = R3/Lm, h = 1/(Lm*Cm)

fignum = 1;

%------------------------------------
% SPICE data is read from a comma-separated values (CSV) file

M = csvread('LSS_MOV_3loops_spice.csv', 1,1); %reads data from
% the csv formatted file starting at row 1 and column 1

time_spice = M(:,2); % takes all rows of column three
I3_spice = M(:,3);  % takes all rows of column four

%------------------------------------
% DE definition for RLC circuit
function xdot = RLC_dot(t, x, flag, params)
[x1 x2 x3 x4 x5] = deal(x(1), x(2), x(3), x(4), x(5));
x1_dot = x2;
x2_dot = -a*x2 - (b-a*c)*x1 - a*d*x3 + a*e*x4;
x3_dot = c*x1 - d*x3 +e*x4;
x4_dot = x5;
x5_dot = -f*x5 + c*g*x1 - g*d*x3 + (e*g-h)*x4;
xdot = [x1_dot; x2_dot; x3_dot; x4_dot; x5_dot]; % return
                                              % column vector
end
```

```
time = [0 1e-5];

tic

% ode15s with variable timesteps
X0(:,1) = [0; Vc0/L1; 0; 0; 0]; flag = ''; params = [];
options = odeset('RelTol', 1e-3);
[t_out, X_out] = ode15s(@RLC_dot, [time(1) time(end)], X0,...
options);
I1 = X_out(:,1); J = X_out(:,2); I2 = X_out(:,3); I3 = X_out(:,4);
K = X_out(:,5);

figure(fignum);
   subplot(211)

   plot(time*1e6, I3, '--r', 'linewidth', 2);hold on;
   plot(time_spice*1e6, I3_spice, '-k', 'linewidth', 1);
   legend('ode15s based simulation','SPICE based simulation');
   axis ([0 10 -0.13 0.4])
   title('(a)')
   xlabel('Time (us)'); ylabel('MOV "off" state current (A)');
   subplot(212)
   plot(diff(time)*1e6,'+-k');
   xlabel('Index number'); ylabel('Time step (us)');
   title('(b)')
toc
end
```

C.6 Development of State Equations (6.18)–(6.23) for the Assumed Fixed States of the MOV Driven by the LSS

In order to solve this 4-loop circuit we will assume that the loop currents are i_1, i_2, i_3, and i_4. The Laplace transformed network representation of Fig. C.3 is given in Fig. C.4.

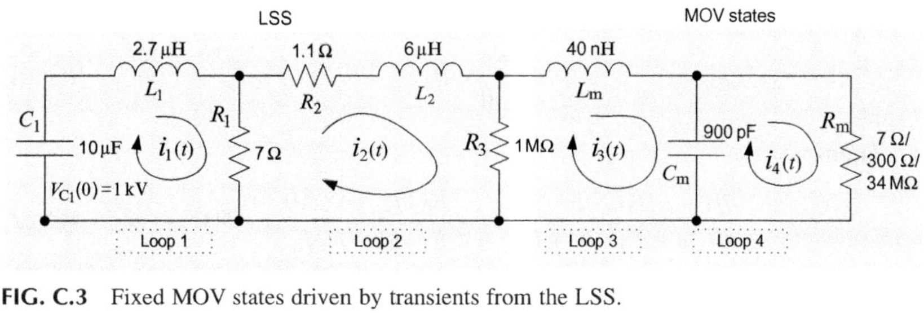

FIG. C.3 Fixed MOV states driven by transients from the LSS.

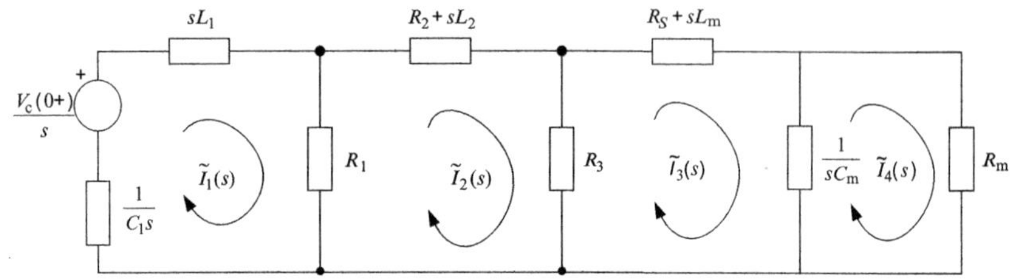

FIG. C.4 Transformed network for the circuit in Fig. C.3.

$$\left(sL_1+R_1+\frac{1}{sC_1}\right)\tilde{I}_1(s)-R_1\tilde{I}_2(s)=-\frac{V_{c1}(0+)}{s} \tag{C.19}$$

$$-R_1\tilde{I}_1(s)+(R_1+R_2+R_3+sL_1)\tilde{I}_2(s)-R_3\tilde{I}_3(s)=0 \tag{C.20}$$

$$\left(R_3+R_S+sL_m+\frac{1}{sC_m}\right)\tilde{I}_3(s)-R_3\tilde{I}_2(s)-\frac{1}{sC_m}\tilde{I}_4(s)=0 \tag{C.21}$$

$$\left(R_m+\frac{1}{sC_m}\right)\tilde{I}_4(s)-\frac{1}{sC_m}\tilde{I}_3(s)=0 \tag{C.22}$$

Let us first consider the first loop Eq. (C.19). Multiplying Eq. (C.19) by sC_1

$$\left(s^2L_1C_1+sR_1C_1+1\right)\tilde{I}_1(s)-(sR_1C_1)\tilde{I}_2(s)=C_1V_{c1}(0+) \tag{C.23}$$

Taking the inverse Laplace transform of Eq. (C.23)

$$\frac{d^2i_1}{dt^2}L_1C_1+\frac{di_1}{dt}R_1C_1+i_1-R_1C_1\frac{di_2}{dt}=C_1V_{c1}(0+)\cdot\delta(t) \tag{C.24}$$

Dividing Eq. (C.24) by L_1C_1, we get

$$\frac{d^2i_1}{dt^2}+\frac{R_1}{L_1}\frac{di_1}{dt}+\frac{1}{L_1C_1}i_1-\frac{R_1}{L_1}\frac{di_2}{dt}=\frac{V_{c1}(0+)}{L_1}\cdot\delta(t) \tag{C.25}$$

In order to solve this second-order DE numerically, we recast it into two coupled first-order DEs. Let

$$\frac{di_1}{dt}=j_1 \tag{C.26}$$

Then

$$\frac{dj_1}{dt}=-\frac{R_1}{L_1}j_1-\frac{1}{L_1C_1}i_1-\frac{R_1}{L_1}\frac{di_2}{dt}+\frac{V_{c1}(0+)}{L_1}\cdot\delta(t) \tag{C.27}$$

Now let us consider the second-loop Eq. (C.20). Taking the inverse Laplace transform of Eq. (C.20), we get

$$i_2(R_1+R_2+R_3)+L_2\frac{di_2}{dt}-i_1R_1-i_3R_3=0 \tag{C.28}$$

Dividing by L_2 and rearranging

$$\frac{di_2}{dt}=\frac{R_1}{L_2}i_1+\frac{R_3}{L_2}i_3-\frac{R_1+R_2+R_3}{L_2}i_2 \tag{C.29}$$

Substituting Eq. (C.29) in Eq. (C.27), we get

$$\frac{dj_1}{dt}=-\frac{R_1}{L_1}j_1-\frac{1}{L_1C_1}i_1-\frac{R_1}{L_1}\left(\frac{R_1}{L_2}i_1+\frac{R_3}{L_2}i_3-\frac{R_1+R_2+R_3}{L_2}i_2\right)+\frac{V_{c1}(0+)}{L_1}\cdot\delta(t) \tag{C.30}$$

Rearranging Eq. (C.30)

$$\frac{dj_1}{dt}=-\frac{R_1}{L_1}j_1-\left(\frac{1}{L_1C_1}-\frac{R_1^2}{L_1L_2}\right)i_1-\frac{R_1(R_1+R_2+R_3)}{L_1L_2}i_2-\frac{R_1R_3}{L_1L_2}i_3+\frac{V_{c1}(0+)}{L_1}\cdot\delta(t) \tag{C.31}$$

Now let us consider the third-loop Eq. (C.21). Multiplying it by $sC_{\rm m}$

$$\widetilde{I}_3(s)\left[1+s(R_3+R_S)C_{\rm m}+s^2L_{\rm m}C_{\rm m}\right]-\widetilde{I}_2(s)sR_3C_{\rm m}-\widetilde{I}_4(s)=0 \tag{C.32}$$

Taking the inverse Laplace transform of Eq. (C.32), we get

$$\frac{d^2i_3}{dt^2}L_{\rm m}C_{\rm m}+\frac{di_3}{dt}(R_3+R_S)C_{\rm m}+i_3-R_3C_{\rm m}\frac{di_2}{dt}-i_4=0 \tag{C.33}$$

Dividing Eq. (C.33) by $L_{\rm m}C_{\rm m}$

$$\frac{d^2i_3}{dt^2}+\frac{(R_3+R_S)}{L_{\rm m}}\frac{di_3}{dt}+\frac{1}{L_{\rm m}C_{\rm m}}i_3-\frac{R_3}{L_{\rm m}}\frac{di_2}{dt}-\frac{1}{L_{\rm m}C_{\rm m}}i_4=0 \tag{C.34}$$

In order to solve the second-order DE, Eq. (C.34) numerically, we recast it into normal form: Let

$$\frac{di_3}{dt}=j_3 \tag{C.35}$$

Then Eq. (C.34) can be written as

$$\frac{dj_3}{dt}=-\frac{(R_3+R_S)}{L_{\rm m}}j_3-\frac{1}{L_{\rm m}C_{\rm m}}i_3+\frac{R_3}{L_{\rm m}}\left(\frac{R_1}{L_2}i_1+\frac{R_1}{L_2}i_3-\frac{R_1+R_2+R_3}{L_2}i_2\right)+\frac{1}{L_{\rm m}C_{\rm m}}i_4 \tag{C.36}$$

Rearranging Eq. (C.36) gives

$$\frac{dj_3}{dt} = -\frac{(R_3+R_S)}{L_m}j_3 + \frac{R_1R_3}{L_2L_m}i_1 - \frac{R_3(R_1+R_2+R_3)}{L_2L_m}i_2 + \left(\frac{R_3^2}{L_2L_m} - \frac{1}{L_mC_m}\right)i_3 + \frac{1}{L_mC_m}i_4 \tag{C.37}$$

Now let us consider the fourth loop Eq. (C.22). Multiplying it by sC_m

$$(sR_mC_m - 1)\tilde{I}_4(s) - \tilde{I}_3(s) = 0 \tag{C.38}$$

Taking the inverse Laplace transform of Eq. (C.32), we get

$$R_mC_m\frac{di_4}{dt} + i_4 - i_3 = 0 \tag{C.39}$$

Dividing Eq. (C.39) by R_mC_m

$$\frac{di_4}{dt} = \frac{1}{R_mC_m}i_3 - \frac{1}{R_mC_m}i_4 \tag{C.40}$$

Now Eqs (C.26), (C.29), (C.31), (C.35), (C.37), and (C.40) make up the set of six state equations required to simulate the working of Fig. C.3 (same as Fig. 6.12). These six equations can be written as state equations (6.18)–(6.23) provided we make the following substitutions:

$$a = R_1/L_1, b = 1/L_1C, c = R_1/L_2, d = (R_1+R_2+R_3)/L_2, e = R_3/L_2, f = g = R_3/L_m, h = 1/L_mC_m, \text{and } k = 1/C_mR_m$$

The impulse response that results from the delta function input in Eq. (C.31) can be taken care of by using the initial condition that arises from the impulse input, and making the input itself equal to zero. Hence the initial conditions would be $i_1(0)=0$, $j_1(0)=V_{C1}(0+)/L$, $i_2(0)=i_3(0)=j_3(0)=i_4(0)=0$, where $V_{C1}(0+)$ is the initial voltage of the LSS capacitor or the amplitude of the surge voltage generated by the LSS.

C.7 Numerical Simulation of an Assumed Fixed State of an MOV Validated With a SPICE-Based Simulation

MATLAB code used to produce Fig. 6.13C

```
%————————————————————————————————————————————————————————————————
function LSS_MOV_four_loops_Rm_7ohms

% Solve for the third loop current (out of 4) thru Rm.
% ODE15s (with jacobian). Works well with RelTol = 1e-3 & 1e-4.
% Change Rm =300 ohms and Rm = 34 Mohms for the other two
% simulations;
% respective SPICE data have to be imported for validation
```

```
R1 = 7;
C = 10e-6;
L1 = 2.7e-6;
R2 = 1.1;
L2 = 6e-6;
R3 = 1e6;
Rs = 0;
Lm = 40e-9;
Cm = 900e-12;
Rm = 7;
Vc0 = 1000;

a = R1/L1; b = 1/(L1*C);
c = R1/L2; d = (R1+R2+R3)/L2;
e = R3/L2; f = (R3+Rs)/Lm;
g = R3/Lm; h = 1/(Lm*Cm);
k = 1/(Cm*Rm);

fignum = 1;

   %------------------------------------------------

% SPICE data is read from a comma-separated values (CSV) file

M=csvread('LSS_MOV_4loops_spice.csv', 1,1); %reads data from the
     % csv formatted file starting at row 1 and column 1
time_spice = M(:,2);       % takes all rows of column three
I3_spice = M(:,3);          % takes all rows of column four
   %------------------------------------------------
   % DE definition for RLC circuit
 function xdot = RLC_dot(t, x, flag, params)
   [x1 x2 x3 x4 x5 x6] = deal(x(1), x(2), x(3), x(4), x(5),...
                                   x(6));
   x1_dot = x2;
   x2_dot = -a*x2 - (b-a*c)*x1 - a*d*x3 + a*e*x4;
   x3_dot = c*x1 - d*x3 +e*x4;
   x4_dot = x5;
   x5_dot = -f*x5 + c*g*x1 - g*d*x3 + (e*g-h)*x4 + h*x6;
   x6_dot = k*x4 - k*x6;
   xdot = [x1_dot; x2_dot; x3_dot; x4_dot; x5_dot;...
      x6_dot]; % return column vector
 end

time = [0 0.2e-4];

% set initial conditions
      X0(:,1) = [0; Vc0/L1; 0; 0; 0; 0]; flag = ''; params = [];
```

```
% define Jacobian matrix
       J = [ 0        1  0    0        0  0
           (a*c-b) -a -a*d  a*e      0  0
            c        0 -d    e        0  0
            0        0  0    0        1  0
            c*g      0 -g*d (e*g-h) -f  h
            0        0  0    k        0 -k];
  options = odeset('RelTol', 1e-4 ,'Stats','on','Jac',J);
tic
 [t_out, X_out] = ode15s(@RLC_dot, [time(1) time(end)], X0,...
                                         options);
toc
  time = t_out;
  I1 = X_out(:,1); J1 = X_out(:,2); I2 = X_out(:,3);
       I3 = X_out(:,4); J3 = X_out(:,5); I4 = X_out(:,6);

figure(fignum);
   plot(time*1e6, I3, '--r', 'linewidth', 2);hold on;
   plot(time_spice*1e6, I3_spice, '-k', 'linewidth', 1);
   legend('ode15s based simulation','SPICE based simulation');
   title('(c)')
   xlabel('Time (us)'); ylabel('Loop 3 / MOV current (A)');

figure(2);
   plot(time(1:end-1)*1e6,diff(time)*1e6,'-+k');
   title('Diff(time) Vs Index - (ode23tb - with Jac:...
            Rm = 500 ohms; RelTol = 1e-4)')
   xlabel('Time (us)'); ylabel('Time difference (us)');
end
```

[By changing R_m to its other instances of 300 Ω and 34 MΩ in the code, the simulations pertaining to Fig. 6.13A and B, respectively, can be produced; the respective SPICE data has to be imported for validation.]

Appendix D

State Equations and MATLAB Code for Chapter 7

D.1 Development of State Equations (7.2)–(7.7) for an MOV Driven by the LSS

The complete equivalent circuit for this situation was given in Fig. 7.3 and is repeated here for convenience as Fig. D.1.

In order to solve this 4-loop circuit we will assume that the loop currents are i_1, i_2, i_3, and i_4.

Considering loop 1 and writing the KVL equation,

$$V_{C1}(0+)\Theta(t) - \frac{1}{C_1}\int i_1 dt - L_1\frac{di_1}{dt} - R_1(i_1 - i_2) = 0 \tag{D.1}$$

where $\Theta(t)$ is the Heaviside function and $V_{C1}(0+)\Theta(t)$ is the initial voltage of the capacitor C_1.

Differentiating Eq. (D.1), we get

$$V_{C1}(0+)\delta(t) - \frac{1}{C_1}i_1 - L_1\frac{d^2i_1}{dt^2} - R_1\left(\frac{di_1}{dt} - \frac{di_2}{dt}\right) = 0 \tag{D.2}$$

Rearranging Eq. (D.2)

$$\frac{d^2i_1}{dt^2} = \frac{V_{C1}(0+)}{L_1}\delta(t) - \frac{1}{L_1C_1}i_1 - \frac{R_1}{L_1}\left(\frac{di_1}{dt} - \frac{di_2}{dt}\right) \tag{D.3}$$

Considering the auxiliary:

$$\frac{di_1}{dt} = j_1 \tag{D.4}$$

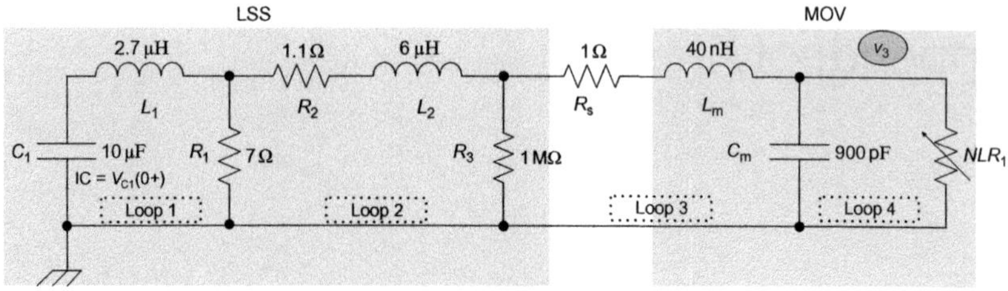

FIG. D.1 Overall equivalent circuit for an MOV driven by an LSS.

Now Eq. (D.3) can be written as

$$\frac{dj_1}{dt} = -\frac{R_1}{L_1}j_1 - \frac{1}{L_1C_1}i_1 - \frac{R_1}{L_1}\frac{di_2}{dt} \tag{D.5}$$

where the exposed δ-function has been eliminated by making the initial condition $j_1(0) = V_{C1}(0+)/L_1$.

Let us consider loop-2 now; taking KVL results in

$$i_2(R_1+R_2+R_3) + L_2\frac{di_2}{dt} - i_1R_1 - i_3R_3 = 0 \tag{D.6}$$

Dividing by L_2 and rearranging

$$\frac{di_2}{dt} = \frac{R_1}{L_2}i_1 - \frac{R_1+R_2+R_3}{L_2}i_2 + \frac{R_3}{L_2}i_3 \tag{D.7}$$

By substituting Eq. (D.7) in Eq. (D.5), we get

$$\frac{dj_1}{dt} = -\frac{R_1}{L_1}j_1 - \frac{1}{L_1C_1}i_1 - \frac{R_1}{L_1}\left(\frac{R_1}{L_2}i_1 - \frac{R_1+R_2+R_3}{L_2}i_2 + \frac{R_3}{L_2}i_3\right) \tag{D.8}$$

Rearranging Eq. (D.8)

$$\frac{dj_1}{dt} = -\frac{R_1}{L_1}j_1 - \left(\frac{1}{L_1C_1} - \frac{R_1^2}{L_1L_2}\right)i_1 - \frac{R_1(R_1+R_2+R_3)}{L_1L_2}i_2 - \frac{R_1R_3}{L_1L_2}i_3 \tag{D.9}$$

Moving on to loop-3, let us take KVL again

$$(R_3+R_S)i_3 + L_m\frac{di_3}{dt} + \frac{1}{C_m}\int i_3dt - R_3i_2 - \frac{1}{C_m}\int i_4dt = 0 \tag{D.10}$$

Differentiating Eq. (D.10), we get

$$\frac{di_3}{dt}(R_3+R_S) + \frac{d^2i_3}{dt^2}L_m + \frac{1}{C_m}i_3 - R_3\frac{di_2}{dt} - \frac{1}{C_m}i_4 = 0 \tag{D.11}$$

Rearranging Eq. (D.11),

$$\frac{d^2i_3}{dt^2} + \frac{(R_3+R_S)}{L_m}\frac{di_3}{dt} + \frac{1}{L_mC_m}i_3 - \frac{R_3}{L_m}\frac{di_2}{dt} - \frac{1}{L_mC_m}i_4 = 0 \tag{D.12}$$

Introducing auxiliary,

$$\frac{di_3}{dt} = j_3 \tag{D.13}$$

Converting Eq. (D.12) into a first-order DE, we get

$$\frac{dj_3}{dt} = -\frac{(R_3 + R_S)}{L_m} j_3 - \frac{1}{L_m C_m} i_3 + \frac{R_3}{L_m} \frac{di_2}{dt} + \frac{1}{L_m C_m} i_4 \tag{D.14}$$

The current i_4 is the current through the nonlinear resistance of the MOV model and is assumed to be governed by Eq. (7.1) which is repeated here as Eq. (D.15).

$$i_4 = K_1 {v_3}^{\alpha_1} + K_2 {v_3}^{\alpha_2} \tag{D.15}$$

By substituting Eqs (D.15), (D.7) in Eq. (D.14), we get

$$\frac{dj_3}{dt} = -\frac{(R_3 + R_S)}{L_m} j_3 + \frac{R_3}{L_m} \left(\frac{R_1}{L_2} i_1 - \frac{R_1 + R_2 + R_3}{L_2} i_2 + \frac{R_3}{L_2} i_3 \right)$$

$$- \frac{1}{L_m C_m} i_3 + \frac{1}{L_m C_m} (K_1 {v_3}^{\alpha_1} + K_2 {v_3}^{\alpha_2}) \tag{D.16}$$

Rearranging Eq. (D.16)

$$\frac{dj_3}{dt} = -\frac{(R_3 + R_S)}{L_m} j_3 + \frac{R_1 R_3}{L_2 L_m} i_1 - \frac{R_3(R_1 + R_2 + R_3)}{L_2 L_m} i_2$$

$$+ \left(\frac{R_3^2}{L_2 L_m} - \frac{1}{L_m C_m} \right) i_3 + \frac{1}{L_m C_m} (K_1 {v_3}^{\alpha_1} + K_2 {v_3}^{\alpha_2}) \tag{D.17}$$

Let us consider the nodal voltage v_4 now; since this is the voltage across the capacitor C_m, we have

$$v_4 = \frac{1}{C_m} \int (i_3 - i_4) dt \tag{D.18}$$

Differentiating Eq. (D.18), we get

$$\frac{dv_4}{dt} = \frac{1}{C_m} (i_3 - i_4) \tag{D.19}$$

Substituting Eq. (D.15) in Eq. (D.19)

$$\frac{dv_4}{dt} = \frac{1}{C_m} (i_3 - K_1 {v_3}^{\alpha_1} - K_2 {v_3}^{\alpha_2}) \tag{D.20}$$

Now Eqs (D.4), (D.9), (D.7), (D.13), (D.17), and (D.20) make up the set of six state equations corresponding to the state variables i_1, j_1, i_2, i_3, j_3, and v_3. This set of six state equations was successfully used to simulate the working of Fig. D.1 (same as Fig. 7.3). These six equations can be written as state equations (7.2)–(7.7) provided we make the following substitutions:

$$a = R_1/L_1, b = 1/L_1C, c = R_1/L_2, d = (R_1 + R_2 + R_3)/L_2, e = R_3/L_2, f = (R_3 + R_S)/L_m, g = R_3/L_m, h = 1/L_mC_m, \text{and } k = 1/C_m.$$

The initial conditions for this simulation are $i_1(0) = 0$, $j_1(0) = V_{C1}(0+)/L$, $i_2(0) = i_3(0) = j_3(0) = v_4(0) = 0$, where $V_{C1}(0+)$ is the initial voltage of the LSS capacitor or the amplitude of the surge voltage generated by the LSS.

D.2 Numerical Simulation and Validations of an MOV Driven by an LSS

MATLAB code (LSS_MOV_with_nonlinearR.m) used to produce figures from Figs 7.4 to 7.7.

```
%——————————————————————————————————————————————
——————————————————————
function LSS_MOV_ODE15S
% Comparison of simulated MOV current and voltage
% (with Lm & Cm & nonlinear R) with those obtained using the LSS.
% MOV used: 275L40C : MOV model used: Power-law based model
% ODE solver used: ode15s (with routine to return jacobian).
R1 = 7;
C = 10e-6;
L1 = 2.7e-6;
R2 = 1.1;
L2 = 6e-6;
R3 = 1e6;
Rs = 1;
Lm = 40e-9;
Cm = 900e-12;
Vc0 = 1000;
a = R1/L1; b = 1/(L1*C);
c = R1/L2; d = (R1+R2+R3)/L2;
e = R3/L2; f = (R3+Rs)/Lm;
g = R3/Lm; h = 1/(Lm*Cm);
k = 1/Cm;
% values from polynomial fit
[k1, alpha] = deal(0, 4.0413);
[k2, beta]  = deal(3.4628*10^-72.65, 26.2396);
% changed to obtain a reduced error in energy absorbed by the MOV
% [k2, beta]= deal(2.4496*10^-50.15, 18.3338);
fignum = 1;
```

```
% DE definition for RLC circuit
function xdot = RLC_dot(t, x, flag, params)...
      [x1 x2 x3 x4 x5 x6] = deal(x(1), x(2), x(3), x(4), x(5),...
                                    x(6));
       x1_dot = x2;
       x2_dot = -a*x2 - (b-a*c)*x1 - a*d*x3 + a*e*x4;
       x3_dot = c*x1 - d*x3 +e*x4;
       x4_dot = x5;
       x5_dot = -f*x5 + c*g*x1 - g*d*x3 + (e*g-h)*x4 + ...
            h*(k1*x6^alpha + k2*x6^beta);
       x6_dot = k*x4 - k*(k1*x6^alpha + k2*x6^beta);
      % return column vector
       xdot = [x1_dot; x2_dot; x3_dot; x4_dot; x5_dot; x6_dot];
 end
 % The function that returns the Jacobian matrix
 function J = FJac(t,x)
     J(6,6)=0;
     J(1,2)=1;
     J(2,1)=(a*c-b);
     J(2,2)=-a;
     J(2,3)=-a*d;
     J(2,4)=a*e;
     J(3,1)=c;
     J(3,3)=-d;
     J(3,4)=e;
     J(4,5)=1;
     J(5,1)=c*g;
     J(5,3)=-g*d;
     J(5,4)=(e*g-h);
     J(5,5)=-f;
     J(5,6)=h*(k1*alpha*x(6)^(alpha-1)+ k2*beta*x(6)^(beta-1));
     J(6,4)=k;
     J(6,6)=-k*(k1*alpha*x(6)^(alpha-1)+ k2*beta*x(6)^(beta-1));
 end
   time = [0 1e-4];
X0(:,1) = [0; Vc0/L1; 0; 0; 0; 0]; flag = ''; params = [];
options = odeset('RelTol', 1e-3,'Stats','on','Jac',@FJac);
tic
[t_out, X_out] = ode15s(@RLC_dot, [time(1) time(end)], X0,...
                        options);
toc
time = t_out;
I1 = X_out(:,1); J1 = X_out(:,2); I2 = X_out(:,3); I3 = X_out
(:,4); J3 = X_out(:,5); V3 = X_out(:,6);
```

```
% LSS data (start)
% reads data from the csv formatted file starting at row 17 and
% column 1
M = csvread('1KV_Surge_275L40C.csv', 17,1);
% just to see the first few lines (prints the first five rows and
% all the columns of the current M)
M(1:5,:)
t = M(:,4);
I = M(:,3);        % takes all rows of column three
V = M(:,2);        % takes all rows of column two
% LSS data (end)
figure(fignum);
   subplot(311)
   plot(time*1e6, I3*10,'--r', 'linewidth', 2);hold on;
   plot (t*1e6,I*10,'-r');hold on;
   plot(time*1e6, (Lm*J3)+V3,'--b', 'linewidth', 2);hold on;
   plot (t*1e6,V,'-b');hold on;
   axis([-5 90 -40 1200])
   legend('I_m_o_v(Simulation)','I_m_o_v(Validation)',...
       'V_m_o_v(Simulation)','V_m_o_v(Validation)')
   title('(a)');
   xlabel('Time (us)'); ylabel('Current (x10)(A) / Voltage (V)');
   subplot(312)
   Pcs = V3.*(I3/1e3);
   plot (time*1e6,Pcs,'--g','linewidth', 2);hold on;
   Plss = V.*(I/1e3);
   plot (t*1e6,Plss,'k');
   legend('P_m_o_v(Simulation)','P_m_o_v(Validation)')
   xlabel('time(us)');
   ylabel('Power (kW)');
   zoom on; axis tight;
   axis([-5 90 0 70]);
   title('(b)');
   subplot(313)
   Ecs = (cumsum(Pcs(1:end-1).*diff(time)))*1e3;
   plot (time(1:end-1)*1e6,Ecs,'--g','linewidth', 2);hold on;
   t1 = M(11,4);
   t2 = M(12,4);
   dt = t2-t1;
   Elss = cumsum(Plss*1e3)*dt;
   plot (t*1e6,Elss,'k');hold on;
   xlabel('time(us)');
   ylabel('Energy (J)');
   zoom on; axis tight;
```

```
    axis([-5 90 0 1.2]);
    title('(c)');
    legend('E_m_o_v(Simulation)','E_m_o_v(Validation)',...
        'Location','SouthEast')
figure(2);
    plot(time(1:end-1)*1e6,diff(time)*1e6,'-+k');
    title('Diff(time) Vs Index - ...
    (ode23t - with Jac routine: RelTol = 1e-3)')
    xlabel('Time (us)'); ylabel('Time difference (us)');
figure(3);
    plot(V, I,'--k', 'linewidth', 2);grid on;hold on;
    plot(V3, I3,'-m', 'linewidth', 1);grid on;hold on;
    plot(V3, I3,'-m', 'linewidth', 2);grid on;hold on;
    legend('Validation','Initial simulation',...
        'Improved simulation','Location','Best');
    xlabel('MOV voltage (V)'); ylabel('MOV current (A)');
end
```

D.3 PSpice Source File for Simulating an MOV Driven by an LSS

```
** MOV output voltage **
** Circuit Description **
C1 1 0 10UF IC=1000V      ; Capacitance of 10 μF with initial
                            voltage of 1000 V
L1 1 2 2.7UH              ; Inductor of 2.7 μH
R1 2 0 7                  ; Resistance of 7 Ω
R2 2 3 1.1                ; Resistance of 1.1 Ω
L2 3 4 6UH                ; Inductor of 6 μH
R3 4 0 1MEG               ; Resistance of 1 MΩ
R4 5 6 1                  ; Resistance 1 ohm
X 6 0 V20E275             ; Call to sub circuit V20E275
** Analysis Requests **
VX 4 5 DC 0V              ; Measures current through R4
** Output Requests **
* Transient analysis from 0 to 100 μs with a 0.1 μs time increment
and using initial conditions (UIC)
.TRAN 0.1US 100US 0US 0.1US UIC
* Print the result of the transient analysis with the voltage at
node 6 and current through R4
.PRINT TRAN V(6) I(VX)
*******************************
*Varistor PSpice Model Netlist*
*******************************
*     Littelfuse Inc.
```

```
*     March 25, 2003
*     Version 4.2
* Representing typical characteristics for the following UltraMOV
  series varistor:
* V20E275
*
.SUBCKT V20E275 1 2 PARAMS: TOL=0
X1      1 2 UMOV PARAMS: T={1+TOL/100} L=12nH C=900pF a1=2.726
+       a2=5.574E-2    a3=1.837E-2    a4=-3.412E-3    a5=1.362E-2
        a6=-3.772E-2
.ENDS V20E275
*
* Part of the code common to all UltraMOV series varistors
.SUBCKT UMOV 1 2 PARAMS: T=1 C=1pF L=1nH a1=1 a2=0 a3=0 a4=0
a5=0
+      a6=100u a7=100u
E_non_lin    3 1    VALUE {T*(
+      10^(
+      +a1+a2*(log10(limit(v(4),a7,1g))-3)
+      +a3*2^(-log10(limit(v(4),a7,1g))+3)
+      +a4*exp(-log10(limit(v(4),a7,1g))+3)
+      +a5*exp(log10(limit(v(4),a7,1g))-3)
+      +a6*2^(log10(limit(v(4),a7,1g))-3)
+      )
+      -10^(
+      a1+a2*(log10(-limit(v(4),-1g,-a7))-3)
+      +a3*2^(-log10(-limit(v(4),-1g,-a7))+3)
+      +a4*exp(-log10(-limit(v(4),-1g,-a7))+3)
+      +a5*exp(log10(-limit(v(4),-1g,-a7))-3)
+      +a6*2^(log10(-limit(v(4),-1g,-a7))-3)
+      )
+      +limit(v(4)/a7*v(8),-v(8),v(8))
+      )}
L_series     5 6  {L}
H_H1         4 0  VH_H1 1k
VH_H1        5 70 0V
R_R2         0 4  1G
R_series     6 2  100u
V_V1         3 70 0V
E_x_zero     8 0  VALUE {10^(
+       a1+a2*(log10(a7/1e3))
+       +a3*2^(-log10(a7/1e3))
+       +a4*exp(-log10(a7/1e3))
+       +a5*exp(log10(a7/1e3))
+       +a6*2^(log10(a7/1e3))
+        )}
```

```
R_x_zero      8 0  1G
C_parallel     1 5 {C}
.ENDS UMOV
*
.END      ; End of circuit file
```

D.4 Development of State Equations (7.9)–(7.15) for a 2-Stage TVSS Driven by an LSS

The complete equivalent circuit for this situation is given in Fig. 7.12 and is repeated here for convenience as Fig. D.2.

We will consider five loop currents i_1, i_2, i_3, i_4, and i_5 for the loops shown in Fig. D.2 and also the nodal voltage v_3 across the nonlinear resistance NLR_1, as the current through it is a function of this voltage.

Since the first two loops of this circuit (the LSS) are identical to the one given in the 4-loop circuit of Fig. D.1, the following three state equations remain unchanged.

$$\frac{di_1}{dt} = j_1 \tag{D.21}$$

$$\frac{dj_1}{dt} = -\frac{R_1}{L_1}j_1 - \left(\frac{1}{L_1C_1} - \frac{R_1^2}{L_1L_2}\right)i_1 - \frac{R_1(R_1+R_2+R_3)}{L_1L_2}i_2 - \frac{R_1R_3}{L_1L_2}i_3 \tag{D.22}$$

$$\frac{di_2}{dt} = \frac{R_1}{L_2}i_1 - \frac{R_1+R_2+R_3}{L_2}i_2 + \frac{R_3}{L_2}i_3 \tag{D.23}$$

Considering loop-3 and writing the KVL equation, we get

$$(R_3+R_S)i_3 + \frac{1}{C_m}\int i_3dt - R_3i_2 - \frac{1}{C_m}\int i_4dt = 0 \tag{D.24}$$

Differentiating Eq. (D.24), we get

$$\frac{di_3}{dt}(R_3+R_S) + \frac{1}{C_m}i_3 - R_3\frac{di_2}{dt} - \frac{1}{C_m}i_4 = 0 \tag{D.25}$$

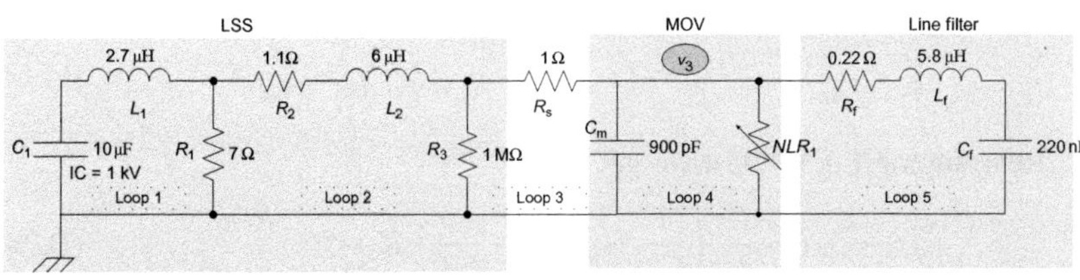

FIG. D.2 LSS driving the first two stages of a category-B protection unit.

Rearranging Eq. (D.25), we get

$$\frac{di_3}{dt}=\frac{R_3}{(R_3+R_S)}\frac{di_2}{dt}-\frac{1}{C_m(R_3+R_S)}i_3+\frac{1}{C_m(R_3+R_S)}i_4 \tag{D.26}$$

The current (i_4-i_5) is the current through the nonlinear resistance of the MOV model and is assumed to be governed by Eq. (7.8), which is repeated here as Eq. (D.27)

$$i_4-i_5=K_2{v_3}^{\alpha_2} \tag{D.27}$$

By substituting Eqs (D.23), (D.27) in Eq. (D.26), we get

$$\frac{di_3}{dt}=\frac{R_3}{(R_3+R_S)}\left(\frac{R_1}{L_2}i_1-\frac{R_1+R_2+R_3}{L_2}i_2+\frac{R_3}{L_2}i_3\right)$$

$$-\frac{1}{C_m(R_3+R_S)}i_3+\frac{1}{C_m(R_3+R_S)}(i_5+K_2{v_3}^{\alpha_2}) \tag{D.28}$$

Rearranging Eq. (D.28), we get

$$\frac{di_3}{dt}=\frac{R_3R_1}{L_2(R_3+R_S)}i_1-\frac{R_3(R_1+R_2+R_3)}{L_2(R_3+R_S)}i_2+\frac{1}{(R_3+R_S)}\left(\frac{R_3^2}{L_2}-\frac{1}{C_m}\right)i_3$$

$$+\frac{K_2}{C_m(R_3+R_S)}{v_3}^{\alpha_2}+\frac{1}{C_m(R_3+R_S)}i_5 \tag{D.29}$$

Let us consider the nodal voltage v_3 now; since this is the voltage across the capacitor C_m, we have

$$v_3=\frac{1}{C_m}\int(i_3-i_4)dt \tag{D.30}$$

Differentiating and substituting for i_4 from Eq. (D.27), we get

$$\frac{dv_3}{dt}=\frac{1}{C_m}(i_3-i_5-K_2{v_3}^{\alpha_2}) \tag{D.31}$$

Considering loop-5 and writing the KVL equation, we get

$$v_3=R_f i_5+L_f\frac{di_5}{dt}+\frac{1}{C_f}\int i_5 dt \tag{D.32}$$

Differentiating and substituting for $\frac{dv_3}{dt}$ from Eq. (D.31), we get

$$\frac{1}{C_m}(i_3-i_5-K_2{v_3}^{\alpha_2})=R_f\frac{di_5}{dt}+L_f\frac{d^2i_5}{dt^2}+\frac{1}{C_f}i_5 \tag{D.33}$$

Rearranging Eq. (D.33), we get

$$\frac{d^2i_5}{dt^2}=\frac{1}{L_fC_m}i_3-\frac{1}{L_f}\left(\frac{1}{C_m}+\frac{1}{C_f}\right)i_5-\frac{1}{L_fC_m}K_2{v_3}^{\alpha_2}-\frac{R_f}{L_f}\frac{di_5}{dt} \tag{D.34}$$

Introducing the auxiliary

$$\frac{di_5}{dt} = j_5 \tag{D.35}$$

Now Eq. (D.34) can be written as

$$\frac{dj_5}{dt} = \frac{1}{L_f C_m} i_3 - \frac{1}{L_f}\left(\frac{1}{C_m} + \frac{1}{C_f}\right) i_5 - \frac{R_f}{L_f} j_5 - \frac{1}{L_f C_m} K_2 {v_3}^{\alpha_2} \tag{D.36}$$

Now Eqs (D.21), (D.22), (D.23), (D.29), (D.31), (D.35), and (D.36) make up the set of seven state equations corresponding to the state variables i_1, j_1, i_2, i_3, v_3, i_5, and j_5 . This set of seven state equations was successfully used to simulate the working of Fig. D.2 (same as Fig. 7.12). These seven equations can be written as state equations (7.9)–(7.15) provided we make the following substitutions:

$$a = R_1/L_1, b = 1/L_1C, c = R_1/L_2, d = (R_1 + R_2 + R_3)/L_2, e = R_3/L_2,$$
$$f = R_3/(R_3 + R_S), g = 1/C_m(R_3 + R_S), h = 1/L_fC_m, k = 1/L_fC_f, l = 1/C_m$$

and $m = R_f/L_f$

The initial conditions for the simulation are $i_1(0) = 0$, $j_1(0) = V_{C1}(0+)/L$, $i_2(0) = i_3(0) = v_3(0) = i_5(0) = j_5(0) = 0$. As before, the value given for $j_1(0)$ results due to the initial voltage $V_{C1}(0+)$ of the capacitor C_1, which is the amplitude of the surge voltage generated by the LSS.

D.5 Numerical Simulation and Validations of a 2-Stage (MOV and a Line Filter) TVSS/SPD Driven by an LSS

MATLAB code (LSS_MOV_LineFilter.m) used to produce figures from Figs 7.13 to 7.15.

```
%-----------------------------------------------------------------------
function LSS_MOV_LineFilter_ODE15S
% Simulated voltage, current, power and energy plots for a TVSS
% with an MOV and line-filter elements (Cf and Lf) compared with
% LSS validations.
% MOV used - 275L40C
% ODE solver used: ode15s (with routine to return jacobian)
% LSS used: Noiseken LSS-6110
R1 = 7;
C1 = 10e-6;
L1 = 2.7e-6;
R2 = 1.1;
L2 = 6e-6;
R3 = 1e6;
Rs = 1;
```

```
Rf = 0.22;
Cm = 900e-12;
Lf = 5.8e-6;
Cf = 0.22e-6;
Vc0 = 1000;
a = R1/L1; b = 1/(L1*C1);
c = R1/L2; d = (R1+R2+R3)/L2;
e = R3/L2; f = R3/(R3+Rs);
g = 1/(Cm*(R3+Rs)); h = 1/(Lf*Cm);
k = 1/(Lf*Cf); l = 1/Cm;
m = Rf/Lf;
[k2, beta]= deal(2.4496*10^-50.15, 18.3338);
fignum = 1;
% DE definition for RLC circuit
      function xdot = RLC_dot(t, x, flag, params)
     [x1 x2 x3 x4 x5 x6 x7] = deal(x(1), x(2), x(3), x(4),...
                x(5), x(6), x(7));
      x1_dot = x2;
      x2_dot = -a*x2 - (b-a*c)*x1 - a*d*x3 + a*e*x4;
      x3_dot = c*x1 - d*x3 + e*x4;
      x4_dot = c*f*x1 - d*f*x3 + (e*f-g)*x4 + g*x6 + g*k2*x5^beta;
      x5_dot = l*x4 - l*x6 - l*k2*x5^beta;
      x6_dot = x7;
      x7_dot = h*x4 - (h+k)*x6 - h*k2*x5^beta - m*x7;
      xdot = [x1_dot; x2_dot; x3_dot; x4_dot; x5_dot; x6_dot;
      x7_dot]; % return column vector
 end
 % The function that returns the Jacobian matrix
     function J = FJac(t,x)
     J(7,7)=0;
     J(1,2)=1;
     J(2,1)=(a*c-b);
     J(2,2)=-a;
     J(2,3)=-a*d;
     J(2,4)=a*e;
     J(3,1)=c;
     J(3,3)=-d;
     J(3,4)=e;
     J(4,1)=c*f;
     J(4,3)=-d*f;
     J(4,4)=e*f-g;
     J(4,5)=g*k2*beta*x(5)^(beta-1);
     J(4,6)=g;
```

```
        J(5,4)=1;
        J(5,5)=-1*k2*beta*x(5)^(beta-1);
        J(5,6)=-1;
        J(6,7)=1;
        J(7,4)=h;
        J(7,5)=-h*k2*beta*x(5)^(beta-1);
        J(7,6)=-(h+k);
        J(7,7)=-m;
  end
    time = [0 0.9e-4];
    X0(:,1) = [0; Vc0/L1; 0; 0; 0; 0; 0]; flag = ''; params = [];
    options = odeset('RelTol', 1e-3,'Stats','on','Jac',@FJac);
tic
    [t_out, X_out]=ode15s(@RLC_dot, [time(1) time(end)], X0,...
                              options);
toc
       time = t_out;
       I1 = X_out(:,1); J1 = X_out(:,2); I2 = X_out(:,3);
       I3 = X_out(:,4); V4 = X_out(:,5); I5 = X_out(:,6);
       J5 = X_out(:,7);
% LSS data (start)
M = csvread('1KV_275L40C_RFIF_6230.csv', 17,1); %reads data from
% the csv formatted file starting at row 17 and column 1
t = M(:,5);        % takes all rows of column 5
I = M(:,4);        % takes all rows of column 4
VCf = M(:,3);      % takes all rows of column 3
Vmov = M(:,2);     % takes all rows of column 2
ICf = M(:,6);      % takes all rows of column 6
VRf = M(:,7);      % takes all rows of column 7
%  LSS data (end)
% plots for MOV
figure(fignum);
   subplot(411)
    plot(time*1e6, V4,'--k', 'linewidth', 2);hold on;
    plot (t*1e6,Vmov,'-k');
    legend('V_m_o_v (MATLAB simulation)','V_m_o_v (validation)');
    xlabel('Time (us)'); ylabel('Voltage (V) ');
    axis([-5 90 -10 750])
    title('(a)');
   subplot(412)
    plot(time*1e6, (I3-I5),'--k', 'linewidth', 2);hold on;
    plot (t*1e6,((I-ICf)-1),'-k','linewidth', 1);hold on;
    legend('I_m_o_v (MATLAB simulation)','I_m_o_v (validation)');
```

```
        xlabel('Time (us)'); ylabel('Current (A)');
        axis([-5 90 -20 200])
        title('(b)');
      subplot(413)
        Pcs = V4.*((I3-I5)/1e3);
        plot (time*1e6,Pcs,'--k','linewidth', 2);hold on;
        Plss = Vmov.*(((I-ICf)-1)/1e3);
        plot (t*1e6,Plss,'k');
        legend('P_m_o_v(MATLAB simulation)','P_m_o_v(Validation)')
        xlabel('time(us)');
        ylabel('Power (kW)');
        zoom on;
        axis([-5 90 -10 130]);
        title('(c)');
      subplot(414)
        Ecs = (cumsum(Pcs(1:end-1).*diff(time)))*1e3;
        plot (time(1:end-1)*1e6,Ecs,'--k','linewidth', 2);hold on;
        t1 = M(11,5);
        t2 = M(12,5);
        dt = t2-t1;
        Elss = cumsum(Plss*1e3)*dt;
        plot (t*1e6,Elss,'k');
        xlabel('time(us)');
        ylabel('Energy (J)');
        zoom on; axis([-5 90 0 1.2])
        title('(c)');
        legend('E_m_o_v(MATLAB simulation)','E_m_o_v(Validation)',...
          'Location','SouthEast')
    % Plots for the line-filter capacitor Cf
    figure(2);
      subplot(411)
        plot(time*1e6, V4-Lf*J5-Rf*I5,'--k', 'linewidth', 2);hold on;
        plot (t*1e6,VCf,'-k');hold on;grid on;
        legend('V_C_f (MATLAB simulation)','V_C_f (validation)')
        xlabel('Time (us)'); ylabel('voltage (V)');
        axis([-5 90 -10 1100])
        title('(a)');
      subplot(412)
        plot(time*1e6, I5,'--k', 'linewidth', 2);hold on;
        plot(t*1e6, ICf,'-k', 'linewidth', 1);hold on;grid on;
        legend('I_C_f (MATLAB simulation)','I_C_f (validation)')
        xlabel('Time (us)'); ylabel('currents (A)');
        axis([-5 90 -80 100])
```

```
    title('(b)');
   subplot(413)
    Pcs = (V4-Lf*J5-Rf*I5).*(I5)/1e3;
    plot (time*1e6,Pcs,'--k','linewidth', 2);hold on;grid on;
    Plss = VCf.*(ICf)/1e3;
    plot (t*1e6,Plss,'k');
    legend('P_C_f(MATLAB simulation)','P_C_f(Validation)')
    xlabel('time(us)');
    ylabel('Power (kW)');
    zoom on; axis([-5 90 -60 65]);
    title('(c)');
   subplot(414)
    Ecs = (0.5*Cf*(V4-Lf*J5-Rf*I5).*(V4-Lf*J5-Rf*I5));
    plot (time*1e6,Ecs,'--k','linewidth', 2);hold on;grid on;
    Elss =0.5*Cf*VCf.*VCf;
    plot (t*1e6,Elss,'k');
    legend('E_C_f(MATLAB simulation)','E_C_f(Validation)',...
      'Location','NorthEast');
    xlabel('time(us)');
    ylabel('Energy (J)');
    zoom on; axis([-5 90 0 0.13])
    title('(d)');
% Plots for line-filter inductor Lf
figure(3);
   subplot(411)
    plot(time*1e6, Lf*J5,'--k', 'linewidth', 2);hold on;
    plot (t*1e6,Vmov-VCf-VRf,'-k');hold on;
    legend('V_L_f (simulation)','V_L_f (validation)')
    xlabel('Time (us)'); ylabel(' voltage (V)');
    axis([-5 90 -425 400]);
    title('(a)');
   subplot(412)
    plot(time*1e6, I5,'--k', 'linewidth', 2);hold on;
    plot(t*1e6, (ICf+1.0),'-k', 'linewidth', 1);hold on;
    legend('I_L_f (simulation)','I_L_f (validation)')
    xlabel('Time (us)'); ylabel('currents (A)');
    axis([-5 90 -75 95]);
    title('(b)');
   subplot(413)
    Pcs = (Lf*J5).*I5/1e3;
    plot (time*1e6,Pcs,'--k','linewidth', 2);hold on;
    Plss = (Vmov-VCf-VRf).*(ICf+1.0)/1e3;
    plot (t*1e6,Plss,'k');
```

```
        legend('P_L_f(Simulation)','P_L_f(Validation)')
        xlabel('time(us)'); ylabel('Power (kW)');
        zoom on; axis([-5 90 -20 20]);
        title('(c)');
       subplot(414)
        Ecs = (0.5*Lf*(I5).*(I5));
        plot (time*1e6,Ecs,'--k','linewidth', 2);hold on;
        Elss = 0.5*Lf*(VRf/Rf).*(VRf/Rf);
        plot (t*1e6,Elss,'k');
        legend('E_L_f(Simulation)','E_L_f(Validation)')
        xlabel('time(us)'); ylabel('Energy (J)');
        zoom on; axis([-5 90 -0.001 0.025])
        title('(d)');
   end
```

D.6 Development of State Equations (7.18)–(7.25) for a 3-Stage (MOV Followed by a Line Filter and a TVS Diode) TVSS Driven by an LSS

The complete equivalent circuit for this situation is given in Fig. 7.17 and is repeated here for convenience as Fig. D.3.

We will consider six loop currents i_1, i_2, i_3, i_4, i_5, and i_6. The first two loop currents i_1 and i_2 are flowing within the LSS equivalent circuit and are not explicitly shown in Fig. D.3. In addition, we will consider the nodal voltages v_3 and v_5 across the nonlinear resistances NLR_1 and NLR_2.

Since the first four loops of this circuit (2 loops in the LSS, loop-3 and loop-4) are identical to the one given in the 5-loop circuit of Fig. D.2, the following five state equations remain unchanged.

$$\frac{di_1}{dt}=j_1 \tag{D.37}$$

$$\frac{dj_1}{dt}=-\frac{R_1}{L_1}j_1-\left(\frac{1}{L_1C_1}-\frac{R_1^2}{L_1L_2}\right)i_1-\frac{R_1(R_1+R_2+R_3)}{L_1L_2}i_2-\frac{R_1R_3}{L_1L_2}i_3 \tag{D.38}$$

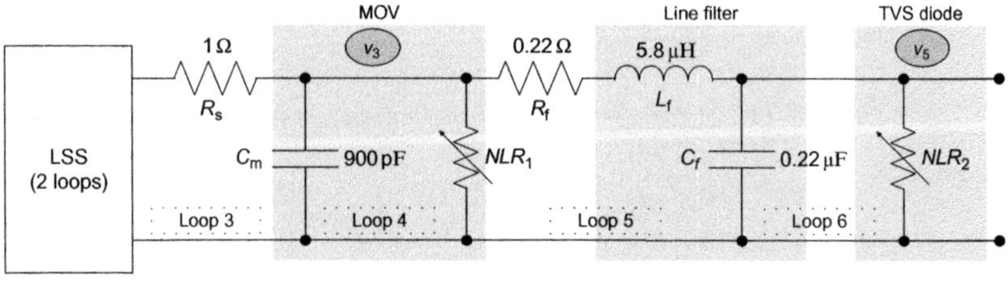

FIG. D.3 Equivalent circuit for the LSS driving a 3-stage 2-level TVSS.

$$\frac{di_2}{dt}=\frac{R_1}{L_2}i_1-\frac{R_1+R_2+R_3}{L_2}i_2+\frac{R_3}{L_2}i_3 \tag{D.39}$$

$$\frac{di_3}{dt}=\frac{R_3R_1}{L_2(R_3+R_S)}i_1-\frac{R_3(R_1+R_2+R_3)}{L_2(R_3+R_S)}i_2+\frac{1}{(R_3+R_S)}\left(\frac{R_3^2}{L_2}-\frac{1}{C_m}\right)i_3 + \frac{K_m}{C_m(R_3+R_S)}{v_3}^{\alpha_m}+\frac{1}{C_m(R_3+R_S)}i_5 \tag{D.40}$$

$$\frac{dv_3}{dt}=l\left(i_3-i_5-k_m v_3^{\alpha_m}\right) \tag{D.41}$$

Considering loop-5 and writing the KVL equation, we get

$$v_3=R_f i_5+L_f\frac{di_5}{dt}+\frac{1}{C_f}\int(i_5-i_6)dt \tag{D.42}$$

Differentiating and substituting for $\frac{dv_3}{dt}$ from Eq. (D.41), we get

$$\frac{1}{C_m}(i_3-i_5-K_m{v_3}^{\alpha_m})=R_f\frac{di_5}{dt}+L_f\frac{d^2i_5}{dt^2}+\frac{1}{C_f}(i_5-i_6) \tag{D.43}$$

Rearranging Eq. (D.43), we get

$$\frac{d^2i_5}{dt^2}=\frac{1}{L_fC_m}i_3-\frac{1}{L_f}\left(\frac{1}{C_m}+\frac{1}{C_f}\right)i_5-\frac{1}{L_fC_m}K_m{v_3}^{\alpha_m}-\frac{R_f}{L_f}\frac{di_5}{dt}+\frac{1}{L_fC_f}i_6 \tag{D.44}$$

Introducing the auxiliary

$$\frac{di_5}{dt}=j_5 \tag{D.45}$$

Now Eq. (D.44) can be written as

$$\frac{dj_5}{dt}=\frac{1}{L_fC_m}i_3-\frac{1}{L_f}\left(\frac{1}{C_m}+\frac{1}{C_f}\right)i_5-\frac{R_f}{L_f}j_5-\frac{1}{L_fC_m}K_m{v_3}^{\alpha_m}+\frac{1}{L_fC_f}i_6 \tag{D.46}$$

Let us consider the nodal voltage v_5 now; since this is the voltage across the capacitor C_f, we have

$$v_5=\frac{1}{C_f}\int(i_5-i_6)dt \tag{D.47}$$

Differentiating Eq. (D.47), we get

$$\frac{dv_5}{dt}=\frac{1}{C_f}(i_5-i_6) \tag{D.48}$$

The current i_6 is the current through the nonlinear resistance of the TVS diode model and is assumed to be governed by Eq. (7.17) which is repeated here as Eq. (D.49).

$$i_6 = K_t v_5^{\alpha_t} \tag{D.49}$$

Substituting Eq. (D.49) in Eqs (D.46), (D.48), we get Eqs. (D.50), (D.51), respectively.

$$\frac{dj_5}{dt} = \frac{1}{L_f C_m} i_3 - \frac{1}{L_f}\left(\frac{1}{C_m} + \frac{1}{C_f}\right) i_5 - \frac{R_f}{L_f} j_5 - \frac{1}{L_f C_m} K_m v_3^{\alpha_m} + \frac{1}{L_f C_f} K_t v_5^{\alpha_t} \tag{D.50}$$

$$\frac{dv_5}{dt} = \frac{1}{C_f}\left(i_5 - K_t v_5^{\alpha_t}\right) \tag{D.51}$$

Now Eqs (D.37), (D.38), (D.39), (D.40), (D.41), (D.45), (D.50), and (D.51) make up the set of eight state equations corresponding to the state variables i_1, $j_1, i_2, \ i_3, \ v_3, \ i_5, \ j_5$, and v_5. This set of eight state equations was successfully used to simulate the working of Fig. D.3 (same as Fig. 7.17). These seven equations can be written as state equations (7.18)–(7.25) provided we make the following substitutions:

$$a = R_1/L_1, b = 1/L_1C_1, c = R_1/L_2, d = (R_1 + R_2 + R_3)/L_2, e = R_3/L_2,$$
$$f = R_3/(R_3 + R_S), g = 1/C_m(R_3 + R_S), h = 1/C_m, k = 1/L_fC_m, l = 1/L_fC_f,$$
$$m = 1/C_f \text{ and } n = R_f/L_f$$

The initial conditions for the simulation are $i_1(0) = 0$, $j_1(0) = V_{C1}(0+)/L$, $i_2(0) = i_3(0) = v_3(0) = i_5(0) = j_5(0) = v_5(0) = 0$. As before, the value given for $j_1(0)$ results due to the initial voltage $V_{C1}(0+)$ of the capacitor C_1, which is the amplitude of the surge voltage generated by the LSS.

D.7 Numerical Simulation and Validations of a 3-Stage (MOV Followed by a Line Filter and a TVS Diode) TVSS/SPD Driven by an LSS

MATLAB code (LSS_MOV_LineFilter_TVS_Diode.m) used to produce figures from Figs 7.18 to 7.19.

```
%---------------------------------------------------------------------
function LSS_TVSS_ODE15S
% Simulated voltage, current, power and energy plots for a TVSS
% with an MOV (for level-1 protection), a line-filter and a TVS
% diode (for level-2 protection, compared with LSS validations.
% LSS used: Noiseken LSS-6110
% MOV used - 275L40C
% TVS diode used - 2 X 1.5KE350CA
% ODE solver used: ode15s (with routine to return jacobian)
R1 = 7;
C1 = 10e-6;
L1 = 2.7e-6;
```

```
R2 = 1.1;
L2 = 6e-6;
R3 = 1e6;
Rs = 1;
Rf = 0.22;
Cm = 900e-12;
Lf = 4.0e-6;
Cf = 0.22e-6;
Vc0 = 1000;
a = R1/L1; b = 1/(L1*C1);
c = R1/L2; d = (R1+R2+R3)/L2;
e = R3/L2; f = R3/(R3+Rs);
g = 1/(Cm*(R3+Rs)); h = 1/Cm;
k = 1/(Lf*Cm);l = 1/(Lf*Cf);
m = 1/Cf; n=Rf/Lf;
% original PL model for 275L40C
[km, alpham] = deal(2.4496*10^-50.15, 18.3838);
% PL model for 2x1.5KE350CA
[kt, alphat] = deal(4.0638e-161, 56.3468);
fignum = 1;
% DE definition for RLC circuit
   function xdot = RLC_dot(t, x, flag, params)
      [x1 x2 x3 x4 x5 x6 x7 x8] = deal(x(1), x(2), x(3), x(4),...
                        x(5), x(6), x(7), x(8));
       x1_dot = x2;
       x2_dot = -a*x2 - (b-a*c)*x1 - a*d*x3 + a*e*x4;
       x3_dot = c*x1 - d*x3 + e*x4;
       x4_dot = c*f*x1 - d*f*x3 + (e*f-g)*x4 + g*x6 +...
                               g*km*x5^alpham;
       x5_dot = h*x4 - h*x6 - h*km*x5^alpham;
       x6_dot = x7;
       x7_dot = k*x4 - (k+l)*x6 - k*km*x5^alpham +...
                         l*kt*x8^alphat-n*x7;
       x8_dot = m*x6 - m*kt*x8^alphat;
       xdot=[x1_dot; x2_dot; x3_dot; x4_dot; x5_dot; x6_dot;...
           x7_dot; x8_dot];    % return column vector
end
% The function that returns the Jacobian matrix
    function J = FJac(t,x)
        J(8,8)=0;
        J(1,2)=1;
        J(2,1)=(a*c-b);
        J(2,2)=-a;
        J(2,3)=-a*d;
```

```
        J(2,4)=a*e;
        J(3,1)=c;
        J(3,3)=-d;
        J(3,4)=e;
        J(4,1)=c*f;
        J(4,3)=-d*f;
        J(4,4)=e*f-g;
        J(4,5)=g*km*alpham*x(5)^(alpham-1);
        J(4,6)=g;
        J(5,4)=h;
        J(5,5)=-h*km*alpham*x(5)^(alpham-1);
        J(5,6)=-h;
        J(6,7)=1;
        J(7,4)=k;
        J(7,5)=-k*km*alpham*x(5)^(alpham-1);
        J(7,6)=-(k+l);
        J(7,8)=l*kt*alphat*x(8)^(alphat-1);
        J(7,7)=-n;
        J(8,6)=m;
        J(8,8)=-m*kt*alphat*x(8)^(alphat-1);
end
time = [0 0.9e-4];
X0(:,1) = [0; Vc0/L1; 0; 0; 0; 0; 0; 0]; flag = ''; params = [];
options = odeset('RelTol', 1e-3,'Stats','on','Jac',@FJac);
tic
[t_out, X_out] = ode23t(@RLC_dot, [time(1) time(end)], X0,...
                                     options);
toc
time = t_out;
I1 = X_out(:,1); J1 = X_out(:,2); I2 = X_out(:,3);...
I3 = X_out(:,4); V4 = X_out(:,5); I5 = X_out(:,6);...
J5 = X_out(:,7); V6 = X_out(:,8);
% LSS data (start)
M   =   csvread('1kV_Surge_275L40CpRFIFp_2x350V_TVSD_031013_0041.
csv',... 17,1);
% reads data from the csv formatted file starting at row 17 and
% column 1
t = M(:,5);
Vmov = M(:,4);           % takes all rows of column 4
Iprot = M(:,3);          % takes all rows of column 3
IRf = M(:,2);             % takes all rows of column 2
Imov= M(:,6);
N   =   csvread('1kV_Surge_275L40CpRFIFp_2x350V_TVSD_081013_0047.
csv',... 17,1);
% reads data from the csv formatted file starting at row 17 and
```

```
% column 1
tt = N(:,4);
Vtvsd = N(:,2);          % takes all rows of column 4
Itvsd = N(:,3);          % takes all rows of column 3
% LSS data (end)
% Plots for the MOV
figure(fignum);
 subplot(411)
    plot(time*1e6, V4,'--k', 'linewidth', 2);hold on;
    plot (t*1e6,Vmov,'-k');
    legend('V_m_o_v (MATLAB simulation)','V_m_o_v (validation)');
    xlabel('Time (us)'); ylabel('Voltage (V) ');
    axis([-5 90 -10 750])
    title('(a)');
 subplot(412)
    plot(time*1e6, km*(V4.^alpham),'--k','linewidth',2);hold on;
    plot (t*1e6,Imov,'-k','linewidth', 1);hold on;
    legend('I_m_o_v (MATLAB simulation)','I_m_o_v (validation)');
    xlabel('Time (us)'); ylabel('Current (A)');
    axis([-5 90 -20 120])
    title('(b)');
 subplot(413)
    Pcs = V4.*(km*(V4.^alpham)/1e3);
    plot (time*1e6,Pcs,'--k','linewidth', 2);hold on;
    Plss = Vmov.*(Imov/1e3);
    plot (t*1e6,Plss,'k');
    legend('P_m_o_v(MATLAB simulation)','P_m_o_v(Validation)')
    xlabel('time(us)');
    ylabel('Power (kW)');
    zoom on;
    axis([-5 90 -10 80]);
    title('(c)');
 subplot(414)
    Ecs = (cumsum(Pcs(1:end-1).*diff(time)))*1e3;
    plot (time(1:end-1)*1e6,Ecs,'--k','linewidth', 2);hold on;
    t1 = M(11,5);
    t2 = M(12,5);
    dt = t2-t1;
    Elss = cumsum(Plss*1e3)*dt;
    plot (t*1e6,Elss,'k');
    xlabel('time(us)');
    ylabel('Energy (J)');
    zoom on; axis([-5 90 0 1.2])
    title('(c)');
```

```
    legend('E_m_o_v(MATLAB...
      simulation)','E_m_o_v(Validation)','Location','SouthEast')
% Plots for the TVS diode
figure(3);
 subplot(411)
    plot(time*1e6, V6,'--k', 'linewidth', 2);hold on;
    plot (tt*1e6,Vtvsd,'-k');
    legend('V_t_v_s_d (MATLAB simulation)','V_t_v_s_d...
             (Validation)','Location','SouthEast');
    xlabel('Time (us)'); ylabel('Voltage (V) ');
    axis([-2 45 -10 800])
    title('(a)');
 subplot(412)
    plot(time*1e6,  kt*(V6.^alphat),'--k',  'linewidth',  2);hold
on;
    plot (tt*1e6,Itvsd,'-k','linewidth', 1);hold on;
    legend('I_t_v_s_d (MATLAB simulation)','I_t_v_s_d...
               (Validation)');
    xlabel('Time (us)'); ylabel('Current (A)');
    axis([-2 45 -10 80])
    title('(b)');
 subplot(413)
    Pcst = V6.*(kt*(V6.^alphat)/1e3);
    plot (time*1e6,Pcst,'--k','linewidth', 2);hold on;
    Plsst = Vtvsd.*(Itvsd/1e3);
    plot (tt*1e6,Plsst,'k');
    legend('P_t_v_s(MATLAB simulation)','P_t_v_s(Validation)')
    xlabel('time(us)');
    ylabel('Power (kW)');
    zoom on; axis([-2 45 -10 60]);
    title('(c)');
 subplot(414)
    Ecst = (cumsum(Pcst(1:end-1).*diff(time)))*1e3;
    plot (time(1:end-1)*1e6,Ecst,'--k','linewidth', 2);hold on;
    tt1 = N(11,4);
    tt2 = N(12,4);
    dtt = tt2-tt1;
    Elsst = cumsum(Plsst*1e3)*dtt;
    plot (tt*1e6,Elsst,'k');
    xlabel('time(us)');
    ylabel('Energy (J)');
    zoom on; axis([-2 45 -0.02 0.2])
    title('(c)');
    legend('E_t_v_s(MATLAB...
```

```
    simulation)','E_t_v_s(Validation)','Location','northeast')
figure(2);
   plot(time(1:end-1)*1e6,diff(time)*1e6,'-+k');
   title('Diff(time) Vs Index - (ode23t - with Jac routine:...
                       RelTol = 1e-3)')
   xlabel('Time (us)'); ylabel('Time difference (us)');
end
```

Index

Note: Page numbers followed by *f* indicate figures and *t* indicate tables.

S

T

U

V

W

X

Y

9780128116647